THE SCIENCE
OF ALLELOPATHY

THE SCIENCE
OF ALLELOPATHY

Edited by

ALAN R. PUTNAM

Michigan State University

and

CHUNG-SHIH TANG

University of Hawaii

A Wiley-Interscience Publication

JOHN WILEY & SONS

New York • Chichester • Brisbane • Toronto • Singapore

Library of Congress Cataloging in Publication Data:

The Science of allelopathy.

   "A Wiley-Interscience publication."
   Bibliography: p.
   Includes index.
   1. Allelopathic agents.   2. Allelopathy.
I. Putnam, Alan R., 1939–   . II. Tang, Chung-Shih,
1938–

QK898.A43S34   1986   581.2′32   86-7822
ISBN 0-471-83027-5

Printed in the United States of America

10 9 8 7 6 5 4 3 2 1

CONTRIBUTORS

JANE P. BARNES, Department of Horticulture and Pesticide Research Center, Michigan State University, East Lansing, Michigan
Present Address:
ARCO Plant Cell Research Institute, Dublin, California

BASIL A. BURKE, ARCO Plant Cell Research Institute, Dublin, California

CHANG-HUNG CHOU, Institute of Botany, Academia Sinica, Taipei, Taiwan, Republic of China

HORACE G. CUTLER, Plant Physiology Unit, United States Department of Agriculture, Agricultural Research Center, Richard B. Russell Research Center, Athens, Georgia

STEVEN O. DUKE, United States Department of Agriculture, Agricultural Research Service, Southern Weed Science Laboratory, Stoneville, Mississippi

FRANK A. EINHELLIG, Department of Biology, University of South Dakota, Vermillion, South Dakota

NIKOLAUS H. FISCHER, Department of Chemistry, Louisiana State University, Baton Rouge, Louisiana

JACOB FRIEDMAN, Department of Biochemistry, Oklahoma State University, Stillwater, Oklahoma
Present Address:
Department of Botany, George S. Wise Faculty of Life Sciences, Tel Aviv University, Tel Aviv, Israel

NURIT FRIEDMAN, Department of Biochemistry, Oklahoma State University, Stillwater, Oklahoma
Present Address:
Department of Botany, George S. Wise faculty of Life Sciences, Tel Aviv University, Tel Aviv, Israel

DURGA KUMARI, Department of Forest Products, University of Minnesota, St. Paul, Minnesota

GERALD R. LEATHER, United States Department of Agriculture, Weed Physiology Laboratory, Frederick, Maryland

JOHN V. LOVETT, Department of Agricultural Science, University of Tasmania, Hobart, Tasmania, Australia

WALTER H. MULLER, Department of Biological Science, University of California, Santa Barbara, Santa Barbara, California

ALAN R. PUTNAM, Department of Horticulture and Pesticide Research Center, Michigan State University, East Lansing, Michigan

ELROY L. RICE, Department of Botany and Microbiology, University of Oklahoma, Norman, Oklahoma

KENNETH L. STEVENS, Western Regional Research Center, United States Department of Agriculture, Agricultural Research Service, Berkeley, California

CHUNG-SHIH TANG, Department of Agricultural Biochemistry, University of Hawaii, Honolulu, Hawaii

GEORGE R. WALLER, Department of Biochemistry, Oklahoma State University, Stillwater, Oklahoma

LESLIE A. WESTON, Department of Horticulture and Pesticide Research Center, Michigan State University, East Lansing, Michigan
Present Address:
Department of Horticulture and Landscape Architecture, University of Kentucky, Lexington, Kentucky

CHIU-CHUNG YOUNG, Department of Soil Science, National Chung Hsing University, Taichung, Taiwan, Republic of China

BAOCHEN ZHANG, Northwest Plateau Institute of Biology, Academia Sinica, Xining, Qinghai, China

PREFACE

In this book, we consider allelopathy to encompass chemical interactions involving all plants, whether they be microbes or higher plants, or inter- or intraspecific (autotoxicity). In this context, we closely follow Molisch's original definition of the term. We feel that microbial (actinomycete, algal, bacterial, fungal) plant products, in particular, must receive serious consideration because of their possible usefulness in agriculture and their potential for manipulation by biotechnologists.

The objective of this book is not only to bring readers up to date on the state of the science of allelopathy, but to assess its impact throughout the world, to look at some new chemistry, and to speculate a bit on the future uses of allelochemicals or allelopathic plants. The introductory chapter appraises the general state of the science, and the following 16 chapters are arranged in three parts. Part 1 offers examples of field observations of allelopathy in both agricultural and natural ecosystems. We are particularly grateful to E. L. Rice (Chapter 2), who reviewed a very neglected aspect, growth stimulation by allelochemicals. Part 2 deals with techniques for the study of allelochemicals and discusses their modes of action. The lack of adequate methodology has long been the major stumbling block in understanding the chemistry and physiology of allelopathy. The use of modern and innovative analytical techniques has enabled researchers to achieve more detailed understanding and to explore new horizons in the potential uses of allelopathy. This part, together with Part 3, on the chemistry and potential uses of allelopathy, offers convincing evidence that the stage has been set for greater advances in the coming decade.

Critics of allelopathy research have raised some important issues which have now strengthened the science. Proof of allelopathy requires not only rigorous protocols, but cooperative efforts by scientists from several disciplines.

At the International Congress of Pacific Basin Societies, held in Hawaii in December 1984, we had the pleasant task of organizing a symposium on allelopathy, which brought together scientists from several countries, representing the disciplines of biochemistry, botany, chemistry, ecology, microbiology, and several agricultural sciences. It became clear that we had gathered a group of re-

PREFACE

searchers dedicated to the advancement of allelopathy—a science which has entered its logarithmic phase of growth during the past decade. We are pleased that many of these experts agreed to contribute a comprehensive chapter for this volume.

Allelopathy has evolved from the usually isolated individual research to an enthusiastic, interacting, maturing science. We feel privileged to witness this rapid transformation, are excited by its broad implications and applications, and humbled by the intriguing and complex nature of the discipline. We realize that the path between maturing and matured science can be winding and treacherous, and it is imperative to carefully examine the past and assess the present, while negotiating toward the future. It is in this spirit we assumed the editorship of *The Science of Allelopathy*.

The editors are indebted to all the contributors for their painstaking efforts to bring this text into realization. Finally, to Sandi and Grace—thank you for your unfailing encouragement and support as well as your capable assistance with the manuscript.

ALAN R. PUTNAM
CHUNG-SHIH TANG

East Lansing, Michigan
Honolulu, Hawaii
September 1986

CONTENTS

PART 2　TECHNIQUES FOR STUDIES OF ALLELOCHEMICALS AND THEIR MODES OF ACTION

PART 3　CHEMISTRY AND POTENTIAL USES OF ALLELOPATHY

THE SCIENCE
OF ALLELOPATHY

1

ALLELOPATHY: STATE OF THE SCIENCE

ALAN R. PUTNAM

*Department of Horticulture
Pesticide Research Center
Michigan State University
East Lansing, Michigan*

CHUNG-SHIH TANG

*Department of Agricultural Biochemistry,
University of Hawaii
Honolulu, Hawaii*

CHEMICAL ECOLOGY AND ALLELOPATHY

In communities of organisms, many species appear to regulate one another by producing and releasing chemical attractants, stimulators, or inhibitors. These phenomena have become classified under the umbrella of chemical ecology. Insects or nematodes may be discouraged from attacking a plant because of the repellant or toxic chemicals that it produces. Plants also produce inhibitors (phytoalexins) in response to attack by pathogenic fungi or bacteria. All of these chemical interactions have been classified (Whittaker and Feeny, 1971) based on their inter- or intraspecific nature and which plant (donor or recipient) receives the ultimate benefit from the chemical release. Allomones, which give adaptive advantage to the producer, include repellants, suppressants, escape substances, venoms, inductants, counteractants, and attractants. Allelopathic chemicals can usually be classified as suppressants.

Allelopathy represents the plant-against-plant aspect of this broader field of chemical ecology. The word allelopathy was coined by Molisch in 1937 to describe the chemical interactions among all plants (microbes and higher plants), including stimulatory as well as inhibitory influences. This has caused some confusion, since allelopathy translates literally as "mutual suffering." Some authors have used the word in a more restricted sense to describe only the harmful effects of one higher plant upon another. As Rice (1984) recently noted, many cases of allelopathy either directly or indirectly involve microbes. Also, chemicals found to inhibit the growth of some species at certain concentrations may stimulate the growth of the same or different species at lower concentrations. For these reasons, the original broad definition of Molisch is probably more appropriate.

To date all cases of alleged allelopathy that have been thoroughly studied appear to involve a complex of chemicals. In no case has one specific phytotoxin been proved to be solely responsible for, or produced as a result of, interference by a neighboring plant.

INTERFERENCE AND ALLELOPATHY

In their communities plants may interact in a positive, negative, or neutral manner. A beneficial association is termed mutualism. For example, some higher plants perform more efficiently in the shade of others, and in other special instances, higher plants and microbial plants form symbiotic associations where benefits are gained by both. It is rare that plants are unaffected by neighbors. However, it may occur when the roots or canopies of higher plants occupy different niches.

It is more common that neighboring plants will interact in a negative manner, where the emergence or growth of one or both is inhibited. The adverse effect of a neighboring plant in an association is termed interference (Muller, 1969). The potential causes of interference include (1) allelospoly—more commonly called competition, which involves depletion of one or more resources required for growth of the association, (2) allelopathy—the addition of toxins by one or more species in the association, and (3) allelomediation—selective harboring of an herbivore that might selectively feed on one species, thus lending advantage to another (Szezepanski, 1977).

Few, if any, field investigations have definitively separated the components of interference because of the complexity of the problem. This was pointed out by Harper (1977), who proposed a more rigorous protocol to search for cause and effect. The term competition has been misused by many (particularly agricultural scientists) to describe interference. Competition always implies resource limitation [e.g., water, space, nutrient(s), or light energy], which we view as a specific mechanism for interference, not the end result. Although many field studies have strongly implicated allelopathy, isolation and identification of the agents (chemicals) requires a rigorous laboratory effort.

Sometimes allelopathy appears to occur among individuals of the same species. Intraspecies toxicity is termed autotoxicity and its ecological purpose has been difficult to interpret. Some have hypothesized that it may function in perennial species particularly as a mechanism to encourage further spreading of vegetative propagules, rather than allowing them to concentrate in one area.

Chemicals that impose allelopathic influences have been called allelochemicals or allelochemics. Grummer (1955) classified inhibitors based upon their source of origin and upon the organism affected by their action (suscept) (Table 1.1). These terms have not been widely used, perhaps because they are either nondescriptive (i.e., koline) or nonspecific (e.g., antibiotic). The terms phytoinhibitins and saproinhibitins have been suggested as appropriate to describe compounds of plant and microbial origin, respectively, which inhibit higher plants (Fuerst and Putnam, 1983).

Chemicals with allelopathic potential are present in virtually all plant tissues, including leaves, flowers, fruits, stems, roots, rhizomes, and seeds. Whether these compounds are released into the environment in sufficient quantities and with enough persistence to affect a neighboring or a succeeding plant remains a critical question in many cases of alleged allelopathy. Allelochemicals are released by such processes as volatilization, root exudation, leaching, and decomposition of plant residues (Rice, 1984; Putnam, 1985). Much of the evidence indicates that several chemicals are released together and may exert toxicities in an additive or synergistic manner.

Allelopathy may provide obvious and sometimes startling effects. For example, tomatoes (*Lycopersicon esculentum*) growing near black walnut (*Juglans nigra*) may suddenly wilt and die. More commonly the effects are subtle and thus much more difficult to assess. For example, what would be the long-term impact of a 20% growth reduction caused by allelochemicals during a one-week growth period early in the life of a plant? Would it's performance in the association be permanently impaired as a result of this setback? We suscept that many allelopathic influences occur in this manner and may have long-term effects.

TABLE 1.1
Terms previously used to describe the allelochemicals involved in
interspecific relationships

Term	Donor (Agent)	Recipient (Suscept)	Originator of Term
Koline	Higher plant	Higher plant	Grummer[a]
Phytoinhibitins	Higher plant	Higher plant	Fuerst and Putnam[b]
Marasmin	Higher plant	Microorganism	Gauman[a]
Phytoncide	Microorganism	Higher plant	Waksman[a]
Saproinhibitins	Microorganisms	Higher plant	Fuerst and Putnam[b]
Antibiotic	Microorganisms	Microorganisms	Grummer[a]

[a]Source. Grummer (1955).
[b]Source. Fuerst and Putnam (1983).

Alleged allelochemicals represent a myriad of chemical compounds from simple hydrocarbons and aliphatic acids to complex polycyclic structures. They represent acids, aldehydes, cyanogenic glycosides, thiocyanates, lactones, coumarins, quinones, flavonoids, tannins, alkaloids, terpenoids, steroids, and a variety of other chemicals. These compounds were recently reviewed by Rice (1984).

PROOF OF ALLELOPATHY

Numerous studies have provided excellent evidence for allelopathy, but seldom have investigators followed a specific protocol (similar to Koch's postulates for proof of disease) to achieve convincing proof (Fuerst and Putnam, 1983). Proof could generally involve the following sequence of studies:

1. Demonstrate interference using suitable controls, describe the symptomology, and quantitate the growth reduction.
2. Isolate, characterize, and assay the chemicals against species that were previously affected. Identification of chemicals that are not artifacts is a key step in proof of allelopathy.
3. Obtain toxicity with similar symptomology when chemical(s) are added back to the system.
4. Monitor release of chemicals from the donor plant and detect them in the environment (soil, air, etc.) around the recipient and, ideally, in the recipient.

Field Studies

Proof of interference by living plants has involved both additive and substitutive experiments. In the first case, the density of one plant (usually a crop) is kept constant while the density of another (usually a weed) is increased, hence the overall plant population becomes a variable (Zimdahl, 1980). These studies demonstrate interference and quantitate growth reduction as a function of density. Their main disadvantage is that they reveal little regarding the mechanism of interference. In substitutive or replacement series experiments, the total plant density is kept constant and ratios of the two (or more) species are varied (DeWit, 1961). These experiments provide more insight into the possible nature of the interference and the relative aggressiveness of the species, and one particular result (mutual inhibition) provides indirect evidence for allelopathy.

When first published, field studies of allelopathy in *Salvia* (Muller et al., 1964) were criticized because they did not consider the possible effects of animals. These problems can be eliminated by fencing the animals or including other proper controls.

Many observations of alleged allelopathy involve the effects of plant residues on the emergence or growth of other plants. In these experiments, poor performance can be attributed to factors other than allelopathy. For example, the plant residues may exert physical influences (reducing light energy or soil temperature), or microbes which degrade them may immobilize nutrients, particularly nitrogen. One effective method of addressing the problem of physical influences is to use an inert substance which can provide similar shading characteristics as a control. For example, Barnes and Putnam (1983) used wood shavings to simulate the physical influences of rye (*Secale cereale*). The material, which resembled rye in color and shape, was added in quantities that provided equal light reduction. The problem of possible nutrient immobilization may be best addressed by providing a superoptimal nutrient supply. It would also be beneficial to periodically monitor the nutrient condition of the soil.

All field studies are subject to interaction by a variety of herbivores, pests, and associated microbes. When plant residues are evaluated, one must be careful to look for insect, nematode, and disease influences. Soil microflora contributed important allelochemicals in a number of studies (McCalla and Haskins, 1964). Their participation in allelopathic phenomena should probably be considered the rule rather than the exception.

Laboratory and Greenhouse Studies

To date, most research activities on allelopathy have been concentrated on the apparent cases that are conspicuous under field conditions. It is also possible that chemical interference among coexisting organisms is a common evolutionary strategy among plants, and that nonapparent allelopathy could be as widespread as the ubiquitous distribution of secondary metabolites in the plant kingdom.

For both apparent and nonapparent allelopathy, further proof by laboratory and greenhouse studies is required. Under controlled conditions, factors in competition may be segregated and it is possible to prove that chemical interactions are either totally or partially responsible for the interference observed. Since the sources and types of allelochemicals vary, different methods have been devised for greenhouse and laboratory verification of their presence. Although it is not the purpose of this chapter to compare and evaluate the individual methods already reported (Rice, 1984), we would like to examine the crucial issues involved in some of the familiar techniques.

For volatile compounds emitted from the above-ground part of plants, air samples may be collected from their close vicinity (Muller, 1965). This is an excellent approach since at no time are the plants disturbed, and the collection of the compounds is a confirmation of their presence under natural conditions. Modern techniques for air pollutant analyses (Sherma, 1979) can be readily adopted. Various types of synthetic adsorbents are available and the samples collected may be used for both bioassay and GC/MS analysis. So far, however, this approach has not been used to its full advantage.

Plant leachates have been collected to support the presence of extracellular bioactive compounds. Under field conditions, leaching may be caused by dew drops, rain, or irrigation water, bringing leachates from leaves and branches to the soil. Roots under flooded conditions may also be considered as being leached by the diluted soil solution. However, leachates do not include intracellular metabolites released because of physical damage inflicted during sample collection. Unfortunately, in many cases it is impossible to judge whether or not damage of the fragile living tissue has occurred, and the sample subsequently collected, in a strict sense, would be of doubtful origin.

Soils collected from the field or greenhouse (e.g., pots) have been frequently used as sources of allelochemicals. Receptor plants can either be planted directly for bioassay (Young and Bartholomew, 1981), or the soil extracted for allelopathic agents (Campbell et al., 1982). While laboratory work on soil appears to be a logical approach to verify field observations, it would be misleading to equate soil extracts with root exudates. This is because (1) recovery or organic compounds from soil is usually difficult, (2) the chance of forming artifacts during extraction is high (Kaminsky and Muller, 1977), and (3) it is difficult to exclude possible contamination by broken root tissues while collecting rhizospheric soil samples. However, soil experiments can be adequate to demonstrate allelopathy due to plant residues. For more discussion on this subject see Chapters 4, 15, and 16.

In devising laboratory and greenhouse studies of allelopathy, efforts have been made to assure that the biological activities obtained are indeed due to the extracellular toxicants of the donor plants. The use of sand cultures to collect nutrient solutions containing allelochemical activities (Abdul-Wahab and Rice, 1967), the stairstep apparatus (Bell and Koeppe, 1972) that allows nutrient effluent from the donor to pass through the acceptor plants, the continuous root exudate extraction system that specifically demonstrates the allelopathic effects of hydrophobic metabolites (Stevens and Tang, 1985), and the use of a system built with PVC pipes and fittings (Pope et al., 1985), were all designed to demonstrate allelopathic effects by the root systems of undisturbed donor plants. With credible methods established for laboratory and greenhouse studies (Rice, 1984), it is the responsibility of the individual researchers to select proper techniques for their particular subject and to execute them with great care.

Chemical Studies

We have recently witnessed a rapid increase in activity on the chemistry of allelopathy. In addition to several international symposia held by chemical societies, the first monograph specifically related to the chemistry of allelopathy was recently published (Thompson, 1985). Increased effort is timely since chemistry has been the Achilles' heel of allelopathy for the following reasons:

1. Allelopathy was not widely recognized in the scientific community until the foundations were laid by the pioneer contributors in the field (*vide infra*) and allelopathy is a relatively young science.

2. Without the powerful modern instrumental methods of analyses it is difficult to solve chemical problems in allelopathy which involve the isolation, identification, and quantitative determination of trace natural products in complex matrices such as soil.

3. It is not possible to study allelopathic chemistry without the combined efforts of several disciplines. A close working relationship has not been evident until recently among the organic chemists, biologists, soil scientists, and ecologists.

It must be stressed that allelopathy is caused by allelochemicals, which are plant metabolites or their products present in the microenvironment, such as the rhizosphere. Unfortunately, current literature often fails to distinguish phytotoxic natural products from allelochemicals. Although a large body of information is available on natural products in plants (Balandrin et al., 1985), these compounds should not be accepted as allelochemicals based on circumstantial evidence. Using tissue extracts for the purpose of studying allelochemicals should not be discouraged; however, we believe that it is essential to assert the difference between the two when assessing the current status of allelopathic chemistry.

Biologists are fully aware of the importance of the chemistry of allelopathy and natural product chemists are joining the biologists with enthusiasm, hence rapid progress in this area can be expected. More refined methods for sample collection (see Chapter 7), improved methods of bioassay (see Chapter 8), and the increased availability of modern instruments for the isolation (e.g., GC, HPLC) and structural elucidation (e.g., GC/MS, NMR, FTIR, see Part 3) of allelochemicals all support our optimism. In addition, the experience accumulated by environmental chemists would be of immense help. After all, allelopathic chemistry may be more appropriately regarded as the microenvironmental chemistry of plants rather than phytochemistry

Additive and Synergistic Effects

In nature the chance that any single allelochemical is the sole responsible agent in the interaction is unlikely. Even in such cases as juglone in black walnut (*Juglans nigra*) (Davis, 1928), where one toxicant appears to dominate, there still exists the possibility that the quantitatively dominating compound may mask the importance of minor but biologically active ones. Furthermore, the soil is such a complicated physical, chemical, and biological entity that organic compounds released into it would be changed rapidly through various mechanisms. Thus the additive and synergistic effects could be commonplace in allelopathy and the study of this subject is as important as it is in the formulation of pesticides and medicines.

Our knowledge regarding additive and synergistic effects is nevertheless very limited. The major information derived from the use of plant phenolics in different combinations was a determination of the magnitude of these effects (Einhellig et al., 1982; Williams and Hoagland, 1982). These model studies would help us to understand the *in situ* significance of these effects once the chemistry of allelopathy—such as the rhizospheric chemistry—gains more solid ground.

Return to the Soil System

To prove allelopathy, the final steps should be to take the isolated compounds and add them back into the system. If these compounds are in fact the allelochemicals, one would hope to produce symptomology similar to that observed in the original interference studies. To succeed it may be necessary to apply all of the active chemicals jointly and at proper concentrations. Barnes and Putnam (1986) were able to reproduce toxicity symptoms similar to those seen from rye residues in the field by adding to soil the same hydroxamic acids isolated from rye tissue. Similar proof has been obtained by introducing the volatile compound ethylene into atmospheric environments.

The frosting on the cake, so to speak, would be to successfully extract or trap the compounds in a field setting. This has seldom been accomplished, but failure to do so would not necessarily negate allelopathy. One problem is the creation of artifacts by harsh extraction techniques (Kaminsky and Muller, 1977). Another is that compounds may be tightly adsorbed to soil colloids. Soil microbes also may rapidly degrade the compounds. In spite of these difficulties, there have been successes, notably the isolation of juglone from soils under walnut trees (Fisher, 1978) and the collection of ethylene from atmospheres surrounding ripening fruits (Molisch, 1937). Perhaps a variation of the trapping system developed by Tang and Young (1982) could prove useful in field situations.

RECENT HISTORY

Major Pioneer Contributors

Although allelopathy was recognized and written about as early as the third century B.C. (Rice, 1984) most of the progress in this science has occurred in the twentieth century. With the risk of offending some who have undoubtedly made major contributions, we wish to emphasize some noteworthy efforts which, in our view, have greatly strengthened the science of allelopathy.

Schreiner and Reed published a series of articles between 1907 and 1909 on toxic compounds produced by plants that could later be extracted from soils. Considering the technology of the time, their research, particularly the chemical characterization work, was remarkable. Molisch (1937) coined the term allelopathy and was a pioneer with studies of ethylene on plants. McCalla et al. published a series of papers from 1948 to 1965 which not only contributed considerable knowledge regarding allelochemicals from plant residues but also brought into focus the importance of microbial products produced by the organisms that degraded those residues.

Two scientists in particular have contributed important research findings and have been responsible for training a large number of the scientists now engaged in the field of allelopathy. C. H. Muller, along with his associates at Santa Barbara, California, published numerous articles on allelopathy by desert and chapparral shrubs. His work to detect volatile inhibitors developed procedures that are still widely accepted today (Muller et al., 1964). E. L. Rice at the University of Oklahoma may well be the outstanding contributor. His pioneering efforts with allelochemicals on nitrifying and nitrogen fixing bacteria (Rice, 1964) have been followed by scores of classical papers on many aspects of allelopathy. His book (Rice, 1984) is the most thorough general reference on the subject. These two scientists greatly helped the science gain the exposure necessary to make it popular.

On a global scale, C. H. Chou of Taiwan has published extensively on allelophathic interactions; A. M. Grodzinsky of the USSR also has published extensively, and in addition has edited a series of books on allelopathy; J. V. Lovett has published numerous papers regarding allelopathic situations on the Australian continent; A. D. Rovira has published several interesting papers on root exudates; and M. Evanari has published many important articles on germination inhibitors.

Notable efforts to determine the mode of action of allelochemicals have been made by F. A. Einhellig, D. E. Koeppe, W. H. Muller, and N. E. Balke. In addition to the pioneer work of Shreiner, Reed, McCalla, and associates, the agricultural implications of allelopathy were also recognized early by K. P. Buchholtz, L. G.Holm, T. Kommedahl, and H. B. Tukey, Jr. Future contributors to our understanding of allelopathy will undoubtedly include natural product chemists who will clarify the qualitative and quantitative aspects of allelochemicals.

Ecological Implications

There has been a wealth of information accumulated demonstrating the role of allelopathy in natural ecosystems. For a comprehensive review, readers should consult the second edition of *Allelopathy* by Rice (1984).

The following sequence of developments is usually found in the establishment of a particular case study. First, evidence from the patterning of an allelopathic plant species shows that competition alone can not explain the sometimes striking phenomena. For example, in intermixed *Salvia leucophylla* and *Artemisia california* invading annual grassland in the Santa Ynez Valley, California, a zone 1- to 2-m wide which is devoid of herbaceous species occurs around the shrub stands (Figure 1.1). Similarly, chamise (*Adenostoma fasiciculatum*) (McPherson and Miller, 1969) and *Eucalyptus camaldulensis* (Del Moral and Muller, 1969) in southern California showed bare zones under and around the trees. Lack of herbaceous plant growth under sycamore (Al-Naib and Rice, 1971), bamboo (Chou and Hou, 1981), and guava (Brown et al., 1983) have similarly been observed. These observations were followed by studies utilizing ex-

Figure 1.1 Zone of inhibition around *Salvia leucophylla* shrubs near Santa Barbara, California. Photograph by C. H. Muller.

clusion methods to show that competition was not the major causative factor. Leachates from the leaves, branches, and roots, extracts from the suspected plant tissues, or leachates or extracts from the soils were often used for the purpose of bioassay, to demonstrate the toxicity of these samples against the germination or seedling growth of the acceptor species. Some of the studies further identified phytotoxic compounds in these samples.

The extensive evidence accumulated on several model cases, such as California chaparral (Chapter 11) and Australian *Eucalyptus* (Chapter 5), contributed to the establishment and popularization of allelopathy. For example, allelopathy has now become a favorite subject at high school science fairs. Of more importance, however, these established cases lead us to speculate that allelopathy is a common ability acquired by the plant kingdom through the course of evolution. What we have been paying attention to could represent only a small fraction of cases in the category of the apparent allelopathy. The nonapparent allelopathy, which may not provoke noticeable symptoms, has been largely unexplored. These interactions may be caused by the marginal but persistent presence of either rhizospheric compounds active against the acceptor species or mycorrhizae and organisms important to the nitrogen cycle. These interactions may have far reaching implications in ecology.

Agricultural Implications of Allelopathy

Detrimental influences of plants on agricultural sites have been observed for centuries. Perhaps the best documented of these are the influences from plant litter or residues. Replant problems are common to agriculture. They are particularly acute with perennial crops (Putnam and Duke, 1978), but also occur with selected annuals under some environmental conditions. The release of toxins from decaying herbage and root tissues has frequently been demonstrated. When plant residues are allowed to remain on the soil surface, as in conservation tillage farming practices, they may reduce the emergence and growth of weeds and/or crops (Putnam and DeFrank, 1983).

There is considerable evidence that a number of weed species may impose interference through allelopathic mechanisms. Aggressive perennials, including quackgrass (*Agropyron repens*), johnsongrass (*Sorghum halepense*), and yellow nutsedge (*Cyperus esculentus*) have all been implicated. Worldwide, about 90 weed species have alleged allelopathic attributes (see Chapter 3).

Weed seeds can lie dormant in the soil for decades and still remain viable (Kivilaan and Bandurski, 1973). During this period they must resist decay by soil microbes. Chemical inhibitors appear to be responsible for both phenomena. Certain lactones, phenolic compounds, flavonoids, and tannins are present in seeds in relatively high concentrations (Rice, 1984).

Weeds may impose stresses on crops through indirect methods. Rice (1984) discovered that a number of weed species can inhibit the growth of the nitrogen fixing bacteria *Azotobacter* and *Rhizobium*. Members of the latter genus are particularly important for nitrogen fixation on a variety of legume crops grown worldwide. More recently, quackgrass has been shown to reduce nodulation and nitrogen fixation on a number of legumes (Weston and Putnam, 1985).

The positive impact of allelochemicals has only been discussed seriously for about the past 10 years. Their primary utility could develop in the pest management area. This approach is discussed in more detail in this chapter and in Chapters 9, 16, and 17.

There are a number of ways that allelopathic plants might be managed. For example, pasture plants could be maintained in proper species composition to provide optimum nutrition for grazing animals. Densities and distributions of forest species might be better managed by understanding their potential for allelopathy or autotoxicity. There is considerable evidence that symbiotic mycorrhizal fungi important to tree species are adversely affected by allelochemicals (Brown and Mikola, 1974).

Toxic compounds from plant residues have often been implicated in poor crop performance. Therefore, a better understanding of which plants release toxins that subsequently inhibit succeeding crops or crop symbionts is needed. This would allow more objective decisions to be made regarding how to rotate crops and manage the residues (tillage vs. no-tillage, etc.). Adsorbants such as activated charcoal might also be utilized to quickly remove soil toxins on smaller areas.

In horticulture, grafting or budding is widely utilized to exploit the benefits of scion, rootstock, or their interaction. For example, apples (*Malus sylvestris*) are now almost exclusively budded to rootstocks that induce dwarfing. Rootstocks can provide resistance to insect, disease, and nematode problems and contribute tolerance to other environmental stresses. Biochemical rejection mechanisms make many interfamily, intergeneric, and even interspecific grafts impossible. Moroz and Popivshchy (1973) suggested that apple rootstocks low in allelochemicals (particularly phlorizin) would overcome much of the autointolerance problem.

Allelopathic plants may also provide a strategy for vegetation management in aquatic systems. The diminutive spikerush (*Eleocharis coloradoensis*) has been reported to displace such vigorous and unwanted aquatic plants as pondweed (*Potamogenton* species) and *Elodea* in canals and drainage ditches. Frank and Dechoretz (1980) attributed this to allelopathic effects.

Allelopathy as a Maturing Scientific Discipline

Although scientific observations of allelopathy began long before Molisch coined the term in 1937 (Rice, 1984), it is in recent years that allelopathy attracted wide attention among scientists. Research articles increased steadily from the early 1960s, and since 1970 the number of scientific papers relating to allelopathy jumped to more than 50 per year on an average basis. The surge of activity included several international meetings, often followed by special journal issues [e.g., *Journal of Chemical Ecology*, 9 (8), 1983] and symposium series, such as *Allelochemicals and Pheromones* (Chou and Waller, 1983) and *The Chemistry of Allelopathy* (Thompson, 1985). In addition, the second edition of Rice's *Allelopathy* was published in 1984, 10 years after the debut of his celebrated first edition.

This flourish of activity suggests that allelopathy is now a maturing scientific discipline. A state of rapid progress may be envisioned based on the existing knowledge and available research techniques, where enrichment of evidence and improvement of techniques, however, are still very much in need. In this stage new research directions should abound, and as a multidisciplinary science, it provides a fertile ground for both biologists and physical scientists to explore new opportunities for relevant research.

We find that as a maturing science allelopathy is already contributing to the solution of practical problems in agriculture and providing explanations for observed plant–plant interactions. We foresee the acceleration of the maturation process through the growing body of scientists joining this area of research.

THE FUTURE OF ALLELOPATHY

Ecosystem Management

Plant succession in old field and cut-over forests has intrigued ecologists for decades. The appearance and disappearance of species and changes in species dominance over time has been attributed to numerous factors including physical changes in the habitat, seed production and dispersal, competition for re-sources, or combinations of these. Rice and co-workers (1984) presented exten-sive evidence that allelopathy may play an important role in the disappearance of the pioneer weeds (those most rapidly invading old fields). Rice and Pancholy (1972) also presented evidence that succession proceeds toward the selection of plants that inhibit nitrification and thus conserve the ammonium form of nitro-gen for their own use. Additional findings in this area could help us manage vegetation more effectively.

Certain reforestation problems have also been linked to allelopathy. There are logged-over sites on the Allegheny Plateau in northwestern Pennsylvania that have remained essentially treeless for up to 80 years (Horsley, 1977). Several herbaceous weed species produce toxins that inhibit the establishment of the black cherry (*Prunus serotina*) seedlings that normally reinfest these sites. Among the more active are goldenrods (*Solidago*) and the *Aster* species.

In many ecosystems, plants tend to pattern themselves as pure stands or as individuals spaced in rather specific densities or configurations. Many desert species show obvious zones of inhibition around which few, if any, alien species are able to invade (Muller et al., 1964). These patterns often cannot be ade-quately explained by competition alone, and are probably caused by a combina-tion of factors including allelopathy. Patterning occurs with herbaceous plants as well as woody shrubs and trees.

Black mustard (*Brassica nigra*) can form almost pure stands after invading the annual grasslands of coastal southern California (Muller, 1969). This was attributed to inhibitors released from the dead stalks and leaves which do not permit the germination and growth of other plants. These observations provide agronomists with hope that almost pure stands of crops (over weeds) could be achieved by the use of an allelopathic mechanism.

Numerous reports (primarily from Eastern Europe) indicate that the growth and yield of certain crops may be increased when they are grown in concert with other species. Although this concept has been virtually ignored in the United States because of large-scale mechanized agriculture, it may have important im-plications for smaller scale or subsistence agricultural systems. A number of le-gumes increase maize (*Zea mays*) growth (Lykhvar and Nazarova, 1970), and the addition of small amounts of white mustard (*Brassica hirta*) or wild helio-trope (*Heliotropium europeaum*) increased the yields of several crops (Melni-chenko et al., 1972; Grechkanev and Rodionov, 1971).

Allelopathic mechanisms are implicated in several of these beneficial associa-tions. Lastuvka (1970) demonstrated that substances secreted by plants can in-fluence ion absorption and accumulation by other plants. Studies with ^{14}C-labeled plants indicated that there is considerable exchange of root excretions between species (Rakhteenko et al., 1973).

Pest Management

The possible use of allelochemicals to defend crop plants against insects, nema-todes, diseases, and weeds has recently received considerable attention. With insects and nematodes one could capitalize on repellants, antifeedants, growth disrupters, or toxicants (Jacobson, 1975). Disease resistance might be achieved through fungistasis or antibiosis (Baker and Cook, 1974). Both living crop plants and crop residues have been screened for inhibition of weed growth (Fay and Duke, 1977; Putnam and DeFrank, 1983).

Initially, one must determine which crop plants have the greatest potential for allelopathy. Perhaps the poorest place to look is in the cultivars that are now grown commercially. The selection of these plants for exceptional seedling vigor and other desirable growth characteristics may have eliminated individuals with the desired allelopathic attributes. The allelopathic potential of our cultivars is not well known, although there are obvious differences in "competitive ability," which in reality has been a measure only of their total interference capability. Allelopathic characteristics are more likely to occur in crop predecessors or "wild types" that have evolved in the presence of allelopathic and competitive influences from other species (Putnam and Duke, 1978).

Recently, a number of weed scientists attempted to directly exploit allelopathy as a weed management strategy. One approach has been to screen for allelopathic types in germ plasm collections of crops. Presumably, this character could be incorporated into a cultivar by conventional plant breeding or other genetic recombination strategies. Superior weed suppressing types have been reported in cucumber (Putnam and Duke, 1974), oats (Fay and Duke, 1977), sunflowers (Leather, 1983), and soybeans (Massantini et al., 1977).

Another approach is the utilization of allelopathic rotational crops or companion plants in annual or perennial cropping systems. Living rye (*Secale cereale*) and its residues have provided good suppression of a variety of weeds (see Chapter 16). Similarly, residues of sorghums, barley, wheat, and oats can provide exceptional suppression of a number of weed species. Some crop plants appear to grow successfully in these residues, whereas others are severely damaged (Figure 1.2; Putnam and DeFrank, 1983). Much more research is needed in this area, which will complement the trend toward conservation tillage production systems.

Two of the major functions of allelopathic compounds in seeds are to prevent seed decay and to control germination. Rice (1984) devoted an entire chapter of his recent book to prevention of seed decay, whereas Evanari (1949) and more recently Taylorson and Hendricks (1977) reviewed germination inhibitors and their role in seed dormancy. Weed seeds may remain viable in the soil for decades. The numbers have been estimated by numerous investigators throughout the world and the reported range is 5.2×10^5 to 3.5×10^9 seeds per hectare (Kropac, 1966). Weed problems in agroecosystems result not only from enormous numbers of propagules, but also from lack of uniformity in their germination.

Major research efforts should be directed toward finding methods to increase weed–seed decay or methods to stimulate or inhibit weed–seed germination. The authors are aware of no successful attempts to enhance weed–seed decay under field conditions, although it can be accomplished in the laboratory by removing

Figure 1.2 Weed suppression under cherry trees in a Michigan orchard using residues of sorghum. The sorghum is grown in the strips betw een the cherry rows and foliage clippings are blown under the tree.

the microbial inhibitors from the seed (Evanari, 1949). The selection of a microorganism able to destroy weed seeds is one possible approach. Another approach is somehow to absorb or inactivate the inhibitor that may be protecting the seed against decay.

Seed dormancy is a complex phenomenon in which several compounds including abscisic acid, gibberellic acid, cytokinins, and ethylene are implicated (Taylorson and Hendricks, 1977). Seeds may germinate as a result of an external chemical stimulus; for example, germination of *Striga asiatica* seeds is enhanced by the compound strigol, which may be exuded by the roots of susceptible host plants. This compound is active at concentrations as low as $10^{-15}M$. Ethylene also enhances germination of *Striga* and is utilized by soil injection to achieve suicidal germination of the weed seed before the host crop plants are present (Eplee, 1975). There are undoubtedly other compounds that could be used in this manner.

Developing Novel Agricultural Chemicals

The modern techniques of agricultural production rely heavily on insecticides, herbicides, and other chemical reagents for crop protection. According to the U.S. Department of Agriculture, our farmers spent 3.5 billion dollars on pesticides in 1982. The increased reliance on chemical control has become an overwhelming economical burden, but perhaps more important, it could pose a serious threat to the public health and the environment. An alarming example has been the contamination of ground water in several states. This situation calls for the development of more environmentally feasible methods of pest control. Among such approaches as integrated pest management and genetic engineering for disease- and pest-resistant crops, the use of natural plant chemicals offers an additional avenue for alleviating the problem. Allelochemicals have been described as the plants' own herbicides (Putnam, 1983). Indeed, if plants have

TABLE 1.2.
Natural products used commercially or under commercial development as pesticides or plant growth regulators

Compound	Use	Source
Avermectin	Nematocide, miticide	*Streptomyces* sp.
Bialaphos	Herbicide	*Streptomyces* sp.
Blasticidin S	Fungicide	*Streptomyces* sp.
Gibberellins	Plant growth regulator	*Gibberella* sp.
Glufosinate	Herbicide	*Streptomyces* sp.
Kasugamycin	Fungicide	*Streptomyces* sp.
Nicotine	Insecticide	*Nicotiana tabacum*
Polyoxin	Fungicide	*Streptomyces* sp.
Quassin	Insecticide	*Quassia amara*
Rethrins	Insecticides	*Chrysanthemum cineraraefolium*
Rotenoids	Insecticides	*Derris* and *Lonchocarpus*
Ryanodine	Insecticide	*Ryania speciosa*
Streptomycin	Fungicide	*Streptomyces* sp.
Tetranactin	Miticide	*Streptomyces* sp.
Triacontanol	Plant growth regulator	Numerous
Unsaturated isobutylamides	Insecticides	*Anacyclus pyrethrum* *Helliopsis longipes*
Validamycin	Fungicide	*Streptomyces* sp.

Sources. Rice (1983) and Misato (1982).

provided us with normal natural products as medicines, it is only reasonable to assume that bioactive chemicals were originally acquired for their own needs.

Through the study of allelopathy, it becomes clear that allelochemical modifications by plants play an important role in natural and agricultural ecology. The allelochemicals, therefore, should serve as the model for future herbicides if environmental compatibility is a required feature. Novel agricultural chemicals may be synthesized based on the structure and characteristics of these compounds. Synergistic or antagonistic compounds may be designed to amend the soil, thus enhancing or suppressing the existing allelopathic interaction in the interest of crop management.

At present the number of natural products utilized as pesticides is limited (Table 1.2) and many are restricted to use in Japan. Allelopathy is a maturing science with its weakness in chemistry, and our knowledge often has not reached the stage of application. In addition, compared to the synthetic agricultural chemicals, the cost of procurring natural products is often prohibitively high. This limitation may be alleviated by advances in biotechnology. Methods for cell and tissue culture would provide continuous production processes for the industrial scale operation, and the plant-derived chemicals can be expected to play an increasingly significant role in agricultural and ecological management.

REFERENCES

Abdul-Wahab, A. S. and E. L. Rice. 1967. *Bull. Torrey Bot. Club.* 94:486.

Al-Naib, F. A. and E. L. Rice, 1971. *Bull. Torrey Bot. Club.* 98:75.

Baker, K. F. and J. R. Cook. 1974. *Biological Control of Plant Pathogens.* Freeman, San Francisco. 483 pp.

Balandrin, M. F., J. A. Klocke, E. S. Wurtele, and W. H. Bollinger. 1985. *Science.* 228:115.

Barnes, J. P. and A. R. Putnam. 1983. *J. Chem. Ecol.* 9:1045.

Barnes, J. P. and A. R. Putnam. 1986. *J. Chem. Ecol.* (In press.)

Bell. D. T. and D. E. Koeppe. 1972. *Agron. J.* 64:321.

Brown, R. L., C. S. Tang, and R. K. Nishimoto. 1983. *Hort. Science.* 18:316.

Brown, R. T. and P. Mikola. 1974. *Acta Forestry Fenn.* 141:1.

Campbell, G, J. D. H. Lambert, T. Arnason, and G. H. N. Towers. 1982. *J. Chem* Ecol. 8:961.

Chou, C. H. and M. H. Hou. 1981. In Annual Report for July, 1980-June, 1981, pp. 10–11 Inst. of Botany, Academia Sinica, Taipei, Taiwan.

Chou, C. H. and G. R. Waller (eds.). 1983. *In Allelochemicals and Pheromones.* Institute of Botany, Academia Sinica, Taipei, Taiwan.

Davis, E. F. 1928. *Am. J. Bot.* 15:620.

Del Moral, R. and C. H. Muller. 1969. *Bull. Torrey Bot. Club.* 96:467.

DeWit, C. T. 1961. *Symp. Soc. Exp. Biol.* 15:314.

Einhellig, F. A., M. K. Schon, and J. A. Rasmussen. 1982. *J. Plant Growth Reg.* 1:251.

Eplee, R. E. 1975. *Weed Sci.* 23:433.

Evanari, M. 1949. *Bot. Rev.* 15:153.

Fay, P. K. and W. B. Duke. 1977. *Weed Sci.* 25:224.

Fisher, R. F. 1978. *Soil Sci. Soc. Am. J.* 42:801.

Frank, P. A. and N. Dechoretz. 1980. *Weed Sci.* 28:499.

Fuerst, E. P. and A. R. Putnam. 1983. *J. Chem. Ecol.* 9:937.

Grechkanev, O. M. and V. I. Rodionov. 1971. *In* A. M. Grodzinsky (ed.), *Biochemical and Physiological Bases for Plant Interaction in Phytocenosis.* Naukova Dumka, Kiev (in Russian with English summaries). 2:94.

Grummer, G. 1955. *Die gegenseitige Beeinflussung hoherer Pflanzen-Allelopathie.* Fischer, Jena. 162 pp.

Harper, J. L. 1977. *Population Biology of Plants.* Academic, New York. 892 pp.

Horsley, S. B. 1977. *Can. J. Forestry.* 7:205.

Jacobson, M. 1975. *Insecticides from Plants. A Review of the Literature.* 1954–1971. ARS Handbook. Washington, D.C. GPO. 461 pp.

Kaminsky, R. and W. H. Muller. 1977. *Soil Sci.* 124:205.

Kivilaan, A. and R. S. Bandurksi. 1973. *Am. J. Bot.* 60:146.

Kropac, Z. 1966. *Pedobiologia* 6:105.

Lastuvka, Z. 1970. *In* A. M. Grodzinsky (ed.), *Biochemical and Physiological Bases for Plant Interaction in Phytocenosis.* Naukova Dumka, Kiev (in Russian with English summaries). 1:37.

Leather, G. R. 1983. *Weed Sci.* 31:37.

Lykhvar, D. F. and N. S. Nazarova. 1970. *In* A. M. Grodzinsky (ed.), *Biochemical and Physiological Bases for Plant Interaction in Phytocenosis.* Naukova Dumka, Kiev (in Russian with English summaries). 1:83.

Massantini, F., F. Caporali, and G. Zellini. 1977. *Symp. Different Methods Weed Control Integration.* 1:23.

McCalla, T. M. and F. A. Haskins. 1964. *Bacteriol. Rev.* 28:181.

McPherson, J. K. and C. H. Muller. 1969. *Ecol. Monograph.* 39:177.

Melnichenko, A. N., O. Grechkanyov, and V. Rodionov. 1972. *In* A. M. Grodzinsky (ed.), *Biochemical and Physiological Bases for Plant Interaction in Phytocenosis.* Naukova Dumka, Kiev (in Russian with English summaries). 3:42.

Misato, T. 1982. *In Pesticide Chemistry: Human Welfare and the Environment.* Proc. 5th Int. Congr. Pest. Chem. Kyoto. 231 pp.

Molisch, H. 1937. *Der Einfluss einer Pflanze auf die andere-Allelopathie.* Fischer, Jena.

Moroz, P. A. and I. I. Popivshchy. 1973. *In* A. M. Grodzinsky (ed.), *Biochemical and Physiological Bases for Plant Interaction in Phytocenosis.* Naukova Dumka, Kiev (in Russian with English summaries). 4:75.

Muller, C. H. 1965. *Bull. Torrey Bot. Club.* 92:38.

Muller, C. H. 1966. *Bull. Torrey Bot. Club.* 93:332.

Muller, C. H. 1969. *Vegetatio.* 18:348.

Muller, C. H., W. H. Muller, and B. L. Haines. 1964. *Science.* 143:471.

Pope, D. F., A. C. Thompson, and A. W. Cole. 1985. *In* A. C. Thompson (ed.), *The Chemistry of Allelopathy.* American Chemical Society, Washington, D.C.

Putnam, A. R. 1983. *Chem. Eng. News.* 61:34.

Putnam, A. R. 1985. Weed Allelopathy. *In* S. O. Duke (ed.), *Weed Physiology.* CRC Press, Boca Raton, FL. 1:131.

Putnam, A. R. and J. DeFrank. 1983. *Crop Protect.* 2:173.

Putnam, A. R. and W. B. Duke. 1974. *Science.* 185:370.

Putnam, A. R. and W. B. Duke. 1978. *Annu. Rev. Phytopathol.* 16:431.

Rakhteenko, I. N., I. A. Kaurov, and I. F. Minko. 1973. *In* A. M. Grodzinsky (ed.), *Biochemical and Physiological Bases for Plant Interaction in Phytocenosis.* Naukova Dumka, Kiev (in Russian with English summaries). 4:19.

Rice, E. L. 1964. *Ecology.* 45:824.

Rice, E. L. 1983. *Pest Control with Nature's Chemicals.* University of Oklahoma Press, Norman. 224 pp.

Rice, E. L. 1984. *Allelopathy,* Second Edition. Academic, Orlando, Fl..

Rice, E. L. and S. K. Pancholy. 1972. *Am. J. Bot.* 59:1033.

Sherma, J. 1979. *Manual of Analytical Quality Control for Pesticides and Related Compounds.* U.S. Environmental Protection Agency, Office of Research and Development, Research Triangle Park NC. pp. 186–188.

Stevens, G. and C. S. Tang. 1985. *J. Chem. Ecol.* 11:1411.

Szezepanski, A. J. 1977. *Aquatic Bot.* 3:193.

Tang, C. S. and C. C. Young. 1982. *Plant Physiol.* 69:155.

Taylorson, R. B. and S. B. Hendricks. 1977. *Ann. Rev. Plant Physiol.* 28:331.

Thompson, A. C. (ed.). 1985. *The Chemistry of Allelopathy.* American Chemical Society, Washington, D.C.

Weston, L. A. and A. R. Putnam. 1985. *Crop Sci.* 25:561.

Whittaker, R. H. and P. P. Feeny. 1971. *Science.* 171:757.

Williams, R. D. and R. E. Hoagland. 1982. *Weed Sci* 30:206.

Young, C. C. and D. P. Bartholomew. 1981. *Crop Sci.* 21:770.

Zimdahl, R. L. 1980. *Weed-crop Competition: A Review.* Intl. Plant Protect. Center, Coravallis, OR. 195 pp.

FIELD OBSERVATIONS OF ALLELOPATHY AND AUTOTOXICITY

2

ALLELOPATHIC GROWTH STIMULATION

ELROY L. RICE

Department of Botany and Microbiology,
University of Oklahoma,
Norman, Oklahoma

When Molisch (1937) coined the word allelopathy, he made it clear in his discussion that he meant the term to include stimulatory as well as inhibitory biochemical effects of one plant or microorganism on another. Unfortunately, few scientists involved in allelopathic research have reported examples of stimulatory effects. However, numerous instances of stimulatory effects of microorganisms on other microorganisms and of plants on microorganisms have been reported by scientists who were not consciously working on allelopathic interactions. Some were working on nitrogen-fixing organisms and many were studying plant diseases.

The goal of this chapter is to review the literature on stimulatory allelopathic effects and to present a new example of a striking stimulatory effect of one plant on another.

MICROBIAL GROWTH STIMULATION BY MICROORGANISMS

Bacterial Stimulation by Bacteria

In spite of the voluminous literature on inhibitory effects of bacteria (including the actinomycetes) on other bacteria, there appear to be few reports of stimulatory effects. Leuck and Rice (1976) investigated the effects of 28 bacterial isolates from the rhizosphere of *Aristida oligantha* on three strains of *Rhizobium* and three of *Azotobacter*. Media in which the isolates grew separately were used to condition media in which the nitrogen fixers were grown. Several of the rhizosphere isolates inhibited growth of the test organisms, but only two stimulated growth of any of the test bacteria. *Bacillus cereus* stimulated growth of *Rhizobium* sp. (ATC 10703) and *B. megaterium* stimulated growth of the same strain and of *R. japonicum* (ATC 10324).

Fungal Stimulation of Bacteria

Mallik and Hussain (1972) isolated 41 fungi, 24 actinomycetes, and 66 bacteria from the rhizosphere of *Melilotus alba* and tested these for inhibitory or stimulatory effects against two isolates of *Rhizobium meliloti*. Many of the rhizosphere organisms were inhibitory to growth of the isolates, but only three species of fungi stimulated their growth (Table 2.1).

Hussain and Mallik (1972) isolated 39 species of fungi and 18 distinct bacteria from the rhizosphere of *Trifolium alexandrinum*. They tested all the bacteria and 28 species of fungi against growth of *Rhizobium trifolii* isolated from the nodules of the *Trifolium*. Six species of fungi stimulated growth of the *Rhizobium*, but no bacteria were stimulatory (Table 2.2).

Algal Stimulation of Bacteria

There have been many reports of bacterial inhibition by allelopathic compounds from algae (Rice, 1984), but few reports have been found of bacterial stimulation by algae. Filtrates from cultures of *Anabaena cylindrica, Chlorella pyrenoidosa, C. vulgaris, Nitzschia palea*, and *Scenedesmus quadricauda* stimulated or inhibited growth of *Staphylococcus aureus* in many tests (Jørgensen and Steemann-Nielsen, 1961). Both growth-accelerating and growth-inhibiting sub-

TABLE 2.1

Stimulation of two isolates of *Rhizobium meliloti* by factors in a culture medium of fungi[a]

Fungi	Optical Density	
	Isolate 1	Isolate 2
Control—no metabolities	0.270	0.270
Aspergillus terreus	0.470	0.500
Mucor sp.	0.380	0.375
Penicillium sp.	0.345	0.350

[a]Source. Mallik and Hussain (1972).

TABLE 2.2
Stimulation of growth of *Rhizobium trifolii* by factors in a culture medium
of several fungi[a]

Fungi	Number of Rhizobial Cells ($\times 10^6$)
Control—no metabolities	77
Aspergillus candidus	99
A. nidulans	87
A. ustus	98
Fusarium sp.	125
Penicillium funiculosum	83
Rhizopus sp.	135

Source. Hussain and Mallik (1972).

stances were present simultaneously in the culture solution. Growth of the soil bacterium *Bacillus* sp. was markedly stimulated by the filtrate of a green alga *Hormotila blennista* at a pH of 6.3 (Monahan and Trainor, 1970). Stimulation did not occur at pH 7.7, however.

Stimulation of Fungi by Bacteria

Phytophthora cinnamomi requires stimulation by soil bacteria (*Pseudomonas* spp., *Chromobacterium violaceum*) for profuse production of zoosporangia (Zentmyer, 1965). The stimulatory effect is eliminated from the soil by aerated steam treatment (40–50°C/10 min) or by Millipore filtration of the soil extract. Ayers (1971) showed that the chemical that stimulates zoosporangial formation is nonvolatile, heat stable, water soluble, and effective in high dilution (10^{-9}).

Bacterial Stimulation of Algae

Two strains of the marine bacterial species *Vibrio anguillarum* enhanced the growth of most of 10 species of phytoplankton algae on an enriched agar medium, but did not stimulate growth in an artificial seawater medium (Ukeles and Bishop, 1975). Subsequent studies suggested that the stimulation on agar might have been due to growth substances released by bacterial hydrolysis of the agar.

Fungal Stimulation of Fungi

Spores of most parasitic fungi remain ungerminated while located in their site of production (Bell, 1977). This can be due to several factors, one of which is production by the spores of fungistatic agents that are excreted into the water around the spores. These self-inhibitors generally assure dispersal of viable ungerminated spores. Endogenous germination stimulators that counteract inhibition by self-inhibitors occur in many spores. Nonanal and 6 methyl-5-hepten-2-one were isolated from uredospores of *Uromyces* and *Puccinia* (French et al., 1977). These compounds stimulate germination of stem rust spores that contain methyl ferulate as their inhibitor, but they do not stimulate germination of spores that contain dimethoxy cinnamate as their inhibitor.

Most parasites have to survive prolonged periods of time apart from the host plant. Consequently, the formation of resting propagules, such as sclerotia, constitutes a critical part of the parasite's life cycle. Several observations indicate that the formation of sclerotia may be stimulated by allelochemics (Chet and Henis, 1975; Willetts, 1972). Brandt and Reese (1964) concluded that *Verticillium dahliae* produces a diffusible morphogenetic factor which stimulates the production of microsclerotia. When low concentrations of the diffusable factor were added to cultures of the pathogen, the hyphae swelled and became constricted, septation was increased, and cell walls became thickened. A morphogenetic factor responsible for sclerotia formation is apparently produced by *Sclerotium rolfsii* also (Willetts, 1972). This factor stimulates branching in lateral positions where sclerotia later develop.

Phenols and phenol oxidases have been implicated in differentiation of reproductive structures in fungi (Willets, 1972). Enhanced phenol oxidase activity has been closely correlated with differentiation of conidiophores, protoperithecia, and basidiocarps in various fungi.

Algal Stimulation of Algae

According to Berglund (1969), various investigators on the Swedish west coast demonstrated that certain green algae grew better when cultivated in sea water taken from the *Fucus-Ascophyllum* zone than when cultivated in surface water collected 100 m from shore. Growth was poor also in sea water taken from a depth of 30 m, even after enrichment with nutrients. Growth was stimulated in free surface water and in deep water if certain living algae (*Ulva, Chorda, Ceramium, Ascophyllum, Chondrus*, and *Fucus*) were previously placed in the water for 24 hours. Thus Berglund decided to determine if a particular green alga, *Enteromorpha linza*, produces substances stimulatory to the same alga and to *Enteromorpha* sp. Axenic *E. linza* was grown in an artificial medium for an unspecified period of time, the alga was removed, and one part of this medium was added to four parts of a fresh nutrient medium. The growth of each species in the amended medium was compared with that in a fresh nutrient medium and was found to be markedly stimulated. Water soluble and water insoluble fractions were isolated from the medium in which *E. linza* had grown and both fractions stimulated growth of both species of *Enteromorpha*. The fractions were not identified.

The filtrate of the green alga *Hormotila blennista* was found to be autostimulatory (Monahan and Trainor, 1970). It was also stimulatory to one strain of *Scenedesmus* at pH 6.3. Monahan and Trainor (1971) reported that acid, basic, and volatile acid filtrate extracts of *H. blennista* reduced the lag time of this species at low concentrations. Whole filtrate did not affect the lag time, however. Stimulatory properties of the filtrate were dialyzable and heat labile. They suggested that heat labile, low molecular weight–organic extracellular products were responsible for the stimulatory property of the filtrate.

Keating (1977) reported convincing evidence of the direct role of algal allelopathy in algal succession in a eutrophic lake, Linsley Pond in Connecticut. She studied the bloom sequence over a period of 3 yr and correlated this with the effects of cell-free filtrates of dominant blue-green algae on both their successors and predecessors. She found an unbroken correspondence between the effects of heat-labile stimulatory and inhibitory filtrates and the rise and fall of bloom populations in situ. Keating used only axenic or unialgal (bacterized) isolates from Linsley Pond in all tests. She also collected water samples from the pond before, during, and after bloom maxima, and found that heat-labile effects in these samples correlated well with data obtained from filtrate studies and with the natural sequence in the pond.

Wolfe and Rice (1979) investigated the possible allelopathic interactions of five planktonic green algae isolated from the Cleveland County, Oklahoma area and one yellow-green obtained commercially. Sterile filtrates of axenic cultures were tested in all possible combinations. Numerous instances of significant stimulation or inhibition were observed. Filtrates of *Cosmarium vexatum* significantly inhibited growth of the other five test species in most experiments, but stimulated growth of *C. vexatum*. This combination is probably important in the common production of waterblooms by *C. vexatum* in ponds and swamps. *Pandorina morum* was stimulated in three sterile filtrates from cultures of *Scenedesmus incrassatulus* var. *mononae* with the oldest filtrate showing the least stimulation.

GROWTH STIMULATION OF PLANTS BY MICROORGANISMS

The bakanae or foolish seedling disease of rice, in which seedlings infected with the fungus *Gibberella fujikuroi* grow rapidly and become much taller than healthy plants, is apparently caused by the gibberellin secreted by the fungus (Agrios, 1969).

Youssef and Mankarios (1974) found that each of the six major rhizosphere fungi of cotton stimulated cotton growth when applied as seed treatments. With broad bean, four of the rhizosphere fungi stimulated growth. Culture filtrates from all the same rhizosphere fungi stimulated growth of both cotton and broadbean (Youssef, 1974).

GROWTH STIMULATION OF MICROORGANISMS BY PLANTS

Stimulation of Bacterial and Fungal Growth

As the root system develops in the soil, organic and inorganic compounds exuded from the roots stimulate growth of bacteria and fungi (Hoagland and Williams, 1985). The zone of soil in which the microflora is influenced by the plant root is called the rhizosphere. Bacteria and fungi associated with the rhizosphere may be either attached to the root surface (rhizoplane) or present in the soil surrounding the root. Mycorrhizal fungi grow on the root surface and also penetrate the cortex of the root and develop a symbiotic relationship with the host plant.

Quantitative and qualitative determinations of fungi at various distances from healthy roots of blue lupine seedlings indicated that certain species were preferentially stimulated by the roots, whereas others were apparently unaffected (Papavizas and Davey, 1961). Numbers of bacteria per gram of soil decreased markedly with increased distance from the roots out to about 80 mm.

Kerr (1956) grew sterile seedlings inside cellophane bags buried in soil inoculated with *Pellicularia filamentosa* and found an intense development of the pathogen on the cellophane opposite the roots of the two susceptible hosts, lettuce and radish, but no stimulation opposite tomato roots, which are not susceptible. Buxton (1957) also demonstrated a definite specificity in relation to the germination of spores of *Fusarium oxysporum* f. *pisi* in the exudates of three pea varieties differing in susceptibility to this pathogen. Exudate from a wilt-resistant variety inhibited spore germination, whereas exudate from a susceptible plant stimulated spore germination. There are many other reports of the stimulation of fungal growth and development by factors in plant root exudates, a few of which are listed in Table 2.3.

In soil that has not had recent additions of plant residue or other organic material, microbial respiration proceeds at a low rate (Menzies and Gilbert, 1967). Moreover, fungi apparently exist mostly as spores in a state of fungistasis. This microflora usually responds to the addition of plant residue by spore germination, increased respiration and growth. Menzies and Gilbert (1967) reported that these responses were induced by a volatile component from alfalfa tops, corn leaves, wheat straw, bluegrass clippings, tea leaves, and tobacco leaves, even when the residue was separated from the soil by a 5-cm air gap. There was a rapid outgrowth of hyphae from the soil surface toward the residue before any growth of fungi could be seen in the plant material. A dense network of hyphae could be detected within 24 hours, with many of the filaments oriented at right angles to the soil surface and reaching almost across the air gap. This development did not occur where the residue was omitted or replaced by moist filter paper.

Vapors from distillates of water extracts of the various plant residues had similar effects on growth of fungi and markedly increased numbers of bacteria and the respiratory rate of microorganisms in soil samples (Menzies and Gilbert, 1967). This fascinating allelopathic phenomenon may be very important in the initial colonization of residue, and thus in decomposition.

Water soluble extracts of the litter of four shrub and three conifer species had variable effects on the growth of four species of ectomycorrhizal fungi (Rose et al., 1982). In general, low concentrations stimulated fungal growth, whereas high concentrations had variable effects.

TABLE 2.3
Factors in root exudates stimulatory to fungal growth

Plant Exudate	Fungus Affected	Stage Stimulated	Reference
Tomato, wheat	*Verticillium albo-atrum*	Microsclerotia germination	Schreiber and Green (1962)
Strawberry	*Phytophthora fragariae*	Zoospore attraction	Goode (1956)
Peas	*P. erythroseptica*	Zoospore attraction	Bywater and Hickman (1959)
Solanaceae	*P. parasitica*	Zoospore attraction	Dukes and Apple (1961)
Potato	*Spongospora subterranea*	Spore germination	White (1954)
Tomato	*Colletotrichum atramentarium*	Spore germination	Ebben and Williams (1956)
Peas	*Aphanomyces eutiches*	Spore germination	Scharen (1960)
Turnip	*Pythium mamillatum*	Spore germination	Barton (1957)
Pea, wheat, lettuce	*Gliocladium roseum, Fusarium* spp., *Paecilomyces marquandii*		Jackson (1957)
Beans	*F. solani* f. *phaseci*	Spore germination	Schroth and Snyder (1961)
Strawberry	*Rhizoctonia* sp.	Mycelium growth	Husain and McKeen (1962)

29

Decapitation of inoculated *Trifolium fragiferum* 2 days after germination almost completely eliminated root hair infection by *Rhizobium trifolii*, whereas decapitation 3 days after germination had no effect (Nutman, 1965). These results indicated that some factor necessary for infection thread initiation comes from the cotyledons or plumule, according to Nutman.

Stimulation of Algal Growth

Small amounts (0.2 g/15 ml of Bristol's Solution) of dried plant parts from several species important in the early stages of old-field succession markedly stimulated the growth of two species of blue-green algae (Parks and Rice, 1969) (Table 2.4). Even greater stimulation resulted from the addition of 0.1 g of dried plant material to 15 ml of Bristol's Solution.

Raindrip (artificial rain) collected from leaves of the same species listed in Table 2.4 markedly stimulated growth of the same two algae in most tests. Root exudates from several plant species important in the early stages of old-field succession also strongly stimulated growth of algae present in a soil inoculum from an old field and from the prairie (Table 2.5).

Van Aller et al. (1985) observed that commercial aquaculture ponds that had dense growths of aquatic plants such as *Najas* sp. and *Eleocharis* sp. usually contained fewer algal cells per liter. Moreover, there was a more normal diversity of green algae and diatoms in such ponds, with blue-green algae present only in small numbers. They subsequently found that the aquatic plants *Potamogeton* sp., *Najas* sp., *Thalassia* sp., and *Ruppia* sp. contained numerous oxygenated fatty acids. They identified two of the most prominent ones and found those fatty acids in pond water also. The oxygenated fatty acids were quite inhibitory to algae and particularly to blue-green algae in relatively low concentrations. However, a mixture of the oxygenated fatty acids from *Eleocharis microcarpa*, in a concentration of less than 1 ppm, stimulated the growth of many algal species.

GROWTH STIMULATION OF PLANTS BY PLANTS

Stimulation of Parasitic Plants

Parasitic plants occur in several families and most, but not all, are photosynthetic and capable of maturing to seed set without a host (Lynn, 1985). It is apparently rare, however, that field collections reveal plants that are devoid of attachments. Germination appears to function as one level of chemical recognition in host selection. The developmental feature, however, that is uniquely common to the parasitic angiosperms is the haustorium (Lynn, 1985). This organ forms the physiological and morphological attachment between host and parasite.

Witchweed (*Striga asiatica*) is an economically important root parasite affecting many warm season grasses, including such important crop plants as corn,

TABLE 2.4
Stimulation of algal growth by dried plant material[a,b]

Helianthus annuus	*Erigeron canadensis*	*Ambrosia psilostachya*	*Sorghum halepense*	*Chenopodium album*	*Aristida oligantha*	*Andropogon scoparius*
Parts Stimulatory to Lyngbya sp. (Indiana Culture Collection #488)						
			Inflorescence Roots	Leaves Roots	Leaves Stems	Roots
Parts Stimulatory to Anabaena sp. (Indiana Culture Collection # B380)						
Stems	Stems	Stems	Inflorescence Leaves Roots	Leaves	Leaves Stems	

[a] ANOVA test.
[b] *Source.* Parks and Rice (1969).

TABLE 2.5
Stimulation of growth of algae from soil inoculum by plant root exudates[a]

Plant Exudate	Chlorophyll a (mg/ml)
Control	6.1
Ambrosia psilostachya	22.2[b]
Andropogon scoparius	21.2[b]
Aristida oligantha	10.6[b]
Chenopodium album	10.5[b]
Erigeron canadensis	16.4[b]
Helianthus annuus	10.4[b]
Rhus glabra	13.5[b]
Sorghum halepense	10.1[b]

[a]*Source*. Parks and Rice (1969).
[b]Difference from control significant below 0.05 level (ANOVA test).

grain sorghum, and sugar cane. The viable witchweed seeds may remain dormant in the soil for many years (Pepperman and Blanchard, 1985). The seeds will usually not germinate unless pretreated in a warm, moist environment for several days before exposure to a chemical compound exuded from the roots of a host plant or some nonhost plants. One such compound, strigol, was isolated from the root exudate of cotton and has proved to be a powerful stimulant of witchweed seed germination.

Johnson, Rosebery, and Parker (1976) reported the synthesis and testing of several analogs of strigol, and some were powerful seed germination stimulants for species for both *Striga* and *Orobanche*. Cotton is not a host plant for *Striga* or *Orobanche*, and it is noteworthy that the structures of the germination stimulants exuded from the roots of the host plants remain unknown (Dailey and Vail, 1985).

The haustoria of the parasites do not form when the plants are grown axenically but are rapidly induced in the presence of the host roots or host root exudates (Lynn, 1985). Several haustorial inducing compounds have now been characterized. Xenognosin A and B were identified in gum tragacanth, an exudate of *Astragalus gummifer*, and soyasapogenol B was identified in roots of *Lespedeza sericea*. Activity of the soyasapogenol B was considerably enhanced by the presence of another weakly active fraction which was not identified.

Stimulation of Nonparasitic Plants

Considering the potential economic importance of this topic, surprisingly few reports have appeared in the literature. Neill and Rice (1971) found that rhizosphere soil from *Ambrosia psilostachya* (western ragweed) markedly stimulated the growth of several species of plants that occurred in the same field (Table

TABLE 2.6
Stimulation of plant growth by rhizosphere soil from *Ambrosia psilostachya*
collected in field in July[a]

	Mean Dry Weight (mg) with Standard Error	
Test Species	Control	Test
Amaranthus retroflexus	42 ± 8.0	95 ± 10.0[b]
Andropogon ternarius	25 ± 1.4	33 ± 2.1[b]
Bromus japonicus	22 ± 1.1	46 ± 3.3[b]
Digitaria sanguinalis	56 ± 6.2	117 ± 7.7[b]
Leptoloma cognatum	20 ± 1.9	36 ± 1.8[b]
Rudbeckia hirta	16 ± 0.8	25 ± 1.6[b]
Tridens flavus	27 ± 1.6	43 + 3.2[b]

[a]The control soil was collected in the same field at least 1 m away from the *A. psilostachya* plants.
Source. Neill and Rice (1971).
[b]Difference from control significant at 0.05 level or better.

2.6). This indicated that some stimulating factor was present in the rhizosphere of western ragweed which did not occur in the soil at least 1 m away from the ragweed. Subsequent experiments indicated that growth of two of 10 test species was stimulated by the root exudate of western ragweed, four of the same test species were stimulated by rain drip (leaf leachate) of western ragweed, and four of the test species were stimulated by dried leaves of western ragweed (2 g/kg soil).

Gajić and her colleagues published a series of papers concerning the allelopathic actions of corn cockle (*Agrostemma githago*) on wheat (Gajić, 1966; Gajić and Vrbaški, 1972; Gajić and Nikočević, 1973; Gajić et al., 1976). Field tests over a period of years demonstrated that wheat grain yields were increased appreciably when grown in mixed stands with corn cockle as compared with pure stands of wheat. Sterile seedlings of corn cockle stimulated the growth of sterile wheat seedlings on agar, thus demonstrating that an excreted compound was involved. One of the stimulatory substances isolated from corn cockle was named agrostemmin, and Gajić et al. (1976) reported that application of agrostemmin to wheat fields at the rate of 1.2 g/ha increased grain yields of wheat on both fertilized and unfertilized areas. Moreover, the free tryptophan content of the grains was increased, and the quality of the resulting flour and bread were improved. Thus the evidence is substantial that the stimulating effect of corn cockle on wheat yield in mixed stands is an allelopathic effect.

Chopped alfalfa added to soil stimulated the growth of tomato, cucumber, lettuce, and several other plants (Ries et al., 1977). The stimulatory allelochemical was identified as triacontanol. Subsequent tests with this compound have given variable results, but addition of calcium or lanthanum salts to the triacontanol solution appears to make the stimulatory activity consistent (Maugh, 1981). A steroid, brassinolide, was isolated from rape and alder (*Alnus*) pollen and 1 ng applied to a bean plant caused significant growth stimulation (Maugh, 1981).

Russian knapweed (*Centaurea repens*) is a perennial herb which is rapidly becoming a major threat in many parts of the United States and Canada. Yellow starthistle (*Centaurea solstitialis*), a noxious annual, has become well established in many western rangelands of the United States and often leaves the rangeland unproductive. Fletcher and Renney (1963) found that the growth of tomato and barley was inhibited in soil naturally infested with knapweed previously or in soil with powdered knapweed residues added. An inhibitor was isolated that was soluble in both water and ether but was not identified.

Stevens and Merrill (1985) subsequently identified numerous sesquiterpene lactones and two chromenes from Russian knapweed and yellow starthistle. Of the sesquiterpene lactones isolated, acroptilin, repin, solstitiolide, and centaurepensin stimulated root elongation of lettuce (Black Simpson) at 10 ppm and inhibited it at higher concentrations. The chromenes had no significant biological activity.

Residues from water and methanol extracts (whole plant apparently) of 90 weed and crop species were dissolved separately in distilled water and diluted to concentrations of 3, 30, and 300 ppm for bioassays (Nicollier, Pope, and Thompson, 1985). The extracts were tested against root and shoot growth of Purple Top turnip. Extracts of six species significantly stimulated root growth of turnip and extracts of 18 species significantly retarded root growth. Active compounds were identified from several of the species found to have allelopathic potential.

Some gardeners have told me that ground-ivy (*Glechoma hederacea*) takes over their lawns and gardens. This suggested that ground-ivy might be allelopathic in addition to being very competitive because of its creeping growth form. I decided, therefore, to investigate the allelopathic potential of this species.

MATERIALS AND METHODS

The reported interference of ground-ivy against lawn and garden plants caused me to select a grass and a garden species as test plants. *Bromus tectorum* (downy brome) was selected as the grass species and *Raphanus sativus* (radish—Burpee's Champion) was chosen as the garden species.

Effects of Volatile Compounds on Test Plants

Ground-ivy has a distinctive odor, as do many species of the mint family; thus the first tests concerned the possible effects of volatile compounds on the test species. Muller's (1966) sponge bioassay method was used. Two test chambers and two control chambers were utilized for each test species, and 50 seeds of either downy brome or radish were placed on filter paper on a moist sponge in each chamber. A beaker with 2 g of fresh ground-ivy leaves was placed in each test chamber and an empty beaker was placed in each control chamber before sealing. The chambers were then placed in an incubator at 24°C. The germination percentage and radicle length were measured after 2 days in radish and after 4 days in downy brome.

Effects of Decaying Leaves

Ten radish or 20 downy brome seeds were planted in glazed pots containing either 2 g air-dried ground-ivy leaves (collected in July after flowering) per kilogram of sandy loam soil (test) or 2 g of cellulose per kilogram of soil in the control pots to keep the organic matter content the same. Eight test and eight control pots were used for each species in this experiment and in the second one concerned with decaying leaves.

The pots were placed in a growth chamber on a 16-hr day at a light intensity of 260μ einsteins/m^2/sec and a temperature of 27°C, and an 8-hr dark period at 18°C. The pots were watered when necessary with distilled water. Germination was recorded after 10 days, at which time the plants were thinned to four per pot in radish and five in downy brome. The plants were allowed to grow for another 15 days, at which time they were harvested, separated into roots and tops, and compared on the basis of oven-dry weights.

The experiment was repeated (No. 2 in tables) with several modifications. Ground-ivy leaves were collected in May during the flowering period and they were much smaller. The same sandy loam soil was used but N, P, and K were added at the rate of 60 lb/a of each, based on the weight of soil to the depth of plowing. Perlite was added to the soil in a ratio of 1:3 (v/v) to help prevent hardening of the soil surface in pots. The growing conditions were the same except the day temperature was increased to 29.5°C and the night temperature to 21°C.

Effects of Root Exudates

U-tubes (Tubbs, 1973) made from Pyrex tubes 2.5 cm in diameter and 56 cm long were painted with aluminum paint to exclude light. Ten test and ten control tubes were used for each test species and were filled with aerated Hoagland's nutrient solution. In the test series, test seedlings of uniform shoot and root length were placed in one end of the U-tubes, one seedling per tube, and a ground-ivy plant was placed in the other end of each tube. In the control series, one test seedling was placed in each end of each U-tube. Roots of all plants were inserted through holes in cork stoppers and the plants were held in place with cotton. The plants were grown in the same conditions as those in the first experiment concerned with the effects of decaying leaves. The nutrient solution was replenished each day in each U-tube. Radish seedlings were allowed to grow for 14 days and downy brome for 17 days after they were placed in the U-tubes. They were then harvested, separated into roots and tops, and compared on the basis of oven-dry weights.

RESULTS

Effects of Volatile Compounds

Volatile compounds from ground-ivy leaves did not significantly affect seed germination or growth in length of radicles of either downy brome or radish (Table 2.7).

Effects of Decaying Leaves

Decaying ground-ivy leaves decreased slightly the numbers of seed of both downy brome and radish that germinated in all experiments (Table 2.8). The only statistically significant reduction, however, was against downy brome in experiment 2.

The decaying ground-ivy leaves had a remarkable stimulatory effect on root and shoot growth of both downy brome and radish (Table 2.9). The stimulation of radish growth was considerably greater in each experiment than that of downy brome. Additionally, growth stimulation of both species was considerably less under the conditions of experiment 2 than under those of experiment 1.

Effects of Root Exudates

Root exudates of ground-ivy significantly stimulated both shoot and root growth of radish, but inhibited shoot and root growth of downy brome (Table 2.10). The effect on growth of the radish roots was striking in that the roots attained table size in only 14 days and split most of the corks through which they were originally inserted.

TABLE 2.7

Effects of volatiles from ground-ivy leaves on seed germination and radicle length of downy brome and radish

	Germination (%)		Mean Radicle Length with SD (mm)	
Test Species[a]	Control	Test	Control	Test[b]
Downy brome	97	99	11.2 ± 10.2	11.3 ± 8.3
Radish	99	99	7.6 ± 4.3	8.1 ± 4.1

[a]One hundred control and 100 test seeds of each species used.
[b]Differences from control not statistically significant.

TABLE 2.8

Effects of decaying ground-ivy leaves on seed germination of downy brome
and radish

Test Species	Experiment Number	Mean Number Seeds Germinated per Pot with SD		Percent Germination	
		Control	Test	Control	Test
Downy brome	1[a]	19.1 ± 1.1	17.1 ± 2.7	96	86
	2[a]	18.4 ± 1.8	15.4 ± 3.4[d]	92	77
Radish	1[b]	6.6 ± 2.4	6.0 ± 0.9	66	60
	2[c]	13.0 ± 1.5	11.2 ± 2.1	87	75

[a]Twenty seeds planted per pot. Eight control and eight test pots planted in all experiments.
[b]Ten seeds planted per pot.
[c]Fifteen seeds planted per pot.
[d]Difference from control significant at 0.05 level.

DISCUSSION

The evidence indicates that ground-ivy is strongly allelopathic to at least some species. Both root exudates and decaying leaves of ground-ivy significantly stimulated growth of radish plants and decaying leaves significantly stimulated growth of downy brome. Additionally, root exudates of ground-ivy inhibited growth of downy brome. Volatile allelochemicals were apparently not involved; thus it appears that water-soluble compounds were responsible for the results. Rainfall leachates of the leaves were not tested for allelopathic activity, but it is possible that such leachates could have significant effects.

The reduced stimulatory effect of decaying ground-ivy leaves is experiment 2 is intriguing. The most important point is, of course, that there was still a highly significant growth stimulation of both test species in spite of numerous changes in conditions. The greatest change in growth in the second experiment was the marked increase in control growth. The experimental changes in the control conditions were a slight increase in temperature; addition of N, P, and K; and addition of Perlite. It is doubtful that the small temperature change or Perlite was responsible for the increased control growth relative to that of the test. It is more likely that the added minerals were responsible for the difference, even though they were added to the test soil also. Obviously, the large growth differential between control and test in experiment 1 was not caused by the addition of minerals in the leaf material of ground-ivy, because only 2 g of this material were added per kilogram of soil in the test and only a very small percentage of the leaf material consisted of minerals. This suggests that a chemical(s) in the ground-ivy might have increased the uptake of minerals by the test plants, because a well-known mechanism of action of some allelopathic compounds is on the up-

TABLE 2.9

Effects of decaying ground-ivy leaves on seedling growth of downy brome and radish

Test Species	Experiment Number	Mean Dry Weight with SD (mg) and Percent Stimulation					
		Shoots			Roots		
		Control	Test	Stimulation (%)	Control	Test	Stimulation (%)
Downy brome[a]	1	6.4 ± 2.4	55.7 ± 23.8[c]	770	8.0 ± 2.2	28.1 ± 15.8[c]	251
	2	41.0 ± 19.1	57.4 ± 16.4[c]	40	21.4 ± 10.4	27.0 ± 7.6[d]	26
Radish[b]	1	23.9 ± 12.9	278.1 ± 123.9[c]	1064	12.7 ± 6.7	184.6 ± 133.3[c]	1354
	2	133.6 ± 37.8	358.7 ± 97.8[c]	168	179.8 ± 56.6	357.4 ± 140.0[c]	99

[a]Each value is average of 40 replicates.
[b]Each value is average of 32 replicates.
[c]Difference from control significant beyond the 0.001 level.
[d]Difference from control significant at 0.01 level.

TABLE 2.10
Effects of root exudates of ground-ivy on seedling growth of downy brome and radish

Mean Dry Weight with SD (mg) and Percent Stimulation or Inhibition

Test Species	Shoots			Roots		
	Control[a]	Test[b]	Stimulation or Inhibition (%)	Control[a]	Test[b]	Stimulation or Inhibition (%)
Downy brome	71.4 ± 16.8	53.6 ± 12.4[c]	25	22.8 ± 9.4	15.1 ± 5.8[c]	34
Radish	239.8 ± 100.0	352.5 ± 118.6[c]	47	206.4 ± 141.2	325.0 ± 147.4[c]	58

[a]Each value is average of 20 replicates.
[b]Each value is average of 10 replicates.
[c]Difference from control significant at 0.05 level or better.

take of minerals. Obviously, these speculations need to be tested with much additional experimentation. Moreover, the allelochemicals responsible for the growth stimulation and inhibition by ground-ivy need to be identified.

CONCLUSIONS

Most research projects in allelopathy have been designed in such a way that only the inhibitory results are considered significant in explaining the biological problem under investigation. Thus any resulting stimulatory effects have been ignored or mentioned only incidentally. Phytopathologists have been very aware, however, of the significance of the chemical stimulation of growth and development of pathogens and parasitic plants by microorganisms and host plants. Hence there have been many reports of allelopathic stimulation of fungi by other fungi or by plants, and many concerning stimulation of seed germination or haustorial formation of parasitic plants by host plants. There have been numerous reports of allelopathic stimulation of algae by other algae in relation to bloom formation and phytoplankton succession. There have been at least a few reports of the chemical stimulation of each type of microorganism (bacteria, fungi, and algae) by the same and other types. Instances of allelopathic stimulation of plants by microorganisms have also been documented. The allelopathic growth stimulation of nonparasitic plants by plants has been grossly ignored, considering its potential economic importance. There have been several cases documented and a new instance is reported here, that of the remarkable growth stimulation of radish and downy brome grass by *Glechoma hederacea* (ground-ivy). Decaying ground-ivy leaves (2 g/kg of soil) stimulated growth of downy brome shoots by as much as 770% and radish shoots by as much as 1064%. Downy brome root growth was stimulated up to 251% and radish root growth up to 1354%. Root exudates of ground-ivy significantly stimulated both shoot and root growth of radish. Volatiles from ground-ivy leaves did not significantly affect seed germination or radicle growth of the two test species.

REFERENCES

Agrios, G. N. 1969. *Plant Pathol.* Academic, New York.

Ayers, W. A. 1971. *Can. J. Microbiol.* 17:1517.

Barton, R. 1957. *Nature.* 180:613.

Bell, A. A. 1977. *In* C. G. McWhorter, A. C. Thompson, and E. W. Hauser (eds.), *Report of the Research Planning Conference on the Role of Secondary Compounds in Plant Interactions (Allelopathy).* USDA, Agricultural Research Service, Tifton, GA, pp. 64–99.

Berglund, H. 1969. *Physiol. Plant.* 22:1069.

Brandt, W. H. and J. E. Reese. 1964. *Am. J. Bot.* 51:922.

Buxton, E. W. 1957. *Trans. Brit. Mycol. Soc.* 40:145.

Bywater, J. and C. J. Hickman. 1959. *Trans. Brit. Mycol. Soc.* 42:513.

Chet, J. and Y. Henis. 1975. *Ann. Rev. Phytopathol.* 13:169.

Dailey, O. D., Jr. and S. L. Vail. 1985. *In* A. C. Thompson (ed.), *The Chemistry of Allelopathy.* American Chemical Society, Washington, D.C., pp. 427–435.

Dukes, P. D. and J. L. Apple. 1961. *Phytopathology.* 51:195.

Ebben, N. H. and P. H. Williams. 1956. *Ann. Appl. Biol.* 44:425.

Fletcher, R. A. and A. J. Renney. 1963. *Can. J. Plant Sci.* 43:475.

French, R. C., C. L. Graham, A. W. Gales, and R. K. Long. 1977. *J. Agric. Food Chem.* 25:84.

Gajić, D. 1966. *J. Sci. Agric. Res.* 19:63.

Gajić, D. and G. Nikočević. 1973. *Fragm. Herb. Jugoslavica* 18:1.

Gajić, D. and M. Vrbaški. 1972. *Fragm. Herb. Croatica* 7:1.

Gajić, D., S. Malenčić, M. Vrbaški, and S. Vrbaški. 1976. *Fragm. Herb. Jugoslavica* 63:121.

Goode, P. M. 1956. *Trans. Brit. Mycol. Soc.* 39:367.

Hoagland, R. E. and R. D. Williams. 1985. *In* A. C. Thompson (ed.), *The Chemistry of Allelopathy.* American Chemical Society, Washington, D.C., pp. 301–325.

Husain, S. S. and W. E. McKeen. 1962. *Phytopathology.* 52:14. (Abstract.)

Hussain, A. and M. A. B. Mallik. 1972. *J. Sci.* 1:139.

Jackson, R. M. 1957. *Nature.* 180:96.

Johnson, A. W., G. Rosebery, and C. Parker. 1976. *Weed Res.* 16:223.

Jørgensen, E. and E. Steemann-Nielsen. 1961. *Physiol. Plant.* 14:896.

Keating, K. I. 1977. *Science.* 196:885.

Kerr, A. 1956. *Aust. J. Biol. Sci.* 9:45.

Leuck, E. E., II and E. L. Rice. 1976. *Bot. Gaz.* 137:160.

Lynn, D. G. 1985. *In* A. C. Thompson (ed.), *The Chemistry of Allelopathy.* American Chemical Society, Washington, D.C., pp. 55–81.

Mallik, M. A. B. and A. Hussain. 1972. *J. Sci.* 1:133.

Maugh, T. H., II. 1981. *Science.* 212:33.

Menzies, J. D. and R. G. Gilbert. 1967. *Soil Sci. Soc. Amer. Proc.* 31:495.

Molisch, H. 1937. *Der Einfluss einer Pflanze auf die andere-Allelopathie.* Fischer, Jena.

Monahan, T. J. and F. R. Trainor. 1970. *J. Phycol.* 6:263.

Monahan, T. J. and F. R. Trainor. 1971. *J. Phycol.* 7:170.

Muller, C. H. 1966. *Bull. Torrey Bot. Club.* 93:332.

Neill, R. L. and E. L. Rice. 1971. *Amer. Midland Natur.* 86:344.

Nicollier, G. F., D. F. Pope, and A. C. Thompson. 1985. *In* A. C. Thompson (ed.), *The Chemistry of Allelopathy.* American Chemical Society, Washington, D.C., pp. 207–218.

Nutman, P. S. 1965. *In* K. F. Baker and W. C. Snyder (eds.), *Ecology of Soil-Borne Plant Pathogens.* University of California Press, Berkeley, pp. 231–246.

Papavizas, G. C. and C. B. Davey. 1961. *Plant Soil.* 14:215.

Parks, J. M. and E. L. Rice. 1969. *Bull. Torrey Bot. Club.* 96:345.

Pepperman, A. B., Jr. and E. J. Blanchard. 1985. *In* A. C. Thompson (ed.), *The Chemistry of Allelopathy.* American Chemical Society, Washington, D.C., pp. 415–425.

Rice, E. L. 1984. *Allelopathy.* 2nd ed. Academic, Orlando, FL.

Ries, S. K., V. Wert, C. C. Sweeley, and R. A. Leavitt. 1977. *Science.* 195:1339.

Rose, S. L., D. A. Perry, D. Pilz, and M. M. Schoeneberger. 1982. *Allelopathic Effects of Litter on the Growth and Colonization of Mycorrhizal Fungi.* North American Symposium on Allelopathy, Nov. 14–17, Urbana-Champaign, IL. (Abstract.)

Scharen, A. L. 1960. *Phytopathology.* 50:274.

Schreiber, L. R. and R. J. Green. 1962. *Phytopathology*. 52:751. (Abstract.)

Schroth, M. N. and W. C. Synder. 1961. *Phytopathology*. 51:389.

Stevens, K. L. and G. B. Merrill. 1985. *In* A. C. Thompson (ed.), *The Chemistry of Allelopathy*. American Chemical Society, Washington, D.C., pp. 83–98.

Tubbs, C. H. 1973. *Forest Sci*. 19:139.

Ukeles, R. and J. Bishop. 1975. *J. Phycol*. 11:142.

Van Aller, R. T., G. F. Pessoney, V. A. Rogers, E. J. Watkins, and H. G. Leggett. 1985. *In* A. C. Thompson (ed.), *The Chemistry of Allelopathy*. American Chemical Society, Washington, D.C., pp. 387–400.

White, N. H. 1954. *Aust. J. Sci*. 17:18.

Willetts, H. J. 1972. *Biol. Rev*. 47:515.

Wolfe, J. M. and E. L. Rice. 1979. *J. Chem. Ecol*. 5:533.

Youssef, Y. A. 1974. *Folia Microbiol*. 9:381.

Youssef, Y. A. and A. T. Mankarios. 1974. *Mycopath. Mycol. Appl*. 54:173.

Zentmyer, G. A. 1965. *Science*. 150:1178.

3

ADVERSE IMPACTS
OF ALLELOPATHY
IN AGRICULTURAL SYSTEMS

ALAN R. PUTNAM AND LESLIE A. WESTON

Department of Horticulture and
Pesticide Research Center
Michigan State University
East Lansing, Michigan

Some of the earliest observations of allelopathy concerned harmful effects of crops upon other crops or weeds. Theophrastus (ca. 300 B.C.) observed that chick pea (*Cicer arietinum*) "exhausts" the ground and destroys weeds (Rice, 1984). Pliny (Plinius Secundus, 1 A.D.) not only reported that a number of crops including chick pea, barley (*Hordeum vulgare*), and bitter vetch (*Vicia ervilia*) "scorch up" cornland, but he also recognized toxicity from walnut (*Juglans regia*) trees. He attributed the toxicity of plants to their scents or juices and indicated that bracken fern (*Pteridium aquilinum*) might even be controlled by breaking young stalks and allowing "the juice trickling down out of the fern itself to kill the roots." DeCandolle (1832) recognized "soil sickness" problems on agricultural lands and suggested this might be caused by chemical exudates from crops. He also made the astute observation that crop rotation might alleviate the problem.

Perhaps the earliest work to clearly demonstrate allelopathy on agricultural sites was an elegant series of experiments conducted in the early 1900s by Schreiner and Reed. These scientists and their associates successfully isolated a number of phytotoxic chemicals from plants and soil (Schreiner and Reed, 1907a and b, 1908; Schreiner and Shorey, 1909; Schreiner and Sullivan, 1909) using chemical techniques which today would be considered primitive.

WEEDS WITH ALLELOPATHIC POTENTIAL

Weeds now impose at least a 10-billion-dollar crop loss in the United States each year. Although the literature abounds with articles on "weed competition," seldom has allelopathy been considered or even mentioned in these studies. In mixed stands of crops and weeds, several mechanisms of interference may be working simultaneously or sequentially. Under field conditions, it has been virtually impossible to separate these mechanisms. Allelopathic potential has now been suggested for about 90 species of weeds (Table 3.1). The evidence for allelopathy by many of these species remains weak and may not hold true in future testing (Putnam, 1985). For some species, however, the evidence is quite convincing.

Allelopathic chemicals may directly influence crop emergence or growth or they may affect the crop symbionts or its pests. The indirect effects may be important in determining the overall efficiency of the crop plants. Allelopathy may be one mechanism through which aggressive weeds species gain dominance on agricultural sites.

One weed species for which considerable evidence supports allelopathy is quackgrass (*Agropyron repens*). This rhizomatous perennial causes severe crop losses in temperate regions of the Northern Hemisphere. It may be useful to review the research done with quackgrass because it could serve as a model for future investigations with other species. Allelopathy by quackgrass involves many facets and demonstrates how various management techniques may influence the toxic potential of a weed.

Ahlgren and Aamodt (1939) were the first to suggest a chemical basis for the interference between quackgrass and legumes, particularly the *Trifolium* species. They could not explain their data based on resource limitation alone. Bandeen and Buchholtz (1967) experienced similar difficulties. Although corn growth was severely inhibited by quackgrass, they could not overcome the inhibition by supplying additional fertilizer and water. Similarly, Kommedahl et al. (1970) found that adding large concentrations of nitrogen only partially corrected yield reductions imposed by previous stands of quackgrass.

Several detrimental effects caused by quackgrass were later shown in greenhouse tests. The addition of dead quackgrass rhizomes to soils reduced alfalfa (*Medicago sativa*) emergence and growth (Kommedahl et al., 1957). Oats and alfalfa were also chlorotic and stunted when grown in the presence of quackgrass shoot residues (Ohman and Kommedahl, 1964). These workers also demonstrated that hot water extracts of quackgrass roots, stems, rhizomes, and leaves decreased alfalfa seedling growth by 60–85%. Leachates from pots of living quackgrass severely decreased seed germination and root growth (Kommedahl, 1957).

TABLE 3.1
Common agroecosystem weeds with alleged allelopathic potential

Scientific Name	Common Name	First Reference[a]
Abutilon theophrasti	Velvetleaf	Gressel and Holm (1964)
Agropyron repens	Quackgrass	Kommedahl et al. (1959)
Agrostemma githago	Corn cockle	Gajić and Nikočević (1973)
Allium vineale	Wild garlic	Osvald (1950)
Amaranthus dubius	Amaranth	Altieri and Doll (1978)
Amaranthus retroflexus	Redroot pigweed	Gressel and Holm (1964)
Amaranthus spinosus	Spiny amaranth	VanderVeen (1935)
Ambrosia artemisiifolia	Common ragweed	Jackson and Willemsen (1976)
Ambrosia cumanensis	—	Anaya and DelAmo (1978)
Ambrosia psilostachya	Western ragweed	Neill and Rice (1971)
Ambrosia trifida L.	Giant ragweed	Letourneau et al. (1956)
Antennaria microphylla	Pussytoes	Selleck (1972)
Artemisia absinthium	Absinth wormwood	Bode (1940)
Artemisia vulgaris	Mugwort	Mann and Barnes (1945)
Asclepias syriaca	Common milkweed	Rasmussen and Einhellig (1975)
Avena fatua	Wild oat	Tinnin and Muller (1971)
Berteroa incana	Hoary alyssum	Bhowmik and Doll (1979)
Bidens pilosa	Beggar-ticks	Stevens and Tang (1985)
Boerhovia diffusa	Spiderling	Sen (1976)
Brassica nigra	Black mustard	Muller (1969)
Bromus japonicus	Japanese brome	Rice (1964)
Bromus tectorum	Downy brome	Rice (1964)
Calluna vulgaris	—	Salas and Vieitez (1972)
Camelina alyssum	Flax weed	Grummer and Beyer (1960)
Camelina sativa	Largeseed falseflax	Grummer and Beyer (1960)
Celosia argentea	—	Pandya (1975)
Cenchrus biflorus	Sandbur	Sen (1976)
Cenchrus pauciflorus	Field sandbur	Rice (1964)
Centaurea diffusa	Diffuse knapweed	Fletcher and Renney (1963)
Centaurea maculosa	Spotted knapweed	Fletcher and Renney (1963)
Centaurea repens	Russian knapweed	Fletcher and Renney (1963)
Chenopodium album	Common lambsquarters	Caussanel and Kunesch (1979)
Cirsium arvense	Canada thistle	Stachon and Zimdahl (1980)
Cirsium discolor	Tall thistle	Letourneau et al. (1956)
Citrullis colocynthis	—	Bhandari and Sen (1971)
Citrullis lavatus	—	Bhandari and Sen (1972)
Cucumis callosus	—	Sen (1976)
Cynodon dactylon	Bermudagrass	VanderVeen (1935)
Cyperus esculentus	Yellow nutsedge	Tames et al. (1973)
Cyperus rotundus	Purple nutsedge	Friedman and Horowitz (1971)
Daboecia polifolia	—	Salas and Vieitez (1972)
Digera arvenis	—	Sarma (1974)
Digitaria sanguinalis	Large crabgrass	Parenti and Rice (1969)
Echinochloa crus-galli	Barnyardgrass	Gressel and Holm (1964)
Eleusine indica	Goosegrass	Altieri and Doll (1978)
Erica scoparia	Heath	Ballester et al. (1977)

(continued)

45

TABLE 3.1 *Continued*

Scientific Name	Common Name	First Reference[a]
Euphorbia corollata	Flowering spurge	Rice (1964)
Euphorbia esula	Leafy spurge	Letourneau and Heggeness (1957)
Euphorbia supina	Prostrate spurge	Brown (1968)
Galium mollugo	Smooth bedstraw	Kohmmedahl (1965)
Helianthus annuus	Sunflower	Rice (1974)
Helianthus mollis	—	Anderson et al. (1978)
Hemarthria altissima	Bigalta limpograss	Tang and Young (1982)
Holcus mollis	Velvetgrass	Mann and Barnes (1947)
Imperata cylindrica	Alang-alang	Eussen (1978)
Indigofera cordifolia	Wild indigo	Sen (1976)
Iva xanthifolia	Marshelder	Letourneau et al. (1956)
Kochia scoparia	Kochia	Wali and Iverson (1978)
Lactuca scariola	Prickly lettuce	Rice (1964)
Lepidium virginicum	Virginia pepperweed	Bieber and Hoveland (1968)
Leptochloa filiformis	Red sprangletop	Altieri and Doll (1978)
Lolium multiforum	Italian ryegrass	Naqvi and Muller (1975)
Lychnis alba	White cockle	Bhowmik and Doll (1979)
Matricaria inodora	Mayweed	Mann and Barnes (1945)
Nepeta cataria	Catnip	Letourneau et al. (1956)
Oenothera biennis	Evening primrose	Bieber and Hoveland (1968)
Panicum dichotomiflorum	Fall panicum	Bhowmik and Doll (1979)
Parthenium hysterophorus	Ragweed parthenium	Sarma et al. (1976)
Plantago purshii	Wooly plantain	Rice (1964)
Poa pratensis	Bluegrass	Alderman and Middleton (1925)
Polygonum aviculare	Prostrate knotweed	Al Saadawi and Rice (1982)
Polygonum orientale	Princesfeather	Datta and Chatterjee (1978)
Polygonum pensylvanicum	Pennsylvania smartweed	Letourneau et al. (1956)
Polygonum persicaria	Ladysthumb	Martin and Rademacher (1960)
Portulaca oleracea	Common purslane	Letourneau et al. (1956)
Rumex crispus	Dock	Einhellig and Rasmussen (1975)
Saccharum spontaneum	Wild cane	Amritphale and Mall (1978)
Salsola kali	Russian thistle	Lodhi (1979)
Salvadora oleoides	—	Mohnat and Soni (1976)
Schinus molle	California peppertree	Anaya and Gomez-Pompa (1971)
Setaria faberi	Giant foxtail	Schreiber and Williams (1967)
Setaria glauca	Yellow foxtail	Gressel and Holm (1964)
Setaria viridis	Green foxtail	Rice (1964)
Solanum surattense	—	Sharma and Sen (1971)
Solidago sp.	Goldenrod	Letourneau et al. (1956)
Sorghum halepense	Johnsongrass	Abdul-Wahab and Rice (1967)
Stellaria media (L.)	Common chickweed	Mann and Barnes (1950)
Tagetes patula	Wild marigold	Altieri and Doll (1978)
Trichodesma amplexicaule	—	Sen (1976)
Xanthium pensylvanicum	Common cocklebur	Rice (1964)

[a]Several other reports may also be available. The reference cited is the earliest report of which we are aware.

Welbank (1963) indicated that decaying quackgrass residues produced water-soluble inhibitors of germination. This work was later confirmed by Toai and Linscott (1979). These investigations indicated that anaerobic decomposition under warmer temperatures increased the toxicity. When quackgrass residues are plowed down they might be expected to undergo fermentation. Penn and Lynch (1982) reported that both quackgrass fermentation products and the pathogen *Fusarium culmorum* contributed to the toxicity occurring in barley (*Hordeum vulgare*) planted into plowed fields.

Quackgrass residues also exert toxicity when they are allowed to remain in place in no-tillage systems. Weston and Putnam found that numerous crop species were injured by surface residues (Table 3.2). The toxicity that occurred was more severe than when the quackgrass residues were plowed down. In bioassay, quackgrass foliage was about twice as toxic as the rhizome tissue (Table 3.3). There was no enhancement of activity by exposure to soil microbes over short time periods. These investigators also found that quackgrass residues reduced nodulation by rhizobia in several legumes species (Weston and Putnam, 1985). The primary site appeared to be inhibition of root hair formation, rather than toxicity to rhizobia per se. This is discussed in more detail later in the chapter.

Several investigators have attempted to isolate and identify the toxic compounds released from quackgrass tissues or those produced during the decomposition of its residues. The results are somewhat confusing and incomplete. Both LeTourneau and Heggeness (1957) and Ohman and Kommedahl (1960) reported that toxins from rhizomes were deactivated by activated charcoal. Two phenolic substances, *p*-hydroxybenzoic acid and vanillic acid were reported to be released by dead roots and rhizomes (Grummer and Beyer, 1960). LeFevre and Clagett (1960) isolated a growth inhibitor from rhizomes and Gabor and

TABLE 3.2

Reduced crop growth in response to quackgrass residues either in no-tillage or conventional tillage plot (plowed and diskharrowed)

| Crop | Plant Weight (Percent of Control[a]) | |
	No-Tillage	Conventional Tillage
Alfalfa	3.7	54.8
Cabbage	20.6	73.3
Carrot	8.9	65.2
Cucumber	35.3	83.1
Oats	50.0	76.2
Pea	40.5	79.9
Sorghum	18.7	87.3
Sweet corn	40.4	88.8

[a]The control was an adjacent conventionally tilled plot.

TABLE 3.3
Toxicity of quackgrass residues in soil amendment experiments[a]

Tissue	Killing Method	Root Length[b] (Percent of Control)
Rhizome	Dried	70.4 a
Rhizome	Glyphosate	68.5 a
Herbage	Dried	33.6 b
Herbage	Glyphosate	32.3 b

[a]The experiments were conducted using 1 g of tissue.
[b]Data from the bioassay species barnyardgrass, curly cress, and lettuce were averaged. Means with different suffix letters are significantly different at $P = 101$.

Veatch (1981) tentatively identified a toxic rhizome glycoside with a molecular weight of 460. In neither case was compound identification positively achieved. Grummer (1961) previously isolated a compound called agropyrene from the rhizomes. It had antimicrobial activity, but activity against higher plants was not documented.

Penn and Lynch (1982) considered organic acids produced by fermentation to be the major allelochemicals from decaying residues. More recently, Weston et al. (1986) implicated the flavone tricin (Figure 3.1) and related compounds as toxic components from quackgrass foliage. What role, if any, these compounds play in soil toxicity in the field remains to be proven. As mentioned in Chapter 1, this aspect is usually the weakest link in studies of allelopathy. The quackgrass research may provide useful insight because it indicates that different toxins and mechanisms of toxicity may arise from the same plant, depending on where the tissues are placed and which compounds are released closest to the suscept plant. The research also indicates that even when quackgrass is destroyed with herbicides, the tissues and toxins must be allowed to degrade prior to the successful establishment of crops.

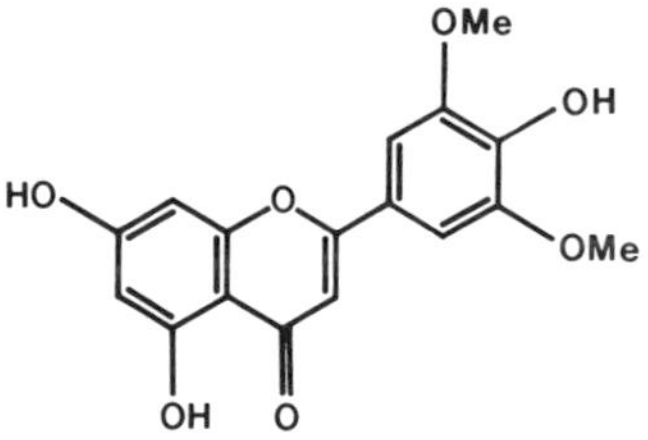

Tricin

Figure 3.1   The flavone tricin isolated from quackgrass herbage (Weston et al., 1986). This compound and related flavonoids are inhibitory to root growth.

TOXICITY FROM CROP RESIDUES

As mentioned, Schreiner and associates reported during the early 1900s, that "soil sickness" due to single cropping might be caused by growth inhibitors added to the soil from the plants. Schreiner and Sullivan (1909) extracted a toxin from soil that had previously grown several crops of cowpeas and found the substance inhibitory to cowpea growth. Furthermore, after the substance was extracted from the soil, the soil was no longer inhibitory.

Many investigations into allelopathy by crop plants have involved the effects of decomposing plant residues. These studies are particularly important now because of the trend toward reduced tillage agriculture which by its nature, purposely maintains plant residues on the soil surface. The toxic influences of straw or other highly carbonaceous plant residues were described in detail by Collison and Conn (1925). They concluded that two separate mechanisms are associated with the toxicity produced by plant residues. The first is toxic chemical agents (allelochemicals), which act quickly and are usually quickly inactivated by colloidal matter. The second involves stimulation of microbial populations, which in turn immobilize much of the nitrogen, making it unavailable to the higher plants.

McCalla and Duley (1948) reported that corn performed poorly under no-tillage (stubble mulch) systems where sweet clover (*Melitotus alba*) or wheat-straw mulch remained at the soil surface. Extracts of sweet clover and microbial products from decaying wheat were both found to be toxic to the corn. Guenzi and McCalla (1962) extracted residues of several crops at the end of the summer and autoclaved half prior to bioassay. Although nonautoclaved extracts were generally more toxic to germination, the autoclaved extracts were more inhibitory to seedling growth.

In later work, Norstadt and McCalla (1963) identified a fungus (*Penicillium urticae*) that produced patulin, a strong inhibitor of seedling growth. Their work, and later studies by McCalla and Haskins (1964) clearly demonstrate that microbial as well as plant toxins contribute to the allelopathic potential of crop residues.

Patrick and Koch (1958) monitored the toxicity of decomposing plant residues using the respiration of tobacco seedlings as an assay. They found the greatest toxicity to occur when decomposition took place under saturated soil conditions. Stage of maturity of the plant residues also affected their toxicity. Residues from young plants were toxic immediately, whereas more mature plants required a longer period to release their toxic substances.

Tang and Waiss (1978) determined that the allelochemicals from decomposing wheat residues were primarily organic acids including acetic, propionic, butyric, and pentanoic acids. Patrick (1971) found these and numerous other compounds in decomposing rye residues (see more details in Chapter 16). Chou and Patrick (1976) identified 18 compounds from decomposing corn residues in soil. These included numerous organic acids, aldehydes, alcohols, and phenolic compounds. Dr. Chou elaborates on these in Chapter 14.

More recently, the impact of residues of several crops used prior to establishing winter wheat in no-tillage planting were assessed (Cochran et al., 1977). Lentils (*Lens culinaris*) and peas (*pisum sativum*) were extremely toxic to fall-planted wheat, causing severe inhibition of root growth. Fall-killed bluegrass (*Poa pratensis*) sod was toxic to wheat planted in the spring.

Although the toxicity of plant residues has usually been considered a problem for subsequent crops, there is now considerable evidence to suggest that crop residues may be managed so as to help suppress certain weed species in agroecosystems (Barnes and Putnam, 1983; DeFrank and Putnam, 1983; Shilling et al., 1985). This approach is considered in more detail in Chapter 16.

In addition to the toxicity exerted by crop residues in annual cropping systems, there are numerous examples of species-specific replant problems with perennial horticultural crops. Perhaps best known is the difficulty encountered when replanting peach trees into areas where old peach orchards were removed. Proebsting and Gilmore (1941) studied this problem intensively. They found that the toxicity was specific to peach and that the bark of old peach roots was particularly toxic to the growth of young peach trees. Amygdalin, a cyanogenic glycoside in peach bark, was believed to break down in the soil thus causing the toxicity. Patrick (1955) further investigated the role of toxins in the peach replant problem. He concluded that microbial breakdown of amygdalin occurred rapidly in areas where peaches had formerly grown, producing high levels of toxic cyanide. In autoclaved soils, amygdalin was not toxic, and its breakdown was much slower in soils that did not contain peach roots. Although nematodes and other factors were implicated in replant difficulties, Patrick et al. (1964) concluded that toxins produced by the breakdown of amygdalin were a primary cause of the peach replant problem.

Replant problems have also occurred in apple nurseries and orchards (Borner, 1959). The toxicity was attributed to compounds released or produced from apple root tissue, including phlorizin, phloretin, *p*-hydroxyhydrocinnamic acid, phloroglucinol, and *p*-hydroxybenzoic acid. These compounds could also be extracted from soil two days after the apple root tissue was added. In later work Borner could only find phlorizin in the apple tissue itself and showed that the other four compounds are produced in the soil by microbial action. Allelopathy or autotoxicity is implicated in numerous other perennial crops. Two in particular, asparagus and coffee, are discussed in detail in Chapters 6 and 15.

INHIBITION OF NITRIFICATION

Since the ammonium form of nitrogen is more tightly adsorbed to soil, it is less susceptible to leaching. The nitrification process converts ammonium to nitrate, which is easily leached; therefore, it would appear that inhibiting nitrification could be advantageous for nitrogen conservation on agricultural soils. One commercial product, nitrapyrin is now marketed specifically as a nitrification inhibitor.

Considerable evidence has been collected to suggest that allelochemicals from living plants or their residues may inhibit nitrification by *Nitrosomonas* in forest (Rice and Pancholy, 1973) and grassland systems (Purchase, 1974). There was little research in agricultural systems after Russell's (1914) initial observation that several crop plants appeared to inhibit nitrification. Moore and Waid (1971) tested the effects of exudates from five crops on nitrification in a clay loam soil near Reading, England. All of the test species initially reduced the rate of nitrification. Although the impact of rape and lettuce exudates were temporary, ryegrass, wheat, and onion root exudates had pronounced and persistent effects. The effect of ryegrass was most pronounced, reducing the rate of nitrification up to 84%. Since neither microbial immobilization nor denitrification appeared to be affected, these workers concluded that the exudates contained inhibitors of nitrification.

Lodhi (1981) monitored the nitrate and ammonium levels, as well as the numbers of *Nitrosomonas* on adjacent sites that either had corn residues removed or maintained on the soil surface. The concentrations of nitrate were much higher than that of ammonium where the residues were removed. The numbers of *Nitrosomonas* were also always higher in the soil without corn residue. Lodhi also monitored phenolic inhibitors released by the corn residues. Two of the inhibitors, ferulic and *p*-coumaric acids were present in much higher concentration in the soil where corn residues had been maintained.

Since loss of nitrate from arable lands is of concern, both because of pollution and conservation of expensive nitrogen, methods for inhibiting nitrification are of great interest. Huber et al. (1977) recently reviewed the literature on nitrification inhibitors for agriculture with particular emphasis on favorable impacts on yield and crop quality.

INHIBITION OF NITROGEN FIXATION

Symbiosis between *Rhizobium* and legume plants is believed to produce about 40% of all biologically fixed nitrogen (Brill, 1977). The soil environment can greatly influence the success of infection and nodulation of legumes by rhizobia. *Rhizobium* survival in soil is influenced by previous plant inhabitants, soil pH, or competitive microflora (Dart, 1974).

Several researchers have shown inhibition of nodulation and nitrogen fixation by allelopathic higher plants (Murthy and Nagodra, 1977; Rice, 1965, 1984; Rice et al., 1981; Jobidon and Thibault, 1982). Pandya and Pota (1978) indicated that root exudates of *Celosia argentea* inhibited proliferation of *Rhizobium* and also reduced radicle elongation and dry weight of bajra seedlings. Rice (1974) indicated that both nitrogen-fixing (*Azotobacter* and *Rhizobium*) and nitrifying bacteria (*Nitrobacter* and *Nitrosomonas*) were severely inhibited by extracts from a number of seed plants. Reduced nodulation occurred on Korean lespedeza (*Lespedeza stipulacea*), white clover (*Trifolium repens*), and red kidney beans (*Phaseolus vulgaris*) after exposure to extracts of *Aristida, Ambrosia, Bromus, Digitaria*, and *Euphorbia* species.

Our recent work (Weston and Putnam, 1985) indicates that legume development and nodulation are both reduced by living or herbicidally killed quackgrass (Table 3.4, Figure 3.2). When quackgrass was killed with glyphosate (a

TABLE 3.4

Influence of various quackgrass regimes on root fresh-weight and nodule number of four legume indicators grown in the greenhouse

Regime	Snap Bean	Navy Bean	Soybean	Pea
Root Weight (g)[a]				
Living quackgrass	2.0 b	1.7 d	4.2 c	0.7 b
Glyphosate killed	3.0 b	2.8 c	6.7 bc	1.0 b
Stifted soil	5.0 a	5.5 a	8.2 ab	2.4 a
Control soil	4.3 ab	4.0 b	10.6 a	2.2 a
Nodule Number[a]				
Living quackgrass	24 c	27 bc	51 b	18
Glyphosate killed	39 c	14 c	82 b	34
Sifted soil	134 b	57 b	66 b	36
Control soil	195 a	107 a	83a	41

[a]Means with different suffix letters within a column are significantly different at $P = 0.05$. These data are adapted from Weston and Putnam (1985).

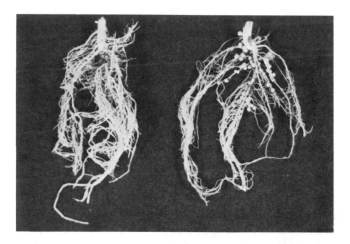

Figure 3.2 Nodulation of snapbean (*Phaseolus vulgaris*) roots in the presence of dead quackgrass residues (left) and in the absence of quackgrass residues (right). Similar results were obtained in the field and greenhouse. Photograph by L. A. Weston.

herbicide with no soil activity), residues of the weed were toxic for about 60 days to legumes and other crops planted through the residue. Aqueous extracts of quackgrass herbage gave similar toxicity and also prevented nodulation. Laboratory studies indicated that compounds from quackgrass leaves are particularly inhibitory to root hair formation, which subsequently prevents infection by the rhizobia. The growth of several *Rhizobium* species on artificial media was not inhibited by these compounds. Tang (personal communication, 1985) observed similar effects when Bigalta limpograss (*Hemarthria altissima*) was present near tropical legumes. Reduced legume efficiency due to inhibitors released by associated weeds my be a worldwide problem. It may be particularly important in areas where cost prohibits the application of synthetic nitrogen sources.

CONCLUSIONS

Allelopathy may adversely impact agricultural systems through various methods. Numerous weed species (now about 90) have alleged allelopathic potential. Although future studies may negate some of these accounts, it appears that sufficient evidence exists to implicate many. In several perennial species, and particularly quackgrass, considerable chemical evidence has been accumulated. The negative impacts of crop residues on subsequent plantings are also well documented. These toxins may be leached from the plant residue, released upon its breakdown, or produced by microbes that use the plant material as a food source. Toxins from surface residues may affect crop performance in no-tillage systems, therefore, more must be learned about their compatibilities. Replanting problems and autotoxicities result from the release of allelochemicals in many perennial cropping systems. These often involve microbes that release toxins upon hydrolysis of conjugates in the plant tissue. Several species release allelochemicals that are inhibitory to nitrogen-fixing bacteria. Others apparently can regulate nitrogen availability through effects on nitrification.

REFERENCES

Abdul-Wahab, A. S. and E. L. Rice. 1967. *Bull. Torrey Bot. Club*. 94:486.

Ahlgren, H. L. and O. S. Aamodt. 1939. *J. Am. Soc. Agron*. 31:982.

Alderman, W. H. and J. A. Middleton. 1925. *Proc. Am. Soc. Hortic. Sci*. 22:307.

Al Saadawi, I. S. and E. L. Rice. 1982. *J. Chem. Ecol*. 8:993.

Altieri, M. A. and J. D. Doll. 1978. *PANS*. 24:495.

Amritphale, D. and L. P. Mall. 1978. *Sci. Cult*. 44:28.

Anaya, A. L. and S. DelAmo. 1978. *J. Chem. Ecol*. 4:289.

Anaya, A. L. and A. Gomez-Pompa. 1971. *Rev. Soc. Mex. Hist. Nat*. 32:99.

Anderson, R. C., A. J. Katz, and M. R. Anderson. 1978. *J. Chem. Ecol*. 4:9.

Ballester, A., J. M. Albo, and E. Vieitez. 1977. *Oecologia*. 30:55.

Bandeen, J. D. and K. P. Buchholtz. 1967. *Weeds*. 15:220.

Barnes, J. P. and A. R. Putnam. 1983. *J. Chem. Ecol.* 9:1045.

Bhandari, M. C. and D. N. Sen. 1971. *Pflanzenphysiol.* 64:466.

Bhandari, M. C. and D. N. Sen. 1972. *Z. Naturforsch.* 27:72.

Bhowmik, P. C. and J. D. Doll. 1979. *Proc. North Cent. Weed Control Conf.* 34:43.

Bieber, G. L. and C. S. Hoveland. 1968. *Agron. J.* 60:185.

Bode, H. R. 1940. *Planta.* 30:567.

Borner, H. 1959. *Contrib. Boyce-Thompson Inst.* 20:39.

Brill, W. J. 1977. *Sci. Am.* 3(March):68.

Brown, D. D. 1968. The possible ecological significance of inhibition by *Euphobia Supina* M.S. Thesis. University of Oklahoma, Norman, OK.

Caussanel, J-P. and G. Kunesch. 1979. *Z. Pflanzenphysiol.* 93:229.

Chou, C. H. and Z. A. Patrick. 1976. *J. Chem. Ecol.* 2:369.

Cochran, V. L., L. F. Elliot, and R. I. Papendizk. 1977. *Soil Sci. Soc. Am. J.* 41:903.

Collision, R. C. and H. J. Conn. 1925. New York State Agr. *Exp. Sta. Tech. Bull.* No. 114. 34 pp.

Dart, P. J. 1974. The infection process. *In* A. Quispel (ed.), *The Biology of Nitrogen Fixation.* North Holland, Amsterdam. 382 pp.

Datta, S. C. and A. K. Chatterjee. 1978. *Indian J. Weed Sci.* 10:23.

DeCandolle, M. A. P. 1832. *Physiologie Vegetale* Tome III. Bechet Jeune, Lib. Fac. Med., Paris, pp. 1474–1475.

DeFrank, J. and A. R. Putnam. 1983. *Crop Prot.* 2:173.

Einhellig, F. A. and J. A. Rasmussen. 1973. *Am. Midl. Nat.* 90:79.

Eussen, J. H. H. (ed.). 1978. *In Studies on the Tropical Weed Imperata Cylindrica (L) Beauv. var Major*, Drukkerig Elinkwijk Bv, Utrecht, Paper No. 7.

Fletcher, R. A. and A. J. Renney. 1963. *Can. J. Plant Sci.* 43:475.

Friedman, T. and M. Horowitz. 1971. *Weed Sci.* 19:398.

Gabor, W. E. and C. Veatch. 1981. *Weed Sci.* 29:155.

Gajić, D. and G. Nikočević. 1973. *Fragm. Herb. Jugoslavica.* 18:1.

Gressel, J. B. and L. G. Holm. 1964. *Weed Res.* 4:44.

Grummer, G. 1961. *Symp. Soc. Exp. Biol.* 15:219.

Grummer, G. and H. Beyer. 1960. *Symp. Brit. Ecol. Soc.* 1:153.

Guenzi, W. and T. McCalla. 1962. *Soil Sci. Soc. Am. Proc.* 26:456.

Huber, D. M., H. L. Warren, D. W. Nelson, and C. Y. Tsai. 1977. *Bioscience.* 27:523.

Jackson, J. R. and R. W. Willemsen. 1976. *Am. J. Bot.* 63:1015.

Jobidon, R. and J. R. Thibault. 1982. *Am. J. Bot.* 69:1213.

Kohmuenzer, S. 1965. *Diss. Pharm.* 17:369.

Kommedahl, T. 1957. *Down to Earth.* 3(Fall):4.

Kommedahl, T., A. J. Linck, and J. V. Bernadini. 1957. *Phytopathology.* 47:526.

Kommedahl, T., J. B. Kotheimer, and J. V. Bernardini. 1959. *Weeds.* 7:1.

Kommedahl, T., K. M. Old, J. H. Ohman, and E. W. Ryan. 1970. *Weed Sci.* 18:29.

LeFevre, C. W. and C. O. Clagett. 1960. *Proc. Northeast Weed Control Conf.* 14:353.

Letourneau, D. and H. G. Heggeness. 1957. *Weeds.* 5:12.

Letourneau, D., G. D. Failes, and H. G. Heggeness. 1956. *Weeds.* 4:363.

Lodhi, M. A. K. 1979. *J. Chem. Ecol.* 5:429.

Lodhi, M. A. K. 1981. *J. Chem. Ecol.* 7:685.

Mann, H. H. and T. W. Barnes. 1945. *Ann. Appl. Biol.* 32:15.

Mann, H. H. and T. W. Barnes. 1947. *Ann. Appl. Biol.* 34:252.

Mann, H. H. and T. W. Barnes. 1950. *Ann. Appl. Biol.* 37:139.

Martin, P. and B. Rademacher. 1960. *Symp. Brit. Ecol. Soc.* 1:143.

McCalla, T. M. and F. L. Duley. 1948. *Science.* 108:163.

McCalla, T. M. and F. A. Haskins. 1964. *Bacteriol. Rev.* 28:181.

Mohnat, K. and S. Soni. 1976. *Comp. Physiol. Ecol.* 1:125.

Moore, D. R. E. and J. S. Waid. 1971. *Soil Biol. Biochem.* 3:69.

Muller, C. H. 1969. *Vegetatio.* 18:348.

Murthy, M. S. and T. Nagodra. 1977. *J. Appl. Ecol.* 14:279.

Naqvi, H. H. and C. H. Muller. 1975. *Pak. J. Bot.* 7:139.

Neill, R. L. and E. L. Rice. 1971. *Am. Midl. Nat.* 86:344.

Norstadt, F. A. and T. M. McCalla. 1963. *Science.* 140:410.

Ohman, J. and T. Kommendahl. 1960. *Weeds.* 8:666.

Ohman, J. and T. Kommendahl. 1964. *Weeds.* 12:222.

Osvald, H. 1950. On antagonism between plants. Proc. 7th Int. Congr. Bot. Stockholm.

Pandya, S. M. 1975. *Geobios* (Jodhpur). 2:125.

Pandya, S. M. and K. B. Pota. 1978. *Natl. Acad. Sci. Lett.* 1:56.

Parenti, R. L. and E. L. Rice. 1969. *Bull Torrey Bot. Club.* 96:70.

Patrick, Z. A. 1955. *Can. J. Bot.* 33:461.

Patrick, Z. A. 1971. *Soil Sci.* 111:13.

Patrick, Z. A. and L. W. Koch. 1958. *Can. J. Bot.* 36:621.

Patrick, Z. A., T. A. Toussoun, and L. W. Koch. 1964. *Annu. Rev. Phytopathol.* 2:267.

Penn, D. J. and J. M. Lynch. 1982. *Plant Pathol.* 31:39.

Plinius Secundus, 1 A.D. *Natural History*, 10 Vol. English translation by H. Rackam, W. H. S. Jones and D. E. Eicholz. Harvard University Press, Cambridge, MA, pp. 1938-1963.

Proebsting, E. L. and A. E. Gilmore. 1941. *Proc. Am. Soc. Hortic. Sci.* 38:21.

Purchase, B. S. 1974. *Plant Soil.* 41:541.

Putnam, A. R. 1985. *Weed Allelopathy. In* S. O. Duke (ed.), *Weed Physiology. Vol. 1.* CRC Press, Boca Raton, FL. 165 pp.

Rasmussen, J. A. and F. A. Einhellig. 1975. *Am. Midl. Nat.* 94:478.

Rice, E. L. 1964. *Ecology.* 45:824.

Rice, E. L. 1965. *Physiol. Plant.* 18:255.

Rice, E. L. 1974. *Allelopathy,* 1st ed, Academic, New York.

Rice, E. L. 1984. *Allelopathy,* 2nd ed. Academic, Orlando, FL. 422 pp.

Rice, E. L. and S. K. Pancholy. 1973. *Am. J. Bot.* 60:691.

Rice, E. L., C. V. Lin, and C. Y. Huang. 1981. *J. Chem. Ecol.* 7:333.

Russell, E. J. 1914. *J. Agric. Sci.* 6:50.

Salas, M. C. and E. Vieitez. 1972. *Ann. Edafal. Agrobiol.* 31:1001.

Sarma, K. K. V. 1974. *Geobios* (Jodhpur). 1:137.

Sarma, K. K. V., G. S. Giri, and K. Subramanyam. 1976. *Trop. Ecol.* 17:76.

Schreiber, M. M. and J. L. Williams, Jr. 1967. *Weeds.* 15:80.

Schreiner, O. and H. S. Reed. 1907a. U.S. Dept. Agric. Bur. Soils Bull. No. 47.

Schreiner, O. and H. S. Reed. 1907b. *Bull. Torrey Bot. Club.* 34:279.

Schreiner, O. and H. S. Reed. 1908. *Bot. Gaz.* 45:73.

Schreiner, O. and E. D. Shorey. 1909. U.S. Dept. Agric. Bur. Soils Bull. No. 53.

Schreiner, O. and M. X. Sullivan. 1909. *J. Biol. Chem.* 6:39.

Selleck, G. W. 1972. *Weed Sci.* 20:189.

Sen, D. N. 1976. Second Progress Report Project No. A7-CR-425, Laboratory of Plant Ecology. University of Jodhpur, Jodhpur, India.

Sharma, K. D. and D. N. Sen. 1971. *Z. Pflanzenphysiol.* 65:458.

Shilling, D. G., R. A. Liebl, and A. D. Worsham. 1985. *In* A. C. Thompson (ed.), *The Chemistry of Allelopathy. American Chemical Society*, Washington, D.C., pp. 243–271.

Stachon, W. J. and R. L. Zimdahl. 1980. *Weed Sci.* 28:83.

Stevens, G. A. and C-S. Tang. 1985. *J. Chem. Ecol.* 11:1411.

Tames, R. S., M. D. V. Gesto, and E. Vieitez. 1973. *Physiol. Plant.* 28:195.

Tang, C. S. and A. C. Waiss, Jr. 1978. *J. Chem. Ecol.* 4:225.

Tang, C. S. and C. C. Young. 1982. *Plant Physiol.* 69:155.

Theophrastus (ca. 300 B.C.) *Enquiry into Plants and Minor Works on Odours and Weather Signs.* 2 Vols. English Translation by A. Hort. W. Neinemaw, London. 1916 pp.

Tinnin, R. and C. H. Muller. 1971. *Bull. Torrey Bot. Club.* 98:243.

Toai, T. V. and D. L. Linscott. 1979. *Weed Sci.* 27:595.

VanderVeen, R. 1935. *Arch. Koffiecult* (Nederland India). 3:65.

Wali, M. K. and L. R. Iverson. 1978. Abstract of the 144th National American Association of Advanced Science Meeting. Washington, D.C., pp. 121–122.

Welbank, P. J. 1963. *Weed Res.* 3:205.

Weston, L. A. and A. R. Putnam. 1985. *Crop Sci.* 25:561.

Weston, L. A. and A. R. Putnam. 1986. *Weed Sci.* 34:366.

Weston, L. A., A. R. Putnam, and B. A. Burke. 1986. *J. Chem. Ecol.* (In press.)

4

THE ROLE OF ALLELOPATHY IN SUBTROPICAL AGROECOSYSTEMS IN TAIWAN

CHANG-HUNG CHOU

Institute of Botany
Academia Sinica
Taipei, Taiwan
Republic of China

In 1937 Molisch used the two Greek words "allelo" and "pathy" to coin allelopathy, which means "reciprocal suffering of two organisms"; thenceforth, in the literature the term usually has been applied to detrimental plant–plant interactions which occur through biochemical effects. However, Molisch meant the term allelopathy to include both detrimental and beneficial effects of one plant upon another. Recently, Rice (1984) defined allelopathy as both beneficial and detrimental chemical interaction among organisms including microorganisms. The confusion was lessened by Muller, who suggested using "interference" in the definition, and by Whittaker, who coined the word "allelochemics" to describe all kinds of chemical interactions among organisms. In 1982 Chou and Waller used "allelochemicals" to mean that both interspecific and intraspecific interactions between organisms were mediated through a chemical process.

Since the 1960s allelopathy has been increasingly recognized as an important ecological mechanism which influences plant dominance, succession, formation of plant communities and climax vegetation, and crop productivity. The allelopathic phenomenon is always related to other environmental parameters and is difficult to isolate from the environmental complex (Muller, 1974). Koeppe et al. (1971, 1976) demonstrated several cases of allelopathic effects in relation to environmental stresses. Putnam and Duke (1974) introduced the concept that allelopathy might be useful in the agricultural practice of selecting crop varieties with high phytotoxic potential to reduce herbicide use. In the last decade, an exponential increase of allelopathic studies occurred in many parts of the world and substantial information was published (Chou, 1983; Chou and Waller, 1983; Muller, 1966, 1971; Rice, 1984). Since 1972 my associates and I have conducted research in the subtropical and tropical areas of Taiwan. This paper describes our work and discusses the role of allelopathy in subtropical agroecosystems in Taiwan.

ALLELOPATHY IN RELATION TO ENVIRONMENTAL FACTORS

Many allelopathic compounds produced by plants are regulated by environmental factors, such as water potential of the environment, temperature, light intensity, soil moisture, nutrients, soil microorganisms, and perhaps others. The compounds are released to the environment by means of volatilization, leaching, decomposition of residues, and root exudation (Muller, 1966, 1974; Chou, 1983; Rice, 1984). First, the terpenoids, such as α-pinene, β-pinene, cineole, and camphor, are released to the environment by volatilization, which is noticeable under drought conditions. The water-borne phenolics and alkaloids are then moved out by rainfall through a leaching process. Next, phytotoxic aglycones such as phenolics are produced during the decomposition of plant residues in soil. Finally, many secondary metabolites, such as scopoletin and hydroquinone, may be released to the surrounding soil through root exudation. The following paragraphs describe phytotoxin production under several environmental regimes in Taiwan.

Plants Under Drought Stress

A study site within a *Leucaena leucocephala* plantation was selected at the Kaoshu village of Pintung county, situated in the southern part of Taiwan. After 3–4 years of culture, there developed an almost total lack of understories except *Leucaena* seedlings underneath the *Leucaena* during the winter season. The absence of weeds was due to a heavy accumulation of *Leucaena* plant residues (such as leaf litter, branches, etc.) which release phytotoxic substances to suppress the growth of many understory species, except *Leucaena* itself. Kuo et al. (1983) found that the aqueous extract of leaf litter produced a significant phytotoxicity on many tested plants, and the phytotoxins were identified as mimosine and seven phytotoxic phenolic acids. The heavy accumulation of the leaf litter in the Kaoshu area was primarily due to drought and heavy winds in the winter season. In contrast, at another *Leucaena* plantation in the Chialin village of Hualean county, where there is heavy rainfall in the winter season, there were luxuriant understory species. In addition, we found little *Leucaena* leaf litter accumulated on the floor. However, at this site we still found bare ground at a particular plot where a dead *Leucaena* tree had fallen, resulting in a relatively heavy accumulation of leaf litter. From the field observations, we concluded that the lack of weeds underneath *Leucaena* trees was primarily due to the phytotoxins released from the leaf litter of *Leucaena* trees, and the accumulation of leaf litter was caused by the droughty winter season or by accidentally fallen trees. It is generally believed that the phenomenon would be more pronounced in older plantations and under severe drought conditions. One interpretation is that the *Leucaena* plants survive drought stress through a strategy whereby their fallen leaves release substantial amounts of phytotoxins to suppress the growth of understories competing with *Leucaena* plants for available water.

Plants Under Water-Logged and Oxygen-Deficient Stress

Many aquatic plants can grow very well under water-logged and oxygen-deficient environments because of their adaptation mechanisms. Although rice plants are not aquatic plants, they grow well in the paddy soils. Patrick and Mikkelsen (1971) indicated that the level of oxygen reached almost zero when it was measured at 25 cm below the soil surface in paddy fields. We obtained similar results in Taiwan paddy soils. In many areas of Taiwan, namely Tsingshui (the central part of Taiwan), Chiatung and Yuanlin (the southern part), and Tungshan (the east coast), paddy fields may either have poor water drainage or high water tables, leading to oxygen deficiency. This is even more pronounced in the second crop season when the monsoon comes. In addition, the farmers in Taiwan have always submerged the rice straw into soil and allowed it to decompose. During the decomposition of rice residues in soil, a significant amount of phytotoxic substances, such as short-chain aliphatic and phenolic acids, were produced (Table 4.1). The amount of these compounds produced reached its maximum at the first month after rice residues were submerged into soil, resulting in the suppression of root growth and panicle initiation followed by decreases in rice yield (Chou and Lin, 1976; Chou and Chiou, 1979; Chou et al., 1977, 1981).

Allelopathy in Relation to Soil Redox Potential (Eh) of Paddy Soil

As mentioned, the oxygen level is nearly zero at 25 cm below the soil surface of paddy fields, and that may cause the reduction of soil redox potential (Eh) (Patrick and Mikkelsen, 1971). Chou and his co-workers measured the soil Eh at the Nankang paddy field and found that the Eh ranged from -100 to 200 mv during the first crop season and from -200 to 100 mv during the second crop season. The same measurement made at the farm of the National Chungshing University of Taichung revealed that the Eh was remarkably low, ranging from -500 to 100 mv during the second crop season. In another pot experiment, we found that the soil Eh was below -300 mv in the treatment of rice straw mixed

TABLE 4.1

Dynamics and distribution of phytotoxins present in extracts of decomposing rice residues in soil using different extraction techniques and decomposed during different time intervals. (Quantitative comparison was made by the order ++++ > +++ > ++ > +)

Compound	Ether Fraction of Aqueous Extract			Ethanol Extract			Alkaline Ethanolic Extract		
	1-Week	2-Week	4-Week	1-Week	2-Week	4-Week	1-Week	2-Week	4-Week
o-Hydroxyphenylacetic acid	++++	++	++	++	++	+			
p-Hydroxybenzoic acid	+	++	++	++	++		++	++	+
cis-p-Coumaric acid				++	++	++	+	+	+
trans-p-Coumaric acid				+	+		++++	+++++	++++
Ferulic acid							+++	+++	++
Vanillic acid							++	++	+
Unknown 1[a,b]							++	++	++
Unknown 2[a,b]							+	+	++
Unknown 3[a,b]							+	+	+

Source. Chou and Lin (1976).

Detection	Unknown 1	Unknown 2	Unknown 3
	0.28 fluorescence	0.33 absorption	0.88 absorption

[a] R_f with 2% acetic acid
[b] Short UV light

60

with soil and was above 100 mv in the treatment of soil alone. Thus the reduction of soil Eh was apparently related to the decomposition of rice residues in soil. The reduction in soil Eh was pronounced at the tillering stage (30–45 days after transplanting) and at the panicling stage (80–90 days after transplanting) (Chou and Chiou, 1979). During this period the growth of rice roots was retarded, the root cells swelled, and many adventitious roots developed. Wu et al. (1976b) interpreted the swelling of root cell to be a kind of adaptive mechanism utilized to obtain more oxygen.

Allelopathy in Relation to Microbial Activity

During the decomposition of plant residues in soil, microbial activities must be involved. Wu et al. (1976b) found that the bacteria *Pseudomonas putida* became dominant in the rhizosphere of the rice paddy, and the populations of *P. putida* were positively correlated to phytotoxin production during the time rice residues were submerged in the saturated soil. They pointed out that in well-drained area of Tsaotune soil, the number of *P. putida* was 701×10^5/g dry soil, while in the poorly drained areas of Lotung, Taan, and Taichung, the number of *P. putida* ranged from 347 to 3412×10^5/g dry soil. It was evident that the number of *P. putida* was exceedingly high in the poorly drained soil, indicating that the organism might use the residues as its carbon source. Wu et al. (1976b) furthermore indicated that the phytotoxic phenolics did not come from the metabolities of this microorganism, but was directly released from the decomposing rice residues. Chou et al. (1977) pointed out that when ammonium sulfate was mixed with rice residues, phytotoxicity was enhanced, indicating that nitrogen fertilizer might favor the growth of decomposer microorganisms and expedite the decomposition rate of rice residues in soil. Wu et al. (1976b) also indicated that the application of ammonium sulfate fertilizer to paddy soil was beneficial to the growth of *P. putida*, and that it might expedite the formation of H_2S, which is toxic to rice growth.

Similarly, the cause of yield decline of sugar cane in Taiwan has been investigated by Wang and his associates (1984) and Wu et al. (1976a). The cause of the reduction is partly due to the phytotoxic effects of decomposing cane residues in soil, and to the microbial activity of *Fusarium oxysporum*, which produces a phytotoxin, fusaric acid, in addition to the phytotoxic phenolics previously mentioned. They found that *F. oxysporum* populations were much greater in the rhizosphere of ratoon sugar cane soil than in unplanted soil. When 10 ppm of fusaric acid were mixed in Murashige and Skoog's medium, the leaves of sugar cane wilted and became chlorotic, indicating that fusaric acid is toxic to sugar cane growth (Wu et al., 1976a).

Phytotoxins in Relation to Nitrogen Availability

Chou et al. (1977) concluded that when more rice stubble was left in the paddy soil, there would be more phytotoxic phenolics and less leachable nitrogen, indicating that the phytotoxins produced may affect nitrogen transformation in soil. They also found that the amount of leachable NH_4-N was about 10 times more than that of NO_3^- N (Chou and Chiou, 1979). Chou and Cheng further designed an experiment using N-isotope tracers incorporated into the soil or soil-rice residues mixture under different temperature regimes and sequences. The results indicated that in the absence of straw, most of the nitrogen fertilizer remained in the mineral forms. Straw enhanced nitrogen immobilization only moderately. The gradual decrease in the proportion of nitrogen fertilizer in the mineral forms was accompanied by a steady increase of nitrogen fertilizer in the amino acid fraction of organic nitrogen. Little accumulation of nitrogen fertilizer in the amino sugar or the insoluble humin fraction was found (Chou et al., 1981). Although the experimental results did not show a distinct trend in relation to temperature variations, the temperature range of 25–30°C tended to favor nitrogen transformation activities.

Interaction of Decomposing Rice Residues with Soil Leachable Cations

During the decomposition of rice residues in soil, the amount of available minerals in soil might be affected and consequently affect plant growth. Chou and Chiou (1979) conducted an experiment to see the effect of rice straw incorporated into soil on the dynamics of some cations in pot soil. The results revealed that the concentrations of K, Cu, and Mn were higher in the first crop season, while those of Na, Ca, Mg, and Zn were higher in the second crop season in Nankang paddy soil, regardless of nitrogen fertilizer application. Most of our findings agree with those of Patrick and Mikkelsen (1971). In the flooded soil the contents of reducible iron and manganese were relatively low. When the pot soil was mixed with rice straw and allowed to decompose, the amount of potassium was significantly higher than that of soil alone, but those of Cu, Fe, Mn, and Zn were average, significantly lower in the soil alone. It is interesting to note that in several poor water-drainage areas of Taiwan, such as Changhwa, Taitung, and Pingtung, zinc deficiency is particularly pronounced during the second crop season.

THE ROLE OF ALLELOPATHY IN AGRICULTURAL PRODUCTIVITY IN TAIWAN

The Allelopathic Mechanism of Low Yield in the Second Crop of Rice

Rice (*Oryza sativa*), the most important crop in Taiwan, is planted twice a year in continuous monoculture systems. For nearly a century the rice yield of the second crop has been generally lower by 25% than that of the first crop (a reduction of about 1000 kg/ha). The reduction of rice productivity has been particularly pronounced in areas of poor water drainage. Although many factors involved in this reduction have been found, their significance has not been fully understood. The cropping system for rice in Taiwan is different from that of other countries. For example, a 3-week fallowing period between the first crop and the second crop is different when compared with a 10-week interval between the second crop and the first crop of the following year. In the first crop (from March to July) the temperature is usually from 15 to 30°C, but in the second crop (August to December) it is from 30 to 15°C. Between two crops, the farmers always leave rice stubble in the field after harvesting, and submerge these residues into soil for decomposition during the fallowing time. During the second crop season, the typhoon (also called monsoon) brings a great amount of rainfall, leading to a high water table in areas where water drainage is poor.

Chou and his associates conducted a series of experiments to elucidate the reason for low yield in the second crop season. The aqueous extracts of paddy soil collected in Nankang were subjected to osmotic concentration measurement and bioassay of phytotoxic properties. The osmotic concentration of the extracts was nearly zero, but the phytotoxicity ranged from 25 to 50%. In pot experiments, a soil–ricestraw mixture (3 kg; 200 g) was saturated with distilled water and allowed to decompose for 1, 2, and 4 weeks under greenhouse conditions. Soil alone was treated in the same manner as a control. At the end of each decomposition time, five rice seedlings (3 weeks old) were transplanted into a pot containing the straw–soil mixture or into the control soil. After 1 month, rice seedlings grown under control conditions were normal and usually about 60 cm tall, whereas the seedlings grew poorly and were only about 36 cm tall in the soil–straw mixture (Table 4.2). The retarded roots were dark brown and the root cell was abnormal and enlarged. Further experimental results showed that when the amount of rice–straw mixed was increased to 100 g/3 kg soil, the phytotoxicity increased with the increase of straw. The toxicity was still persistent after 16

TABLE 4.2
Inhibition of growth of rice seedlings affected by adding different amounts of rice straw decomposed in 3 kg soil for 2 weeks[a]

Amount of Straw Mixed (g)	Length of Seedling (mm)	Inhibition (%)	Dry Weight	
			Grams	Decrease (%)
0	59	0	0.51	0
25	52	12	0.30	41[c]
50	44	25[b]	0.23	55[c]
75	39	34[c]	0.20	61[c]
100	37	37[c]	0.12	76[c]

Source. Chou and Lin (1976).
[a]After decomposition, 3-week old rice seedlings were transplanted into pots; 1 month later, the length and dry weight of seedlings were obtained from the means of 5 seedlings in triplicate.
[b]Statistical significance at 5% level using the analysis of variance.
[c]Statistical significance at 1% level using the analysis of variance.

weeks of decomposition, reflecting a long existence of phytotoxicity during the rice–straw decomposition. The decomposing rice–straw was extracted with ethanol and reextracted with ethyl ether, then the phytotoxins present in the ether fraction of the extract were identified by chromatography. The identified compounds were *p*-hydroxybenzoic, *p*-coumaric, syringic, vanillic, *o*-hydroxyphenylacetic, and ferulic acids (Table 4.1), and propionic, acetic, and butyric acids (Wu et al., 1976b). In particular, *o*-hydroxyphenylacetic acid, the first phytotoxin reported by us, was toxic to rice growth at the concentration of 25 ppm. We found that the concentration of *o*-hydroxyphenylacetic acid reached about 10^{-2} mM in the decomposing rice residues in soil. At the early stage of second crop season, the daytime temperature is usually 30°C, which could expedite the decomposition of rice stubble remaining in soil and release large amounts of phytotoxins. As mentioned, the phytotoxicity of decomposing rice residues could persist for over 4 months; thus the tillering and the panicling of rice plants would be retarded significantly, resulting in a decreased rice yield (Chou and Lin, 1976, Chou et al., 1977; Chou and Chiou, 1979).

The Cause of Low Yield of Monoculture Sugar Cane

The germination and growth of ratoon cane are the two major problems in the farms of the Taiwan Sugarcane Corporation (TSC). The yield of monoculture sugar cane has declined in many sugar cane fields. The causes of this yield reduction have been investigated, yet no single factor causing the reduction can be found. Wang et al. (1967) conducted field and laboratory experiments and concluded that the phytotoxic effect is one of the important factors involved. Five phenolic acids, such as *p*-hydroxybenzoic, *p*-coumaric, syringic, ferulic, vanillic, formic, acetic, oxalic, malonic, tartaric, and malic acids were identified in the decomposing sugar cane leaves in soil under water-logged condition. At 50 ppm of these phenolic acids in water culture, the growth of young sugar cane root was inhibited. On the other hand, the aforementioned aliphatic acids can inhibit the growth of ratoon sugar cane at 10^{-3} M. Furthermore, Wu et al. (1976a) found that the population of *Fusarium oxysporum* in the rhizosphere soil of poor ratoon cane roots was much greater than that of good growing ratoon or of newly-planted sugar cane roots. They found that fusaric acid, the secondary metabolite of the organism, was toxic to the growth of young sugar cane plant *in vitro* (Wu et al., 1976a).

Phytotoxic Effect of Cover Crops on Orchard Plants

Wu et al. (1975) compared the phytotoxic effects of some cover crops, namely *Centrocema, Indigofera*, and Bahia grass on the growth of several receiver crops, namely peas, mustard, cucumber, cauliflower, rape, Chinese cabbage, mungbean, water melon, tomato, and rice. They found that rape was the most sensitive crop to the extracts of these cover crops. Among them, *Centrocema* and *Indigofera* exhibited significant phytotoxic effect, and Bahia grass revealed less toxicity. In addition, the growth of banana was inhibited by the leachate of *Centrocema*. More recently, several cover crops, such as *Bromus catharticus, Pennisetum clandestinum, Lolium multiflorum* (chromosomes 4X and 2X), *Paspalum notatum*, and clover are now under investigation for their allelopathic effect on the productivity of apple and peach plantations in the Lishan area of the Central mountain. A vast area of apple plantation has been situated on the hillsides of the Central mountain since the 1960s. The productivity of the apple plantation was exceedingly high in the first decade after planting, but has gradually decreased in recent years. In fact, this problem has occurred in many European countries and North America as well. Börner (1960, 1971) and Patrick (1955) studied the mechanism for the apple and peach replanting problems, respectively. They reported that phlorizin and amygdalin were respectively involved in each replanting problem. After hydrolysis these two compounds became phytotoxic to plant growth; phlorizin is converted into *p*-hydroxybenzoic acid and its derivatives plus glucose, and amygdalin is hydrolyzed into HCN, benzaldehyde, and glucose. These compounds may contribute to the autointoxication of orchard plantation and the decline of apple and peach productivity observed in Taiwan.

Competitive Exclusion Between Bamboo and Conifer Forests

In many hillsides of the mountainous district in Taiwan, there is a vast area of forest plantations, and in the Chitou area we often find *Cryptomeria japonica* and *Phyllostachys edulis* growing adjacent to one another. However, the *P. edulis* often encroaches to the *C. japonica* area, resulting in a gradual decline in productivity and the eventual death of the *C. japonica*. Chou and Yang (1982) found that the litter of *P. edulis* possesses phytotoxic phenolics, which suppress the growth of its understories. The floristic composition of the two communities showed that the understory species are different. For example, five predominant species of understory in the *P. edulis* community are *Ageratum conyzoides, Commelina undulata, Pilea funkikensis, Pratia nummuaria*, and *Tetrastigma*; in the *C. japonica* community they are *Ficus pumila, Pellionia scabra, Pilea funkikensis, Piper arboriola*, and *Urtica thunbergiana*. These species respond to either different light intensity or to the phytotoxic leachates, resulting in a differential distribution of species density and biomass under the canopy of the two vegetations. The total number of dry weight of seedlings per square meter were much higher in the conifer community than in the bamboo forest, although the light intensity, soil moisture, and nutrient contents are significantly higher in the bamboo habitat. Further experimental results indicated that the aqueous extracts and leachates of bamboo leaves produced more phytotoxicity than that of conifer leaves, reflecting that allelopathy plays a significant role in the regulation of species diversity and production under the canopy of two forests. Nevertheless, the different potential and exclusion between the two forests may be partly due to certain anatomic characteristics, such as rhizomes. There are two types of rhizomes, sympodial rhizocauls and horizontal rhizomes with lateral culms. *P. edulis* has the latter type, and its rhizomes grow rapidly. Thus the invasion of *P. edulis* to *C. japonica* may be caused by (1) the fast growth of rhizomes, which may release phytotoxic root exudates, and (2) allelopathic substances produced by the bamboo leaves and the decomposing litter. The continuous release of water-soluble phytotoxins from *P. edulis* and the accumulation of the compounds in the soil may result in suppression of the understory or in elimination of neighboring plants. The distribution gradient of phytotoxins in the two vegetation zones and their ecotone indicated that the distribution of understory species was controlled by phototoxins. It is also possible that physical factors may interact synergistically with allelopathic substances to produce a more complex interaction in the field.

ALLELOPATHY IN AGRICULTURE PRACTICE IN TAIWAN

Selection of Weed-Controlling Grasses for Pastures

An increased amount of allelopathic research on grassland species has been conducted in many parts of the world during the past decade (Rice, 1984; Chou, 1983). Most of the studies have been concerned with the interpretation of allelopathic phenomena in the field. Only a few studies have employed the allelopathic concept as a practical means of directly controlling weeds. Putnam and Duke (1974) assayed 526 accessions of *Cucumis sativus* L. and 12 accessions of eight related *Cucumis* species representing 41 nations of origin and found that several accessions exhibited strong allelopathic nature, suppressing the growth of some weedy species.

In Taiwan, many grasses have been introduced into pastures, but only a few varieties can be established as forage grasses. Chou and Young (1975) studied 12 subtropical grasses in Taiwan and found that several species revealed great allelopathic potential. Among them, pangola (*Digitaria decumbens*) exhibited the highest toxic effect on test species. Under sufficient nitrogen fertilizer application, pangola grass forms a pure stand and almost no weeds can grow in the stand. In addition, we also found that different varieties of pangola revealed different growth performance and competitive abilities. Liang et al. (1983) thus selected 8 varieties of pangola and conducted field trials and laboratory assays. We found that the invasion ability of cultivars A65, A255, and A254 were highest in the Hsinhwa, Hengchun, and Hwalien stations, respectively; cultivars A79 and A80 were inferior in all stations. Cultivars A84, A254, and A255 revealed the highest toxicity, and phytotoxicity was correlated to the production of toxins, from which nine phytotoxic phenolics were identified. The interference of grasses in the field is very complicated and we believe that allelopathy can not solely explain the interference. Further field and laboratory experiments need to be performed to clarify the role of allelopathy in a grassland ecosystem.

Forest-Pasture Intercropping Systems

Taiwan is an island with two-thirds of the land occupied by forested mountains, the forests being extremely important for water conservation. The limited agricultural land available for crops and pastures forces the agricultural activities moving upward to hillsides or higher elevation areas. A forest and pasture intercropping system may be a way to increase livestock productivity. Recently, we conducted several experiments in the forest area of Houshe Experiment Station of the National Taiwan University located at an elevation of about 2000 m. About 1 hectare of farm was deforested and part of the area was cleaned by removing the leaf litter of the conifer tree (*Cunninghamia lanceolata*) and part was retained unclean to serve as a control. The cleaned and unclean plots were planted with kikuyu grass (*Pennisetum clandestinum*) or left open. The experiment was performed to find out the reciprocal interaction of fir litter and kikuyu grass and to evaluate the allelopathic potential of the two plants on weed growth under natural conditions. The field results indicated that the biomass of kikuyu grass in cleaned plots was significantly higher than that in the unclean control plots. In addition, the number of weeds grown in the plots planted with kikuyu grass was lower than that in the control plots, indicating that the kikuyu grass interferes with weeds. The seedlings of fir regenerated from the deforested area grew well and seemed not to be affected by the neighboring newly planted kikuyu grass. In contrast, the growth of kikuyu grass seemed to be suppressed to a certain degree by the fir litter. Furthermore, we brought fir litter back to the laboratory and evaluated it for phytotoxicity. The bioassay of extracts from kikuyu grass and fir litter showed that the fir extract exhibited higher phytotoxicity than the kikuyu grass. Nevertheless, the kikuyu grass seems to grow well in the plots with fir litter under natural conditions (Chou et al., 1984; unpublished data). It is too early to promulgate the experimental results with certainty, but it appears that the forest–pasture intercropping system may be possible and that some dominant grasses may suppress weed growth, but would not be harmful to forest regeneration.

Forest Intercropping System

On the hillsides of the mountainous district of Taiwan, there is an increasing area of deforestation. The regeneration of the area is important to ecological conservation. Many highly valuable forest species have been planted in forest intercropping systems, such as bamboo, conifers, *Acacia confusa*, *Leucaena leucocephala*, *Liquidembar formosana*, *Casuarina glauca*, *Alnus formosana*, and *Pinus taiwanensis*. We have evaluated the possibility of intercropping with the aforementioned species. The first project was conducted by planting *L. leucocephala* intercropped with other species mentioned above. The results indicated that *Pinus taiwanesis* and *Miscanthus floridulus* grew very well and could tolerate the leachates of *L. leucocephala*, but the remaining species were suppressed by the leachates. As mentioned earlier, we found several phytotoxic phenolics and mimosine, produced by *L. leucocephala*. It is interesting to note that the growth of *Mimosa pudica* was suppressed by *Leucaena* leaf leachates, although the leaf juices of *M. pudica* contain relatively high amounts of mimosine. Among 84 seedlings of *M. pudica* tested, only two survived, reflecting that mimosine can be economically useful to control notorious weeds such as *M. pudica* in the field.

ELIMINATION OF PHYTOTOXIC EFFECTS IN FIELDS TO IMPROVE CROP PRODUCTIVITY

Improvement of Water Drainage

The cause of reduced rice yield in the second crop season in Taiwan is partly due to the phytotoxic substances produced during the decomposition of rice residues in soil in areas of poor water drainage. In addition, the denitrified organism *Pseudomonas putida* was the dominant microbe during the decomposition of rice residues in soil. To eliminate the phytotoxins in paddy soil, a large-scale experiment of improving the drainage system was conducted in Chiatung, where the water table is relatively high and water drainage is poor. The results showed that the rice yield has been increased by at least 30% since the system was improved. We analyzed the soils from poor water drainage and improved water drainage for phytotoxicity and phytotoxins present. There was significantly higher phytotoxicity and a higher amount of phytotoxins present in the poorly drained soil than in soils with improved water drainage (Chou et al., 1984; unpublished data).

Removal of Soil Phytotoxins by Water Flooding

Monocultures of sugar cane exhibited decline problems similar to rice plants. To improve the microbial balance of sick soil and to eliminate soil phytotoxins, water was flooded on the sugar cane soil for a certain period of time, depending upon the nature of farms. The populations of *Fusarium oxysporum* were exceedingly high in the sick soil before the water flooding, but populations were much decreased after flooding. In addition, the amount of phytotoxin, fusaric acid produced by *F. oxysporum*, was significantly lower after flooding treatment, indicating that phytotoxins were leachated out by water flooding (Wang et al., 1984).

Crop Rotation

Many monoculture systems often lead to a soil-sickness problem, which is assumably due to the unbalance of soil microorganisms, the accumulation of soil toxins, mineral deficiency, or abnormal soil pH, which leads to decreased crop productivity. Crop rotation is therefore a good control method to eliminate the cause of the problem. In a natural ecosystem, one plant may replace another by means of allogenetic and autogenic succession (Daubenmire, 1968), which involves allelopathy or autointoxication mechanisms. These successions may take years for one stage and even require several decades or hundreds of years to reach a stable community, called climax. However, a man-made agroecosystem does need crop succession instead of natural vegetation succession. We can shorten the time of succession by introducing a new crop to replace the preceding crop, or by rotating crops. There are many cases of crop rotation throughout the world, and in such areas crop productivity is always high. A typical example is a tobacco–ryegrass–corn rotating system used in Ontario, Canada. Patrick and Koch (1963) found that continuous monoculture of tobacco plants may result in a root-rot disease caused by a soil-borne pathogen, *Thielaviopsis basicola*. However, when the tobacco was rotated to corn or ryegrass, the damage from root rot was much less. When we plant ryegrass or corn plants before tobacco, the decomposition of their residues may produce fungitoxic substances which are able to inhibit the germination of conidia or chlamydospore of *T. basicola*, thus the populations of *T. basicola* can be reduced and controlled to achieve a soil microbial balance (Chou and Patrick, 1972; upublished data).

Although many successful examples of crop rotation have been reported, only a few cases were investigated in relation to allelopathic effects. In Taiwan, pangola grass (*Digitaria decumbens*) provides a high productivity pasture which is stable in many fields. However, after several years of growth, the productivity declines, which has been attributed to autointoxication of the grass. Chou and his associates found that pangola grass produced a certain amount of phytotoxins that could suppress its own growth (Chou et al., unpublished data). The decline of productivity of this grass has been particularly pronounced at the Hengchun Experiment Station of the Taiwan Livestock Research Institute. During the winter season, the Hengchun area is under a severe drought condition, and pangola grass grows poorly. Therefore, we suggested a crop rotation using pangola grass and watermelon sequentially. After the watermelon crop the field was replanted with pangola grass and the yield of pangola grass was increased about 40%. We have not conducted detailed experiments to investigate the reason, but we believe that increased grass production occurs because the phytotoxic effect of prior pangola grass is eliminated. Further experiments are necessary to clarify the role of allelopathy in crop rotation.

Detoxification of Soil Phytotoxins by Nutrient Dressing

Wu and her co-workers conducted field experiments on poorly drained paddy soils of Lotung by applying ammonium sulfate, lime, and green manure to eliminate the phytotoxic effects of rice residues decomposing in soil. Lime produced significantly higher yields than any other treatments (Wu et al., 1976b). They concluded that the increase of rice yield was simply due to a detoxification mechanism of phytotoxins, and calcium may bind with some phytotoxic substances, converting them into nontoxic moieties. In addition, Chou and Chiou (1979) conducted experiments with different forms of nitrogen fertilizer. Ammonium sulfate nitrogen fertilizer gave higher rice yields than that of nitrate nitrogen fertilizer, suggesting that ammonium sulfate nitrogen fertilizer may overcome the phytotoxic effect of decomposing rice residues in soil. This finding agreed with that of Chandrasekaran and Yoshida (1973), who concluded that ammonium sulfate effectively eliminated the injury caused by phytotoxins. We also found that the root system of rice plants was healthy and well developed under the ammonium sulfate dressing as compared to that under the nitrate fertilizer dressing. Hence some nutrients may play a chelating role in detoxifying the phytotoxins in soil, and consequently give a better yield of crops.

Possible Detoxification Mechanism of Phytotoxins by Humic Acid Complex

It has been reported that many phytotoxic substances can be bound with clay minerals or other organic compounds to decrease their phytotoxicity (Wang et al., 1971, 1983; Rice 1984). Wang et al. (1978) found that protocatechuic acid, one of the phytotoxins related to *trans-p*-coumaric acid, can be polymerized into humic acid by using clay minerals as heterogenous catalysts. In fact, the humic acid is a natural substance which can polymerize many kinds of chemicals, including amino acids, flavonoids, terpenoids, aliphatic acids, and nitrogen containing compounds, thus keeping the soil in a fertile state. However, it is also possible that the polymerized phytotoxins fixed into a humic complex can be deploymerized under certain environmental conditions, and release free phenolic compounds that will exert a phytotoxic effect on nearby susceptible plants. If the polymerization of humic substances is a natural phenomenon, the phytotoxic substances may easily be fixed into a humic complex; thus the natural device of an organo-mineral complex of humic acid would be a pool of detoxification for many kinds of toxic substances produced by plants.

ALLELOPATHIC STRATEGY IN PLANT ADAPTATION AND EVOLUTION

Darwin's hypothesis concluded that species evolution is primarily due to competition for survival under natural selection. Of course, during his life the term allelopathy did not exist, but I believe that the allelopathic phenomenon is congruent with Darwin's hypothesis. Even today, many biologists still think that the allelopathic phenomenon is a part of plant competition. Muller (1971) defined competition to mean that one plant utilizes necessary environmental factors, resulting in a shortage which is harmful to another plant sharing the same habitat. Allelopathy, however, means that one plant adds a chemical to the environment, resulting in an effect upon another plant sharing the same habitat. Muller intended to use "interference" instead of "competition" to lessen the confusion that the word competition has generated. In the last decade, progress in bio-chemical ecological research has provided substantial information concerning species evolution in which many ecological phenomena can be easily interpre-tated through chemical mechanisms. Whittaker and Feeny (1971) stated that chemical agents are of major significance in the adaptation of species and orga-nization of communities. Thus, it is certain that allelopathy plays an important role in species evolution and adaptation.

Newman (1978) stated that allelopathy can not be an adaptation but an acci-dent in ecological processes. Muller (1974) strongly emphasized that allelopathy is never a single ecological factor, but should be regarded as one of a number in the environmental complex. However, in many cases, it could play a major role in determining plant diversity, dominance, succession, climax formation, and plant productivity. Newman did not support the idea that allelopathy is an adap-tation strategy and may provide benefit for the producer species, simply because toxins could be harmful to itself as well as to other plants growing nearby. I cannot fully agree with Newman's viewpoint, but agree with Whittaker and Feeny's that some autointoxicate species can survive and adapt for many genera-tions because the organisms possess adaptive autoinhibitors that can limit the population and alleviate excessive crowding.

We have evidence to show that rice plants produce phytotoxic substances that can reduce the productivity of the second crop yield in Taiwan; nevertheless, the rice plants were not killed by autoinhibitors. The adaptive autoinhibition of rice plants can be a significant method of natural selection. For example, we found that both wild rice, *Oryza perennis*, and its associated community of *Leersia hexandra* have autointoxication and allelopathy, and that both species interact with one another in the field. Although both species exhibit allelopathic poten-tial, neither causes the death of the other, although *L. hexandra* sometimes showed higher phytotoxicity than *O. perennis* (Chou et al., 1984). The autoin-toxication of *O. sativa* as well as *O. perennis* and *L. hexandra* does not imply a harmful effect on the cells and tissue producing and containing toxic metaboli-ties, but it will suppress the sprouting of juvenile tissues (i.e., tillers from the mother plant, seed germination, and seedling growth). The self-regulation of population density by self-thinning and phenotypic plasticity is an important adaptive mechanism of plants. There may be conditions in which natural selec-tion works in the direction of reducing autointoxication and increasing allelo-pathic potential. The selective advantage of allelopathy would change according to coexisting plants and community structure, rendering the mode of selection diffuse and disruptive. This may have resulted in the complexity of response patterns we have observed in many of our studies (Chou et al., 1984).

Other evidence also indicates that allelopathy phenomena are more pro-nounced in the area under environmental stresses. This is also of particular im-portance to plant adaptation. Koeppe et al. (1976) indicated that *Helianthus annus* produces significantly higher amounts of phytotoxic substances, such as chlorogenic and neochlorogenic acids, under a phosphorus deficiency condition than under normal conditions. Perhaps the production of phytotoxins can be interpreted as an adaptation strategy aimed at suppressing the growth of its competitor, which utilizes the same nutrient. In addition, plants may produce high amounts of toxic alkaloids in poor nitrogen soil which may be an adaptive strategy to compete with its neighboring plant for nitrogen or autointoxification to limit its own population to insure survival. Many examples of autointoxication or allelopathy as described above may well demonstrate that allelopathy is beneficial to the producer and has an adaptive significance. Allelopathy may also play an appreciable role in natural selection. This topic requires considerably more research.

ACKNOWLEDGEMENT

The author greatly appreciates the help of his former assistants and graduate students from which this is an outgrowth. This study was supported by the Academia Sinica, the National Science Council, and the Council of Agriculture of the Republic of China.

REFERENCES

Börner, H. 1960. *Bot. Rev.* 26:393.

Börner, H. 1971. German research on allelopathy. *In Biochemical Interactions Among Plants*. National Academy of Science, USA, pp. 52-56.

Chandrasekaran, S. and T. Yoshida. 1973. Soil Sci. *Plant. Nutr., Tokyo* 19:39.

Chou, C. H. 1983. Allelopathy in agroecosystems in Taiwan. *In* C H. Chou and G. R. Waller (eds.), *Allelochemicals and Pheromones*. Institute of Botany, Academia Sinica Monograph Series No. 5, Taipei, pp. 27-64.

Chou, C. H. and S. J. Chiou. 1979. *J. Chem. Ecol.* 5:839.

Chou, C. H. and H. J. Lin. 1976. *J. Chem. Ecol.* 2:353.

Chou, C. H. and Z. A. Patrick. 1976. *J. Chem. Ecol.* 2:369.

Chou, C. H. and G. R. Waller. (eds.) 1983. *Allelochemicals and Pheromones*. Institute of Botany, Academia Sinica Monograph Series No. 5. Taipei. 316 pp.

Chou, C. H. and C. H. Yang. 1982. *J. Chem. Ecol.* 8(12):1489.

Chou, C. H. and C. C. Young. 1975. *J. Chem. Ecol.* 1:183.

Chou, C. H., T. J. Lin, and C. I. Kao. 1977. *Bot. Bull. Academia Sinica.* 18:45.

Chou, C. H., Y. C. Chiang, and H. H. Cheng. 1981. *J. Chem. Ecol.* 7:741.

Chou, C. H., M. L. Lee, and H. I. Oka. 1984. *Bot. Bull. Academia Sinica.* 25:1.

Daubenmire, R. 1968. *Plant Communities*, Harper & Row, New York.

Koeppe, D. E., L. M. Rohrbaugh, and S. H. Wender. 1971. *In Biochemical Interactions Among Plants*. National Academy of Science, Washington, D.C., pp. 102-112.

Koeppe, D. E., L. M. Southwick, and J. E. Bittell. 1976. *Can. J. Bot.* 54:593.

Kuo, Y. L., C. H. Chou, and T. W. Hu. 1983. Allelopathic potential of *Leucaena leucocephala*. *In* C. H. Chou and G. R. Waller (eds.), *Allelochemicals and Pheromones*. Institute of Botany, Academia Sinica Monograph Series No. 5., Taipei, pp. 107-119.

Liang, J. C., S. S. Sheen, and C. H. Chou. 1983. Competitive allelopathic interaction among some subtropical pastures. *In* C. H. Chou and G. R. Waller (eds.), *Allelochemicals and Pheromones*. Institute of Botany, Academia Sinica Monograph Series No. 5, Taipei, pp. 121-133.

Molisch, H. 1937. *Der Einfluss einer Pflanza auf dei andere-Allelopathie.* Fischer, Jena.

Muller, C. H. 1966. *Bull. Torrey Bot. Club.* 93:332.

Muller, C. H. 1971. *In Biochemical Interactions Among Plants.* National Academy of Science, Washington, D.C., pp. 64–72.

Muller, C. H. 1974. *In* B. R. Strain and W. D. Billings (eds.), *Handbook of Vegetation Science Part VI: Vegetation and Environment.* Dr. W. Junk B. V., The Hague, pp. 73–85.

Newman, E. I. 1978. *In* J. B. Harborne (ed.), *Biochemical Aspects of Plant and Animal Coevolution.* Academic, London, pp. 327–341.

Patrick, W. H., Jr. and D. S. Mikkelsen. 1971. *In Fertilizer Technology and Use.* 2nd Ed. Soil Science Society of America, Madison, WI, pp. 187–215.

Patrick, Z. A. 1955. *Can. J. Bot.* 33:461.

Patrick, Z. A. 1971. *Soil Sci.* 111:13.

Patrick, Z. A. and L. W. Koch. 1963. *Can. J. Bot.* 41:747.

Putnam, A. R. and W. B. Duke. 1974. *Science.* 185:370.

Rice, E. L. 1984. *Allelopathy,* 2nd ed. Academic, Orlando, FL., 422 pp.

Wang, T. S. C., T. K. Yang, and T. T. Chuang. 1967. *Soil Sci.* 103:239.

Wang, T. S. C., K. L. Yeh, S. Y. Cheng, and T. K. Yang. 1971. *In Biochemical Interactions Among Plants.* National Academy of Sciences, Washington, D.C.

Wang, T. S. C., S. W. Li, and Y. L. Ferng. 1978. *Soil Sci.* 126:16.

Wang, T. S. C., M. C. Wang, and P. M. Huang. 1983. *Soil Sci.* 136:226.

Wang, T. S. C., M. M. Kao, and S. W. Li. 1984. *In* C. H. Chou (ed.), *Tropical Plants.* Institute of Botany, Academia Sinica Monograph Series No. 5, Taipei, pp. 1–9.

Whittaker, R. H. and P. P. Feeny. 1971. *Science.* 171(3973):757.

Wu, M. M. H., S. W. Shieh, C. L. Liu, and C. C. Chao. 1975. *J. Agric. Assoc. China New Series.* 90:54.

Wu, M. M. H., C. L. Liu, and C. C. Chao. 1976a. *J. Chin. Agric. Chem. Soc.* 14(3,4):160.

Wu, M. M. H., C. L. Liu, C. C. Chao, S. W. Shieh, and M. S. Lin. 1976b. *J. Agric. Assoc. China New Series.* 96:16.

5

ALLELOPATHY: THE AUSTRALIAN EXPERIENCE

Department of Agricultural Science
University of Tasmania
Hobart, Tasmania
Australia

Studies of allelopathy in Australia constitute a small but growing proportion of the total chemical and biological research effort. Most examples are associated with plants of agriculture and it is worthy of note that Australia represents the last major land mass on earth to which the major crop and pasture species have been introduced. While ongoing for 200 years, this process is not yet complete. Although stringent plant quarantine procedures are now enforced, in earlier times importations of useful agricultural plants brought with them many of the common insect, disease, and weed species with which they are associated elsewhere. Thus it is convenient to divide this chapter into studies of allelopathy that involve (1) introduced crop, pasture, and weed species and (2) species native to Australia.

ALLELOPATHY BETWEEN INTRODUCED PASTURE AND WEED SPECIES

In the south of mainland Australia annual pastures, regenerating from seed, are an important feature of pastoral agriculture. Some 15 years ago large germination failures in annual medic (*Medicago polymorpha* L., *M. truncatula* Gaertn., and *M. littoralis* Loisel.) pastures occurred in various parts of South Australia. The failure was associated with paddocks in which the cosmopolitan species wireweed (*Polygonum aviculare* L.: Polygonaceae) was dominant (Kloot and Boyce, 1982). Laboratory and field experiments indicated that a water-soluble substance could be leached from aboveground parts of wireweed plants and that this substance inhibited the germination of medic seeds. Removal of aboveground material improved the establishment of medic relative to field quadrats in which wireweed remained in situ (Table 5.1). The increase of the "tops and soil removed" treatment over "tops removed" and "no wireweed" control quadrats was ascribed to removal of the top 1 cm of soil, which contained most of the allelopathic agent, and to compaction of soil in the control area.

Kloot and Boyce (1982) give no indication as to the nature of the allelopathic agent but, from morphological observations, assert that its primary activity appears to be on cell division in the meristem, with a consequent secondary manifestation of radicle deformation.

Stevia eupatoria (Spreng) Willd. (Asteraceae), native to Central America, was introduced into eastern Australia some time before 1933 (the date of the first herbarium specimen held at the Royal Botanic Gardens and National Herbarium, Sydney, New South Wales). *Eupatorium adenophorum* Spreng. (Crofton weed), a member of the same tribe, was also introduced and has become a serious weed in parts of New South Wales. *Stevia eupatoria*, however, whilst maintaining a vigorous stand of approximately 2.5 ha in area near Glen Innes, northern New South Wales, has not spread beyond this area over a period of more than half a century. This is surprising since the weed has effective vegetative and sexual reproduction and, under controlled conditions, shows allelopathy toward the principal introduced legume (*Trifolium repens* L., white clover) in the area it has infested. Leachates of all aboveground parts of the plant are inhibitory to white clover germination and radicle extension (Table 5.2). Air-dried material releases sufficient volatiles to decrease the radicle length of white clover seeds germinating in a closed system but physically separated from the weed material (Lovett, 1982a). Further studies of this species are in progress.

Read (personal communication) considers that allelopathy contributes to the

TABLE 5.1
Percentage establishment of medic, 1 month after sowing, influenced by various soil treatments. (Mean of three plots ± standard error.)

	Wireweed	Tops Removed	Tops and Soil Removed	No Wireweed
Percentage establishment	6.6 ± 1.63	12.2 ± 2.20	28.1 ± 9.42	16.5 ± 4.33

Source. Kloot and Boyce (1982).

TABLE 5.2

Effects of leachates of *Stevia eupatoria* plant parts on germination and early growth of white clover (Means of five replicates.[a])

	Hours from Sowing			
	Germination (x/25)			Radicle Length (mm),
	24	48	72	72
Control	6.80a	16.00a	17.00a	15.2a
Leaves	0.00b	0.20b	1.20b	3.5b
Stems	1.00b	2.40c	4.20c	8.1c
Flowers	0.60b	2.60c	4.00c	6.0bc

Source. Lovett (1982a).

[a]Treatment means identified by the same letter are not significantly different at the 5% level, Studentized Range Test.

competitive ability of *Pennisetum clandestinum* Chiov. (kikuyu grass: Poaceae), an introduction from East Africa which is used as a pasture grass in eastern Australia, particularly in coastal areas. Like *S. eupatoria, P. clandestinum* reproduces vigorously both by seed and vegetatively. Phytochemicals are produced particularly during the decay of stoloniferous material and inhibit the growth of other grass and legume species. Read's observations are supported by preliminary experiments conducted in our laboratory.

Imperata cylindrica (L.) Beauv. (blady grass: Poaceae), native to the old world (Holm et al., 1977) is widely distributed as a vigorous weed of pastures in tropical and warm temperate areas of Australia. Following observation of its allelopathic potential when interfering with the growth of several crop plants (Eussen et al., 1976), Eussen (1978) found several phenolic compounds in fractions of leaf, rhizome, and root extracts and identified vanillic acid in the active fraction of root extracts using paper chromotography. Subsequently, Eussen and Niemann (1981) identified *p*- and *o*-coumaric acid, gentisic acid, vanillic acid, benzoic acid, *p*-hydroxybenzoic acid, vanillin, and *p*-hydroxybenzaldehyde as growth inhibitors in leaf material of *I. cylindrica*. During 1984 Amartalingam (personal communication) identified ferulic, vanillic, *p*-coumaric, and syringic acids from aqueous extracts of both decaying rhizomes and soil from pots in which *I. cylindrica* plants had been grown.

Cenchrus ciliaris L. (buffel grass: Poaceae) is a valued pasture species deliberately, for the most part, introduced to Australia from eastern and southern Africa. The exception is cv. West Australian which Marriott (1955) suggests was introduced between 1870 and 1880 in camel harnesses from India. Cheam (1984) published results of experiments which indicate that *C. ciliaris* excretes a phytotoxic compound that acts as a natural herbicide against the weed species *Calotropis procera* (Willd.) R. Br. (calotrope). The compound has not yet been isolated and identified but it is released from living buffel grass roots, is confined mainly to the top 20 cm of the soil profile, and is "very potent" in its effects on *C. procera* seedlings. In light of other studies (Lovett, 1982b), it is surprising that buffel grass residues do not have any toxic effect on *C. procera* seedlings. This work will be extended to examine the effects of the natural herbicide on other species with a view to developing a commercially valuable compound.

The only report of a pasture weed not only affecting the growth of grasses and a legume but also of other pasture weeds under Australian conditions is provided by Bendall (1975). Working in southern Tasmania, he observed that the growth of annual thistles was restricted to areas not colonized by *Cirsium arvense* (L.) Scop. (creeping thistle: Asteraceae), a perennial probably of European origin. Within a distance of only 2 m a monospecific *C. arvense* population could change to a mixed population of annual thistles. Fresh foliage and roots of *C. arvense* were extracted using either alcohol or water. In bioassays under controlled conditions three "annual" thistles [*Silybum marianum* (L.) Gaertn. variegated thistle, *Carduus pycnocephalus* L. slender thistle, and *Cirsium vulgare* (Savi) ten spear thistle], two grasses (*Hordeum distichon* L. barley and *Lolium perenne* L. perennial ryegrass), one legume (*Trifolium subterraneum* L. subterranean clover) and *C. arvense* itself were used as test species. Both water and ethanol extracts of *C. arvense* roots inhibited germination of its own seeds and of *T. subterraneum*. Water extracts of the foliage showed no phytotoxicity to germination. The extracts had no effect on the germination of any of the other test species but tended to reduce radicle length.

Dried and milled residues of foliage and roots of senescent *C. arvense*, when mixed into a 75:25 sand–peat mixture at 3% plant material by weight (a level of contamination within the range found in the top 15–20 cm of soil under field conditions), also proved to be inhibitory to the growth of all species tested (Table 5.3). Bendall (1975) considered that the demonstration of phytotoxic potential

TABLE 5.3

The effect of *Cirsium arvense* residues on the growth of six test species (Mean dry weights of six replicates, mg, after 4 weeks growth.)

	Residue (percent by weight)				
	Control	3% Roots	3% Foliage	$P < 0.05$	$P < 0.01$
Silybum marianum	327	117	151	77	110
Cirsium vulgare	26	13	14	7	10
Carduus pycnocephalus	208	112	144	38	55
Hordeum distichon	498	249	376	92	131
Lolium perenne	77	37	58	18	26
Trifolium subterraneum	66	53	48	13	18

Source. Bendall (1975).

was indicative of an important role of allelopathy in the aggressiveness of *C. arvense* towards crop, pasture, and annual weed species in crop and pasture associations. Autotoxicity was felt to be important in regulating the growth and longevity of *C. arvense* populations.

Thus it is apparent that although there is interest in allelopathy in Australian pasture systems, the work is largely at a preliminary stage with identification of the compounds concerned made only in the case of *I. cylindrica* and with no evidence of the primary mode of activity in any of the reports cited.

ALLELOPATHY BETWEEN INTRODUCED CROP AND WEED SPECIES

To date, all the published work concerning allelopathy between weed and crop plants in Australia has emanated from our laboratory. In studies of three associations of weeds with crops we have adhered to the principles that (1) aqueous leachates should be prepared with minimal damage to the donor plant, (2) the identity of the allelochemicals should be established, (3) observations of secondary manifestations should be supported by investigation of the primary effects responsible for such manifestations, and (4) field validation of controlled environment observations should be the logical objective of the work.

Grümmer and Beyer (1960) indicated that the yield of *Linum usitatissimum* L. (flax/linseed) could be significantly reduced in the presence of *Camelina sativa* (L.) Crantz (false flax: Brassicaceae) in the field, providing that rain fell at a critical time of year. The adverse effect of the weed on the crop plant was attributed to phytochemicals leached from the *C. sativa* foliage. *C. sativa* is a weed of cultivation in the states of Western Australia, South Australia, Victoria and Tasmania, and was probably introduced from Europe.

In a series of experiments comencing in 1977 we developed a more complete understanding of this system (Figure 5.1). A critical feature of the relationship between *C. sativa* and linseed is the necessity for the presence of bacteria in the *C. sativa* phyllosphere. At least two gram-negative rods, *Enterobacter cloacae* (Jordan) Hormaeche and Edwards and *Pseudomonas fluorescens* (Trevisan) Migula are capable of carrying out the chemical modification, apparently using organic acids leached from the *C. sativa* leaves to fuel their multiplication and metabolic activity. Benzylamine has been identified as an allelochemical in this system. Early growth of Linum is consistently inhibited by this compound, which appears to induce cellular dysfunction, possibly involving a reduced ability of the seedling to metabolize food reserves. There have also been indications that benzylamine can induce hydrophobic conditions in soil that may also be disadvantageous to plant growth (Lovett, 1982b). We take these findings as indications that, under field conditions, benzylamine might have adverse effects on germination and, possibly, other periods of intense metabolic activity. Work in progress seeks to validate this hypothesis and to identify other allelochemicals that may be present in the system.

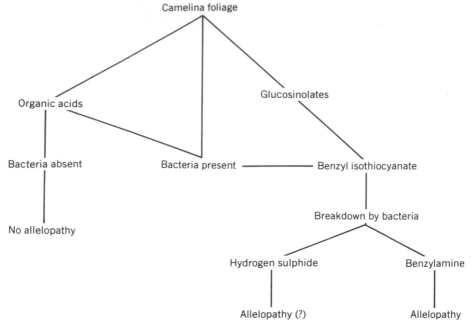

Figure 5.1 *Camelina sativa* and allelopathy.

The bacteria involved in the *C. sativa*/Linum association are cosmopolitan and can function as facultative anaerobes both in the phyllosphere or in soil. Further investigations of their roles in these milieux are in progress.

The role of trichomes in plant defense has been documented by Levin (1973) and other workers. In the association of *Salvia reflexa* Hornem (mintweed: Lamiaceae), an introduction from North America, and *Triticum aestivum* (L.) (wheat), trichomes act as repositories for allelochemicals. The association is common in northwestern New South Wales and southeastern Queensland, where *S. reflexa* is regarded as a weed of significance. Volatiles released from *S. reflexa* foliage retard the germination and early growth of wheat in a closed system through which air is continually circulated. The scent of *S. reflexa* is enhanced when the foliage is wet and it appears that rainfall collapses the trichomes with consequent release of their contents. Thus aqueous washings of foliage show greater inhibitory properties than does the vapor phase.

To date several monoterpenes, including α-pinene, β-pinene, and cineole have been identified from both vapor and leaf washings. An unidentified but potent compound is also present in the washings. Several other workers have identified monoterpenes that manifest a self-defense capability in plants, for example, Muller (1966) in his studies of aromatic shrubs of southern California (which included *S. leucophylla* Greene), Selander et al. (1974) in the association of beetles with Scots pine; Sheehy and Winward (1981) in assessing the relative palatability of *Artemesia* taxa to herbivores, and White et al. (1982) in the grazing behavior of rabbits.

Field validation of the *C. sativa* and *S. reflexa* examples is in progress. A more complete study involves the association of *Datura stramonium* L. (common thornapple: Solanaceae) and *Helianthus annuus* L. (sunflower). *Datura stramonium* is an important weed, present in all Australian states. Both the origin and source of introduction of this species to Australia are obscure. It has been reported as adversely affecting the growth of irrigated summer crops in areas such as northwestern New South Wales (Felton, 1979). *Datura stramonium* contains a number of alkaloids, principally scopolamine and hyoscyamine (Mothes, 1955). Washings of seeds included high concentrations of these alkaloids, both of which were active inhibitors in bioassay (Lovett et al., 1981). The possibility of synergism between the alkaloids remains to be investigated.

In field soils under controlled conditions leachings of the average contents (4.95 g) of one seed capsule of *D. stramonium* significantly reduced the radicle length of sunflower; leachings of 250 g seed for 24 hours gave a further, significant, reduction. This study indicated that the type and amount of clay, the organic matter content, and the particle size of the soil modified allelopathic activity (Table 5.4).

In a lateritic podzolic soil kept under controlled conditions, alkaloids from *D. stramonium* seeds remained at concentrations that were toxic to germinating sunflower seeds for more than 35 weeks. In similar soil in the field the phytotoxicity persisted for 15 weeks.

In several areas of Australia in which *D. stramonium* is a significant weed species double cropping, that is, following a summer crop with a winter crop in the same field, is frequently practiced. The longevity of the *D. stramonium* alkaloids under field conditions suggests that they may remain in soil sufficiently long to pose difficulties in such situations.

During the course of our experiments we were invited to examine soils on the property of an irrigation farmer who had experienced failures of cotton, soy-

TABLE 5.4

Effect of *D. stramonium* seeds and seed washings on radicle growth of *H. annuus* in black earth and lateritic podzolic soils

	H. annuus (72-h Radicle Length, mm)	
Treatment	Black Earth	Lateritic Podzolic
Sterile water	60.7	55.8
D. stramonium seeds	51.0	32.0
D. stramonium seed washings	42.8	22.5
(0.1% LSD = 6.80)		

Source. Levitt and Lovett (1984).

bean, and grain sorghum crops over three summer seasons in fields that were heavily infested with *D. stramonium*. Thin-layer chromatography was used to identify the alkaloids scopolamine and hyoscyamine from the soils of these fields (Table 5.5), where they had been present for up to 7 months. After this relatively lengthy period total alkaloid concentration in the soil was estimated at 0.02% by volume (Levitt and Lovett, 1984).

The presence of hyoscyamine and scopolamine in the seed coat of *D. stramonium* points to an elegant ecological mechanism. In order for the seed to germinate, the alkaloids must be leached from the seed coat but, having been leached, they constitute a powerful barrier to the germination of other species in the vicinity (Lovett et al., 1981). Friedman and Waller (1983) list 18 further species that induce similar phenomena. It was reported (Felton, 1979) that *D. stramonium* was becoming increasingly serious as a weed of irrigated cropping, suggesting that the practice of preirrigation in the affected areas may leach out the alkaloids, permit the *D. stramonium* seed to germinate, and inhibit competition during the vital early phases of seedling growth and development. The observed inhibition of early growth of one potential competitor, sunflower, has been identified with the retarded metabolism of food reserves (Levitt et al., 1984).

On the basis of field experiments carried out during three seasons in central New South Wales, Kemp et al. (1983) argued that the time course of exploitation of space in crop communities is likely to be of greater importance in yield determination than is allelopathy. This may be the case, but the consistent identification of germination and early seedling growth as being vulnerable to allelochemicals, in the work here reviewed as elsewhere, suggests that the significance of allelopathic phenomena should not be underestimated.

TABLE 5.5

Effect of *D. stramonium* infestation on phytotoxicity of black earth soil to *H. annuus*

Sample	H. annuus Radicle Length (mm) (means of 80)[a]		Percentage Reduction	Probability
	Control	Infested		
1	40.5d	15.9ab	60	<0.001
2	35.3cd	20.2b	42	<0.001
3	33.7cd	16.8ab	50	<0.001
4	29.9c	13.7a	54	<0.001

Source. Levitt and Lovett (1984).

[a]Means followed by the same letter are not significantly different at the 5% level, Studentized Range Test.

ALLELOPATHY BETWEEN INTRODUCED CROP SPECIES

Reports of crop plants employing allelopathy to their advantage remain few, the work of Putnam and Duke (1974) with *Cucumis sativus* L. (cucumber: Cucurbitaceae) and Fay and Duke (1977) with *Avena* spp. (oats: Poaceae) being well-known examples. In Australia the only published study is of sunflower, a species with which allelopathy has been identified in American indigenous material (Rice, 1974).

For a period of 3 years little evidence of allelopathy in naturalized (road-side) sunflowers was found in New South Wales (Lovett et al., 1982). However, an unreleased hybrid and its male parent exhibited allelopathic activity against wheat in bioassay (Table 5.6), suggesting the possibility of selecting for allelopathy in cultivated sunflower, if this was perceived as a desirable stratagem.

ALLELOPATHY AND NATIVE AUSTRALIAN PLANTS

Tree Species

There are few reports of studies in Australia of allelopathy by *Eucalyptus* species (family Myrtaceae), perhaps the most characteristic genus of the Australian flora. Water-soluble phenolics, volatile terpenes, and other plant-growth inhibitors are found in eucalypt leaves, bark, and roots (Silander et al., 1983). Conversely, examples of allelopathy by eucalypts in other countries, for example, *E. globulus* ssp. and *E. camaldulensis* Dehnh. in America (Del Moral and Muller 1969, 1970) and *E. microtheca* F. Muell. in Iraq (Al-Mousawi and Al-Naib, 1975), are well known.

Studies in coastal heath vegetation in southern Victoria (Del Moral et al., 1978) indicated that allelopathy was a significant factor in suppressing heath species beneath the eucalypt canopy. *Leptospermum myrsinoides* Schldl. (Myrtaceae) and *Casuarina pusilla* E. D. Macklin (Casuarinaceae) dominate the veg-

TABLE 5.6
The effects of washings of green sunflower foliage on shoot height and root length of wheat after 120-h incubation

Plant Part	Height or Length (mm)[a]			P
	Control	Hybrid	Parent	
Shoot height	47.0a	41.0b	40.0b	<0.001
Longest seminal root	68.3a	61.2b	59.8b	<0.001
Shortest seminal root	24.7a	29.4b	27.7b	<0.05

[a]Means identified by a common letter are not significantly different at the 5% level, Studentized Range Test.

etation in the absence of *E. baxteri* ssp., but are absent or severely suppressed in its presence. Characteristics of the vegetation surrounding isolated individuals of *E. baxteri* are shown in Figure 5.2.

Investigations of total interference between *E. baxteri* and the suppressed species suggested that competition for moisture and nutrients were not of major importance in the relationship. Similarly, because of the sparse leaf canopy of the eucalypt, competition for light appeared to be insignificant. Del Moral et al. (1978) concluded that the suppression was associated with allelochemicals leached from the leaf canopy. Foliar leachates of *E. baxteri* contained gentisic and ellagic acids and large quantities of phenolic glycosides. The extracts were demonstrated to be toxic in bioassays. The litter of *E. baxteri* proved to be readily leached of phenolic acids and other substances and extracts from soil beneath *E. baxteri* were more inhibitory than those from soil beneath heath. The soil was rich in acidic resins and, possibly, terpenoids but free phenolic acids were not obtained, suggesting that chemical change occurred when phenolics in leaf leachates entered the soil but that this did not remove their toxicity.

Fire is an important factor in the ecology of the heath vegetation and the findings of Del Moral et al. (1978) parallel those of Muller (1966) in California, inasmuch as *E. baxteri* establishing after disturbance by fire does not show suppression of understory vegetation until it has exceeded a height of 2 m, when it gradually begins to exert dominance.

Ashton (1981), in discussing the regeneration of eucalypt-dominated vegetation in Australia, comments that fire "is a powerful stimulant to regeneration due to favorable seedbed, soil enrichment and the relief from overstory and understory competition." In unburned forest allelopathic effects on seedlings may act synergistically, involving soil lipids, root exudates, or microbiological antagonisms.

In the absence of fire, the germination and survival rate of *E. regnans* F. Muell. in mature forests of southern Victoria is low (Ashton and Willis, 1982). Seedlings are apparently weakened by allelopathic agents liberated by understory species such as *Cassinia aculeata* (Labill.) R.Br. (Asteraceae) and *Pittosporum undulatum* Vent. (Pittosporaceae). Foliar leachates of *E. regnans* itself

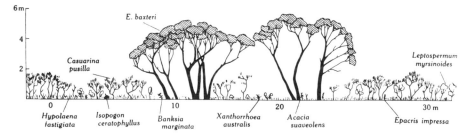

Figure 5.2   A vegetation profile across a well-developed copse of *Eucalyptus baxteri*. In this dense heath vegetation *Casuarina pusilla* is dominant and *Leptospermum myrsinoides* subdominant. Both are absent where the canopy is closed. After Del Moral et al. (1978).

showed no inhibitory activity in bioassay except when the leaf leachates were concentrated well beyond the limits likely to occur in nature (Willis, 1980). The apparent stress induced by allelochemicals of other species on *E. regnans* seedlings apparently predisposed them to further damage by factors such as rhizosphere microorganisms. Of the rhizosphere microflora, *Cylindrocarpon destructans* (Zins.) Scholten was implicated because of its ability to act both as a weak parasite and to exude a powerful toxin and antibiotic. The closed canopy of mature *E. regnans* forests may aid the development of this organism since it favors cool soil conditions.

Unlike *E. regnans*, *E. delegatensis* R. T. Bak. demonstrates autotoxicity in some Tasmanian forests where the effects of fire have also been documented. In a preliminary experiment carried out by Goodwin (personal communication), the area surrounding a large, vigorous, mature individual of *E. delegatensis* was divided into four sectors. Treatments applied were combinations of burning (a hot fire of 2–3 hours duration), trenching (excavating to a depth of 1 m around the sector), and the absence of these treatments. Seedlings of *E. delegatensis* were introduced into each treatment area. Trenching appeared to confer some advantage relative to the unburned/untrenched control, but the burned and trenched sector showed mean increments of seedling growth over a 22-month period which were as much as eight times higher than in the control. Bioassays using *L. usitatissimum* as the test species, with aqueous leachates of soils from the burned and trenched treatments, produced significant differences in radicle growth relative to a water control. Distance from the mature individual tree also affected Linum response (Figure 5.3). The general stimulation of radicle growth

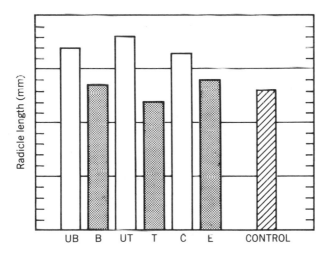

Figure 5.3 Responses of Linum seedlings to leachates of *Eucalyptus delegatensis* forest soils. B, burned; UB, unburned; $P < 0.001$. T, trenched; UT, untrenched; $P < 0.001$. C, close to tree; E, edge of tree canopy, $P < 0.05$.

may indicate the presence of a relatively low concentration of allelochemicals (Lovett, 1982a).

In a second experiment a mature tree was felled and seedlings of *E. delegatensis* were planted on burned and unburned transects running from the stump. As in the previous experiment there were large differences in seedling growth, with burning conferring a great advantage. Soil analyses indicated no gross difference between burned and unburned areas except in organic matter content. The probability that allelochemicals from the crown of the felled tree were adsorbed onto the organic matter fraction and remained toxic in the surface soil is the subject of further investigation.

From these experiments fire emerges as a feature of considerable significance in "natural" Australian ecosystems. This finding is in agreement with the conclusion that allelopathy is often a feature of ecosystems formed under conditions of repeated disturbance (Muller, 1966), especially fire (Rabotnov, 1981). In reviewing the effects of fire on the soil environment, Clinnick (1984) notes that these are primarily on the topmost 5 cm of soil. Various chemical and structural changes result. There is a tendency to increase soil pH and the availability of major elements such as calcium, phosphorus, and potassium. Nitrogen is lost by volatilization, but biotic nitrogen fixation tends to increase following fire. Low-intensity fires have a minimal effect on soil organisms, but the reestablishment of such organisms following high-intensity fires may vary from a few days to several years, depending on site. It is estimated that, in terms of its ability to reduce soil loss, 18 months is the average time span for the reestablishment of vegetation cover. It is during this period that workers such as Muller (1966) have noted the reestablishment of a diverse flora in areas where allelopathy is noted in older, shrub-dominated stands. As in the example of Goodwin, the implications are that volatile and other compounds adsorbed in the organic matter fraction are largely dissipated owing to the effects of fire, permitting growth of species that are otherwise inhibited.

E. delegatensis is a valuable forest tree which is widely used in the timber industries of Tasmania. At a second location in northeastern Tasmania, some 200 m higher than that used by Goodwin, the species is in decline, apparently as a result of interference by rain-forest species. The phenomenon has been investigated for more than 20 years. Ellis (1964) reported the dieback of *E. delegatensis*, noting that the health of the eucalypts tended to decrease with increasing time since burning. Burning was a management technique employed by aboriginals for hundreds of years to regenerate grazing in the eucalypt forests in this area. It has been absent for the greater part of European settlement and particularly since the forests have been managed for timber production. Ellis (1964) concluded that the health of the eucalypts deteriorated with increasing age of the understory, that is, with decreasing frequency of burning. The age of the understory at which dieback commenced decreased with increasing altitude, for given soil drainage conditions. The relatively high rainfall in the area concerned and the presence of rain-forest species in nonmanaged areas suggested that the climax vegetation was rain-forest and that the eucalypts were maintained in a sub-climax stage by fire.

Ellis (1971) carried out further studies and reported an inverse relationship between the incidence of eucalypt dieback and the annual average soil tempera-ture. The rain-forest understory species were implicated through their intercep-tion of solar radiation. It was noted that they probably also increased precipita-tion since microphyllous rain-forest understory species were likely to be more efficient in condensing fog particles (a common feature of the area) than would be a relatively broad-leaved eucalypt stand. It was also noted that organic-mat-ter content increased with the development of a rain-forest understory in the eucalypt forest. Overall, Ellis (1971) concluded that the development of the rain-forest understory in the absence of burning led to a fall in soil temperature of 2–3°C. Such a loss was detrimental to the ability of eucalypt roots to supply the crowns with water and nutrients and this was expressed in the dieback phenome-non.

Ellis et al. (1980) reported on attempts to rejuvenate *E. delegatensis* stands by felling the understory; felling and burning the understory once; and felling the understory and burning repeatedly, as compared with an untreated control. As in the experiments of Goodwin, higher rates of growth were observed on burned plots, the best results being obtained where dieback was most severe as com-pared with control and felled-only plots (Table 5.7). Because felling conferred no improvement in rate of growth it was concluded that competition for moisture or nutrients by the living understory [a complex community in which *Leptosper-mum lanigerum* (Ait) Sm. was often a significant component] was not an impor-tant factor. Attention was focused on the accumulation of organic debris on the forest floor and in the topsoil and it was noted that removal of this debris by fire was associated with the prompt establishment of healthy eucalypt regrowth. El-lis et al. (1980) canvassed the possibility that allelopathy may be involved in the

TABLE 5.7

Stimulation of growth due to burning the understory in a Eucalypt Forest (Ratio of adjusted mean annual growth in tree diameter on burned plots to that on unburned plots, 1963/64-1976)

	Block			
	East (No dieback)	West (Slight Dieback)	South (Moderate Dieback)	North (Severe Dieback)
Burned treatments / Unburned treatments	1.06	1.36	1.46	1.75

Source. Modified from Ellis et al. (1980).

dieback phenomenon. Analyzing the evidence presented in earlier papers (Ellis 1964, 1971) it appeared that lower summer temperatures, higher precipitation, and accumulation of organic matter under the rain-forest understory could create conditions that were relatively unfavorable to the breakdown of humus-accumulated phytotoxins by microorganisms.

Our research group is now cooperating with Ellis in attempts to validate this hypothesis. Studies of allelopathy involving soil have, to date, been limited by the availability of suitable techniques for isolating and monitoring possible soil-trapped allelochemicals. One approach is to utilize trapping media to isolate the compounds involved. Tang and Young (1982) and Brown et al. (1983) have described work in which columns containing XAD-4 resin were used to isolate potential allelochemicals from solutions cycled through root systems growing in inert media. This technique could, presumably, be extended to soil. A complementary technique involves drawing headspace samples through short columns packed with Tenax GC. Detection is achieved by thermal desorption directly into a GC/MS system to enable conclusive identification of components by their GC retention times and mass spectra. Lundgren and Stenhagen (1982) describe the successful use of Tenax GC for the adsorption of wound-emitted leaf volatiles in some species of *Thymus*. We have successfully used Tenax GC traps to monitor monoterpenes emanating from soils collected from beneath a *E. delegatensis* forest. Results, to date, indicate significant differences in the patterns of chemicals emitted under forest maintained in a healthy condition by burning and in stands invaded by understory rain-forest species. Confirmation that allelopathy is a significant factor in this community would lend support to the observation that in natural ecosystems dominant or climax vegetation has a relatively greater capacity to produce inhibitory substances than have competing species (Rabotnov, 1981).

It is intriguing that in two of the small number of Australian studies of eucalypts and allelopathy, *E. delegatensis* appears as both the perpetrator and the victim of allelopathic activity.

Leptospermum lanigerum was noted as being an important component of *E. delegatensis* understory. It is of interest that Di Stefano and Fisher (1983) associate allelopathic activity with invasion of three types of "native" plant communities in southern Florida by *Melaleuca quinquenervia* (Cav.) S. T. Blake, a plant that shares the family (Myrtaceae) and subfamily (Leptospermoideae) of *Leptospermum*. *M. quinquenervia* appears to enjoy the advantages of freedom from organisms that maintain it in equilibrium in its native habitat. It is ironic that an Australian native species should display similar characteristics in America to those that have contributed to the success of the many weeds introduced into Australia during the past 200 years.

Allelopathy is implicit in the so-called "halo effects" which are a common feature of native vegetation in Australia (Lange and Reynolds, 1981). They may be expressions of enhanced or inhibited growth of the vegetation beneath species of *Acacia*, (Mimosaceae), *Casuarina*, and *Eucalyptus*.

Story (1967) identifies the dominant, tree, species as the cause of the halos in grassland (Table 5.8), which disappear if the trees are felled. Explanations for the effect include climatic modification, channelling of water down the stem of the dominant species, the effects of stock congregating beneath the species for shade or shelter, and the influence of tree roots. In his studies, Story (1967) concluded that the symmetry of trees and halos indicated that root effects were the most likely explanation of the phenomenon. Soil-moisture profiles obtained using a neutron moisture meter demonstrated that competition for water was not a satisfactory explanation; competition for plant nutrients by tree roots was also considered but, noting that the halo effect was absent in high rainfall areas where any depletion of nutrients might be expected to be more pronounced, Story (1967) concluded that the halo effect was most likely associated with soil microorganisms or root exudates. In a more recent study (Lange and Reynolds, 1981) similar reasons for the halo effect are advanced, but the authors indicate that no single factor alone provides an adequate explanation for all the characteristics noted in field observations of the halo effect.

The allelopathic potential of a native plant being enhanced by an intermediary is documented in the work of Silander et al. (1983), who associate the bare areas which surround many species of eucalypt with the effects of allelochemicals concentrated in insect frass. Working with *Eucalyptus globulus* ssp. *bicostata* (Maiden et al.) Kirkp., Trenbath and Fox (1976) showed that the frass of insects feeding on the eucalypt foliage caused an inhibition of germinating seeds of *Sinapis alba* L. (white mustard) that was much greater than that due to any of several preparations derived from leaves of the eucalypt. Water leachates of the leaves had, in fact, a slight promotory effect. Direct and indirect estimates suggest that total frass fall below eucalypts may vary from less than 200 to 2500 or more kilograms per hectare per annum (Silander et al., 1983). Variation in frass tolerance was observed among species associated with the eucalypt. *Trifolium repens* and *Themeda australis* (R.Br.) Stapf. (kangaroo grass) were significantly suppressed by frass, whereas the colonizing species *Hypochoeris radicata* L. (catsear) and *Stipa falcata* Hughes (speargrass) showed no such effect (Figure 5.4). Silander et al. (1983) concluded that frass applied to soil in amounts likely

TABLE 5.8
Evidence for the halo effect in the Hunter Valley of New South Wales

Native Grass Species	Halos (mean per transect)	Grassland (mean per transect)	5% LSD	1% LSD
Bothriochloa	2.31	4.87		2.27
Eragrostis	0.27	2.64	2.34	
Sporobolus	0.15	1.33	1.16	

Source. Modified from Story (1967).

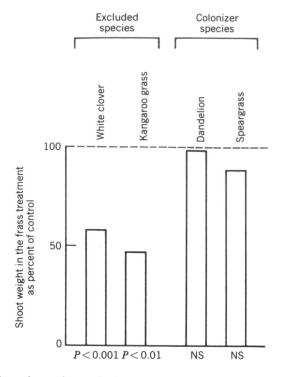

Figure 5.4   Effects of insect frass on herbaceous species. After Trenbath and Silander (1979).

to be found in the field accurately mimicked field observations of excluded and colonizing species.

Eberhard et al. (1975) found that koala bears modify the secondary compounds of eucalypts as the compounds pass through their gut. Preliminary chromatographic analyses revealed some differences and some similarities between extracts of *E. globulus* ssp. *bicostata* leaves and insect frass. Thus Silander et al. (1983) conclude that insects may also modify secondary compounds as they pass through the gut. If so, this example would join the relatively small number of indirect allelopathic effects documented to date, for example, Patrick (1955), Grodzinski (1978), and Lovett and Sagar (1978).

A chrysomelid beetle was identified by Silander et al. (1983) as being responsible for the modification of allelochemicals. De Little and Madden (1975) demonstrated that the larvae of two such eucalypt defoliators showed distinct host preference, both growth and development of the beetle being enhanced when the preferred host was available as a food source.

The synthesis of defensive chemicals from eucalypt foliage ingested by chrysomelid larvae was reported by Moore (1967). Among the compounds produced way hydrogen cyanide, to which the larvae themselves were very resistant. Moore (1967) observed that secretions from these larvae could knock down predatory ants within 20 seconds and kill them within 2 minutes. It was observed that the larvae everted vesicles on the hindbody when threatened and exuded drops of defensive secretion.

The reverse chronology of the examples described above is indicative of a general reluctance among research workers to recognize and accept complex defensive behavior involving plants, directly or indirectly.

Shrub Species

Beeston and Webb (1977) discuss the ecology and control of *Eremophila mitchellii* Benth. (Myoporaceae), one of several shrub species that is invading semi-arid and arid grazing lands in the inland of eastern Australia. They conclude that disturbance of natural communities by land clearing and subsequent overgrazing is the initial cause of the problem. Then *E. mitchelli* becomes an important woody and inedible weed species accounting for between 8 and 22% loss of grazing capacity on properties surveyed (Booth and Barker, 1981). These workers consider that factors important in initiating shrub invasion include reduction in the incidence of fire, overgrazing by rabbits and domestic stock, and the occurrence of a series of abnormally wet years. Harrington (personal communication) suggests that efficiency in competition for moisture accounts for the dominance of the shrub but, on the basis of recent evidence, Walker (personal communication) believes that allelopathy is at least partly responsible. Given that Muller (1966) suggested that allelopathy is most likely to manifest itself in a dry situation, and that these communities have similarities in behavior with those that he documented in California, further research into these Australian species is warranted.

Other Species

Clarke and Hannon (1967) noted that sharply defined vegetation zones often occurred in mangrove swamp and salt-marsh communities of the Sydney region. The possibility that the zonation occurred as a result of species interactions involving phytotoxic exudates, shading, or other negative effects between associated species was canvassed. Although leachates and macerates of *Avicennia* (mangrove: Avicenniaceae) and *Casuarina*, dominants in their respective zones, were used, Clarke and Hannon (1971) were unable to find convincing evidence for allelopathy by these species. Adam (1981) regards this zonation as "a static zonation of a complex environmental gradient."

In what is claimed to be the first demonstration of the presence of allelochemicals in marine habitats Coll et al., (1982) describe an apparatus designed for in situ sampling of substances released from sessile marine organisms. Using C18 reversed-phase SEP-PAK cartridges for collection of compounds and thin-layer chromatography to compare these compounds with extractives of the soft coral tissue from which they had been released, Coll et al. (1982) found that two toxic terpenoids were present in the coral tissue and in surrounding sea water.

The concentration of terpenoids in sea water was estimated at 1–5 mg/l and the authors considered that they could be of significance in aggression between species of marine taxa.

ALLELOPATHY AND THE RHIZOSPHERE

For more than 30 years Dr. A. D. Rovira has been associated with studies of the rhizosphere/rhizoplane. Much of his work falls within the ambit of allelopathy, although he makes infrequent use of the term. Several interrelated strands of investigation emerge when his contribution to the literature is considered consecutively.

Among his early studies, Rovira (1956) published the results of paper chromatography work in which 27 amino-acid compounds were found to be excreted by roots of peas and oats. He concluded that, at least for the first 3 weeks of growth of both peas and oats, actual excretions formed the greater part of the material liberated from the roots and that such excretions must be very important in stimulating the growth of microorganisms in the rhizosphere. He also recognized that some of the material he identified was liberated from sloughed off root cells, that it was difficult to separate this from other components, and that the type and amount of amino compound liberated changed with age. In subsequent studies it was established that rhizosphere and rhizoplane microorganism populations can influence many plant characteristics, including (1) root morphology, (2) root/shoot weight ratios, (3) uptake of calcium and rubidium, (4) uptake of phosphorus and sulphur, (5) mineral content, (6) rate of development and onset of flowering, and (7) crop yield and physiological processes (Rovira, 1965). The mechanisms involved may include the production by microorganisms of metal chelating compounds, growth regulating substances, and protective agents against root pathogens.

Rovira and McDougall (1967) identified a wide range of compounds in wheat root exudates (Table 5.9). This and other work was reviewed by Rovira (1969), who concluded that although the quantities of organic compounds exuded from roots is small (seldom exceeding 0.4% of carbon synthesized), these compounds exert very strong influences on soil microorganisms and may be significant in affecting the availability of nutrients. Some plants may produce exudates that are toxic to neighbors or to germinating seed. In this paper he points to the need for multidisciplinary approaches to resolve outstanding questions and the need to use then novel techniques such as C^{14}-labeling in such studies.

In 1972 Dr. Rovira was awarded the Australian Medal of Agricultural Science. In his address on the occasion of the award (Rovira, 1972) he reported on the use of $^{14}CO_2$ to show three forms of exudation from seedling roots. These were soluble compounds (which he had already studied), water-insoluble material (polysaccharides and sloughed-off root cells), and volatiles (70–80% of which were CO_2). He also presented an electron micrograph of the wheat rhizosphere, thus demonstrating the usefulness of another novel technique. The pos-

TABLE 5.9
Compounds reported in wheat root exudate

Sugars	Amino Acids	Organic Acids	Nucleotides Flavonones	Enzymes
Glucose	Leucine and isoleucine	Oxalic	Flavonone	Invertase
Fructose	Valine	Malic	Adenine	Amylase
Maltose	y-Aminobutyric acid	Acetic	Guanine	Protease
Galactose	Glutamine	Propionic		
Ribose	α-Alanine	Butyric		
Xylose	β-Alanine	Valeric		
Rhamnose	Asparagine	Citric		
Arabinose	Serine	Succinic		
Raffinose	Glutamic acid	Fumaric		
Oligosaccharides	Aspartic acid	Glycolic		
	Glycine			
	Phenylalanine			
	Threonine			
	Tyrosine			
	Lysine			
	Proline			
	Methionine			
	Cystathionine			

Source. Rovira and McDougall (1967).

sibility of manipulating rhizosphere microorganisms as an alternative to direct attack on pathogens using systemic fungicides was also discussed. It was considered that foliar sprays, when leaked from roots, could stimulate bacterial antagonists to root pathogens in the rhizosphere. Such phenomena could be termed allelopathic (*sensu* Molisch, 1937).

In the address several areas for future study were identified. Thus Rovira et al. (1974) applied direct microscopy to the roots of eight grassland species to estimate numbers of bacteria present. These were approximately 10 times greater than those indicated by the conventional plate-counting method; however, although the technique was precise, it was very demanding on the observer. Rovira and Campbell (1974) then demonstrated the potential of scanning electron microscopy for rhizosphere studies, using the technique to describe microorganisms on the roots of wheat. Subsequently, transmission electron microscopy has been applied to study the ultrastructure of the wheat rhizosphere (Foster and Rovira, 1976) and, most recently, the ultrastructure of the root–soil interface (Foster et al., 1983). These studies have produced much new information on the type and location of the microflora of roots and have given indications as to the site and method of root exudation.

Concurrently, the possible role of root exudates in plant protection has been considered. Nematodes, for example, show specificity in their attraction to roots and some nematodes are sensitive to root exudates (Rovira, 1969). Harnessing such substances to give a degree of natural biological control could be most beneficial since organisms such as the cereal cyst nematode (*Heterodera avenae* Woll.) cause substantial losses to cereal production. Rovira et al. (1981) demonstrated that the use of synthetic nematocides can give yield increases of wheat up to 0.9 t/ha in South Australia. Damage caused by the nematode occurs early in the life cycle of the wheat (Simon and Rovira, 1982), but in this study there is no indication as to the likelihood of developing a natural nematocide, as described by Sayre et al. (1965), as an alternative to the synthetic ones currently employed.

Cook and Rovira (1976) report on natural biological control of the take-all fungus of wheat [*Gaeumannomyces graminis* (Sacc.) Arx and Olivier var. *tritici* Walker]. Soils can become suppressive to the fungus following such treatments as the addition of organic matter, elevation of soil temperature, or minimal tillage. Alternatively, there is the development of "take-all decline," where suppression develops over 2–3 yr of wheat monoculture with severe take-all infection. This soil becomes immune to subsequent outbreaks of the fungus if cropped exclusively with wheat or barley. The phenomenon is described as a "specific antagonism" and the evidence of Cook and Rovira (1976) strongly suggests that specific strains of *Pseudomonas fluorescens*, an organism identified with modification of allelochemicals in the phyllosphere (Lovett and Jackson, 1980), are responsible.

Newman et al. (1977) describe experiments in which fumigation and irradiation to sterilize soil showed that soils that were similar chemically could be maintained with different microbial populations under glasshouse conditions. These populations were associated with differences in the growth of higher plants and in the balance between mixtures of higher plant species. For example, fumigation altered the balance, in mixture, in favor of *Lolium perenne* L. (perennial ryegrass) against *Rumex acetosa* L. (sorrel) or *Trifolium repens*, and in favor of *R. acetosa* against *T. repens*. Although Newman and Rovira (1975) found that field distribution patterns of these and five other pasture components in the United Kingdom could possibly be attributed to allelopathy, the term is not used in the context of Newman et al. (1977).

ALLELOCHEMICALS AND PLANT RESIDUES

An area in which soil microorganisms are intimately involved is that of allelochemicals produced after death and during decay of plant residues. Interest in this area has been stimulated by the widespread adoption of stubble retention/reduced cultivation systems in recent times.

Kimber (1967; 1973a,b), working in Western Australia, was among the earliest of the current generation of workers in this field. His findings that cold aqueous extracts of wheat straw produced inhibitors of early seedling growth, that the degree of inhibition was related to the period of decomposition, that prolonged weathering reduced inhibition, and that roots of species used in bioassays were more sensitive to inhibitory chemicals than were shoots have been amplified in succeeding years, notably by J. M. Lynch and his co-workers in the United Kingdom.

In Australia, however, interest in residue phytotoxins has been limited. Lovett and Jessop (1982) and Jessop and Lovett (1982) studied a range of 12 crop plants (four cereals, five legumes, and three oilseeds) and found that all produced residues which had the potential to significantly affect the early growth of wheat under controlled and field conditions. The incorporation of crop residues into soil enhanced phytotoxicity, as observed by Lynch and Cannell (1980).

Subsequently, apparent selectivity in weed control by crop residues has been observed (Purvis et al., 1985). In particular, residues of *Sorghum bicolor* (L.) Moench s.lat. selectively reduced establishment and dry weight of *Echinochloa* spp. in the field, but the broad-leaved weed *Hibiscus trionum* L. (bladder ketmia) was unaffected (Lovett, 1983). This finding is in accordance with that of Putnam and DeFrank (1979), who reported a 90% reduction in weed biomass under a sorghum mulch.

The possibility of double cropping in parts of Australia, where a second crop may follow the first within a few weeks or even days, indicates that there is merit in further examination of these phenomena, particularly if selective weed control may be attained using the same practice.

CONCLUSIONS

From the range of studies reviewed in this chapter (the locations of which are shown in Figure 5.5), a number of topics offer exciting possibilities for further investigation. Soil type and surface litter are important modifiers of allelochemicals; microorganisms may bring about important changes in allelochemicals, in the phyllosphere, in the rhizosphere, and in soil more generally; other organisms, such as insects, may also play an important modifying role; synergism between allelochemicals is a largely neglected field of research but is of potential significance when numbers of chemicals are active simultaneously.

In addition, the basic deficiencies in much research concerning allelopathy remain. Too few chemicals are identified; even fewer are quantified. Too many phenomena are described; too few rigorous investigations of the primary effects of allelochemicals are carried out. Serious limitations to the acceptance of allelopathy as an ecological factor of significance will remain until these shortcomings are rectified.

ACKNOWLEDGMENTS

The cooperation of Dr. W. C. Potts, Dr. P. Adam, Mr. R. Amartalingam, Mr. C. A. Booth, Dr. R. C. Ellis, and Mr. A. Goodwin in providing information for inclusion in this chapter is gratefully acknowledged. In addition Drs. Adam and Ellis and Mr. Booth generously made available photographs.

Figure 5.5 Location of studies of allelopathy in Australia.

REFERENCES

Adam, P. 1981. *Wetlands*. 1:8.

Al-Mousawi, A. H. and F. A. G. Al-Naib. 1975. *J. Univ. Kuwait (Sci.)*. 2:59.

Ashton, D. H. 1981. Mode of regeneration of eucalypt dominated vegetation in Australia. *Proc. 13th Bot. Congr*. Sydney, pp. 21–28.

Ashton, D. H. and E. J. Willis. 1982. *In* E. I. Newman (ed.), *The Plant Community as a Working Mechanism*. Blackwell Scientific Publications, Oxford, UK, pp. 113–128.

Beeston, G. R. and A. A. Webb. 1977. The ecology and control of *Eremophila mitchellii*. *Technical Bulletin No. 2*. Botany Branch, Department of Primary Industries, Brisbane, 84 pp.

Bendall, G. M. 1975. *Weed Res*. 15:77.

Booth, C. A. and P. J. Barker. 1981. *J. Soil Conserv. Serv*. New South Wales. 37:65.

Brown, R. L., C. S. Tang, and R. K. Nishimoto. 1983. *Hort Sci*. 18:316.

Cheam, A. H. 1984. *J. Agr. West. Aust*. 12:69.

Clarke, L. D. and N. J. Hannon. 1967. *J. Ecol*. 55:753.

Clarke, L. D. and N. J. Hannon. 1971. *J. Ecol*. 59:535.

Clinnick, P. F. 1984. A summary-review of the effects of fire on the soil environment. *Technical Report Series*. Soil Conservation Authority of Victoria. 24 pp.

Coll, J. C., B. F. Bowden, and D. M. Tapiolas. 1982. *J. Exp. Mar. Biol. Ecol*. 60:293.

Cook, R. J. and A. D. Rovira. 1976. *Soil Biol. Biochem*. 8:269.

De Little, D. V. and J. L. Madden. 1975. *J. Aust. Entomol. Soc*. 14:387.

Del Moral, R. and C. H. Muller. 1969. *Bull. Torrey Bot. Club*. 96:467.

Del Moral, R. and C. H. Muller. 1970. *Am. Midl. Nat*. 83:254.

Del Moral, R., R. J. Willis, and D. H. Ashton. 1978. *Aust. J. Bot*. 26:203.

Di Stefano, J. F. and R. F. Fisher. 1983. *Forest Ecol. Manage*. 7:133.

Eberhard, H. J., J. McNamara, R. J. Pearse, and I. A. Southwell. 1975. *Aust. J. Zool*. 23:169.

Ellis, R. C. 1964. *Aust. Forestry*. 28:75.

Ellis, R. C. 1971. *Aust. Forestry*. 35:152.

Ellis, R. C., A. B. Mount, and J. P. Mattay. 1980. *Aust. Forestry*. 43:29.

Eussen, J. H. H. 1978. *Studies on the Tropical Weed Imperata cylindrica (L.) Beauv. var. major*. Regional Centre for Tropical Biology, Bogor. p. 5.

Eussen, J. H. H. and G. J. Neimann. 1981. *Zeitschrift für Pflanzenphysiol*. 102:263.

Eussen, J. H. H., S. Slamet, and D. Soeroto. 1976. *Biotrop. Bull*. 10:24.

Fay, P. K. and W. B. Duke. 1977. *Weed Sci*. 25:224.

Felton, W. L. 1979. The competitive effect of *Datura* species in five irrigated summer crops. *Proc. 7th Asian-Pacific Weed Sci. Soc. Conf*., Sydney, pp. 99–104.

Foster, R. C. and A. D. Rovira. 1976. *New Phytol*. 76:343.

Foster, R. C., A. D. Rovira, and T. W. Cock. 1983. *Ultrastructure of the Root-Soil Interface*. The American Phytopathology Society, St. Paul, MN. 157 pp.

Friedman, J. and G. R. Waller. 1983. *J. Chem. Ecol*. 9:1107.

Grodzinski, A. M. 1978. *In* A. M. Grodzinski (ed.), *Problemy Allelopatii*, Naukova Dumka, Kiev.

Grümmer, G. and H. Beyer. 1960. *In* J. L. Harper. (ed.), *The Biology of Weeds*, Symposium of the British Ecological Society, Blackwell, Oxford, pp. 153–157.

Holm, L. G., D. L. Plucknett, J. V. Pancho, and J. P. Herberger. 1977. *The World's Worst Weeds*. University Press of Hawaii, East-West Centre, Honolulu, HI. 609 pp.

Jessop, R. S. and J. V. Lovett. 1982. The effect of stubble residues on the germination and early growth of wheat. *Proc. 2nd Aust. Agron. Conf*., Wagga Wagga, p. 245.

Kemp, D. R., B. A. Auld, and R. W. Medd. 1983. *Agr. Sys.* 12:31.

Kimber, R. W. L. 1967. *Aust. J. Agr. Res.* 18:361.

Kimber, R. W. L. 1973a. *Plant Soil.* 38:347.

Kimber, R. W. L. 1973b. *Plant Soil.* 38:543.

Kloot, P. M. and K. G. Boyce. 1982. *Aust. Weeds.* 1(3):11.

Lange, R. T. and T. Reynolds. 1981. *Trans. R. Soc. South Aust.* 105:213.

Levin, D. A. 1973. *Q. Rev. Biol.* 48:3.

Levitt, J. and J. V. Lovett. 1984. *Plant Soil.* 79:181.

Levitt, J., J. V. Lovett, and P. R. Garlick. 1984. *New Phytol.* 97:213.

Lovett, J. V. 1982a. *Aust. Weeds.* 2:33.

Lovett, J. V. 1982b. *In* J. S. McLaren (ed.), *Chemical Manipulation of Crop Growth and Development*. Butterworth, London, pp. 93–110.

Lovett, J. V. 1983. Allelopathy and weed management in cropping systems. *Proc. 9th Conf. Asian-Pacific Weed Sci. Soc.* Manila, pp. 31–46.

Lovett, J. V. and H. F. Jackson. 1980. *New Phytol.* 86:273.

Lovett, J. V. and R. S. Jessop. 1982. *Aust. J. Agr. Res.* 33:909.

Lovett, J. V. and G. R. Sagar. 1978. *New Phytol.* 81:617.

Lovett, J. V., J. Levitt, A. M. Duffield, and N. G. Smith. 1981. *Weed Res.* 21:165.

Lovett, J. V., S. A. Fraser, and A. M. Duffield. 1982. Allelopathic activity of cultivated sunflowers. *Proc. 10th Intl. Sunflower Conf.* Surfers Paradise, pp. 198–201.

Lundgren, L. and G. Stenhagen. 1982. *Nord. J. Bot.* 2:445.

Lynch, J. M. and R. Q. Cannell. 1980. *Plant Residues*. Agriculture Department and Advisory Service Reference Book Number 321, pp. 26–37.

Marriott, S. J. 1955. *J. Aust. Inst. Agr. Sci.* 21:277.

Molisch, H. 1937. *Der Einfluss einer Pflanze auf die andere—Allelopathie*. Fischer, Jena.

Moore, B. P. 1967. *J. Aust. Entomol. Soc.* 6:36.

Mothes, K. 1955. *Annu. Rev. Plant Physiol.* 6:393.

Muller, C. H. 1966. *Bull. Torrey Bot. Club.* 93:332.

Newman, E. I. and A. D. Rovira. 1975. *J. Ecol.* 63:727.

Newman, E. I., R. Campbell, and A. D. Rovira. 1977. *New Phytol.* 79:107.

Patrick, A. A. 1955. *Can. J. Bot.* 33:461.

Purvis, C. E., R. S. Jessor, and J. V. Lovett. 1985. Weed response to crop stubbles. *Proc. 3rd Aust. Agr. Conf.*, Hobart.

Putnam, A. R. and J. DeFrank. 1979. Use of allelopathic cover crops to inhibit weeds. *Proc. IXth Int. Congr. Plant Protect.*, pp. 580–582.

Putnam, A. R. and W. B. Duke. 1974. *Science.* 185:370.

Rabotnov, T. A. 1981. *Sov. J. Ecol.* 12(3):127.

Rice, E. L. 1974. *Allelopathy*. Academic, New York.

Rovira, A. D. 1956. *Plant Soil.* 7:178.

Rovira, A. D. 1965. *Annu. Rev. Microbiol.* 19:241.

Rovira, A. D. 1969. *Bot. Rev.* 35:35.

Rovira, A. D. 1972. *J. Aust. Inst. Agr. Sci.* 38:91.

Rovira, A. D. and R. Campbell. 1974. *Microb. Ecol.* 1:15.

Rovira, A. D. and B. M. McDougall. 1967. *In* A. D. McLaren and G. F. Peterson (eds.), *Soil Biochemistry*. Marcel Dekker, New York, pp. 417–463.

Rovira, A. D., E. I. Newman, H. J. Bowen, and R. Campbell. 1974. *Soil Biol. Biochem.* 6:211.

Rovira, A. D., P. G. Brisbane, A. Simon, D. G. Whitehead, and R. L. Correll. 1981. *Aust. J. Exp. Agr. Anim. Husb.* 21:516.

Sayre, R. M., Z. A. Patrick, and H. J. Thorpe. 1965. *Nematologica.* 11:263.

Selander, J., O. Kalo, E. Kangas, and V. Pertunnen. 1974. *Ann. Entomol. Fennici.* 40:108.

Sheehy, D. P. and A. H. Winward. 1981. *J. Range Manage.* 34:397.

Silander, J. A., B. R. Trenbath, and L. R. Fox. 1983. *Oecologia.* 58:415.

Simon, A. and A. D. Rovira. 1982. *Aust. J. Exp. Agr. Anim. Husb.* 22:201.

Story, R. 1967. *Aust. J. Bot.* 15:175.

Tang, C. S. and C. C. Young. 1979. *Plant Physiol.* 63(5):supplement p. 105.

Tang, C. S. and C. C. Young. 1982. *Plant Physiol.* 69:155.

Trenbath, B. R. and L. R. Fox. 1976. *Aust. Seed Sci. Newslett.* 2:34.

White, S. M., B. L. Welch, and J. T. Flinders. 1982. *J. Range Manage.* 35:107.

Willis, R. J. 1980. Allelopathy and its role in forests of *Eucalyptus regnans* F. Muell. Ph.D. Thesis, University of Melbourne, Australia.

6

AUTOINTOXICATION OF
ASPARAGUS OFFICINALIS L.

CHIU-CHUNG YOUNG

Department of Soil Science
National Chung Hsing University
Taichung, Taiwan
Republic of China

A reduction in the yield and quality of *Asparagus officinalis* L. commonly occurs in old plantations (Chen, 1978; Hung, 1974; Yang, 1982). The death of plants in continuously cropped asparagus fields is a problem which has also been reported (Endo and Burkholder, 1971; Hung, 1974; Grogan and Kimble, 1959). For example, the percentage of missing plants in replanted fields was on the average 29%, and occasionally went up to 40%. However, the reasons for yield declines and missing plants are not clearly understood. Fungicide treatment of asparagus seeds and fumigation of old asparagus fields have been used to control some pathogens, but stunting and wilting of seedlings still occurs (Lacy, 1979; Weibe, 1967). Some observers suggest that factors other than pathogens could be responsible for these problems.

Soil sickness is a complex natural phenomenon in which a plant degenerates when subsequently grown in the same soil. Replant problems in continuous cropping or rotation systems suggest a chemical effect of substances from previous crops or their residues in the soil. The complex phenomenon is not well understood. Among the possible causes, self-allelopathy or autointoxication has often been suggested (Dommergues, 1978). Autointoxication is a condition in which organisms are inhibited by excessive accumulation of their own waste products. It is a well-recognized phenomenon in woody plants (Muller, 1966), but little is known of the extent of autotoxic mechanisms in field crops. Autointoxication could be particularly important in some continuously cropped systems and pure stands of perennial crops.

The purpose of this chapter is to show autointoxication by root exudates, residues, and soil extracts of *Asparagus officinalis*. The greenhouse experiments included bioassays of root exudates, residues, and soil extracts. The possible reasons for asparagus yield decline and replanting problems in the tropical zone are discussed.

AUTOTOXICITY FROM ROOT EXUDATES

Exudation of chemicals by plant roots is a common phenomenon. Root exudates are substances synthesized in the plant and released into the surrounding soil by healthy and intact living plant roots (Rovira, 1969). The term "root exudate" is here used to describe all organic substances exuded from the roots by any mechanism. The role of root exudates on plant–microbial and plant–plant interactions is not fully understood. There is some evidence indicating that plant exudates can inhibit or stimulate the growth of other plants and microorganisms (Rice, 1974 and 1979; Hale et al., 1978), but little is known of autotoxicity by root exudates.

The very extensive root system of asparagus, a perennial crop, contains storage and absorption roots. Field observations revealed that older plants had fewer absorption roots than younger plants. Extensive death of storage roots on older plants commonly occurs in the older asparagus plantings in Taiwan. A study of root exudates seems justified. Since the biological and chemical complexity of the soil system makes it difficult to clearly interpret results, a simplified system might better assess the impact of root exudates on other plants.

A simplified vermiculite-nutrient culture was used to determine whether root exudates of asparagus affected other asparagus plants (Young, 1984a). The results indicated that the growth of asparagus seedlings receiving root exudates from three asparagus varieties (Mary Washington, California 309, and California 711) was significantly lower than that of controls during the early growth stages (Table 6.1). There were no significant differences in the toxicity of root exudates from these three varieties. The growth of both tops and roots was significantly inhibited by asparagus root exudates. The differences in N, P, and K contents in the tops of asparagus plants among all treatments were small. The results indicated that the inhibition of the growth of asparagus seedlings by asparagus root exudates was not due to a deficiency of mineral nutrients or to the inhibition of uptake and transport of N, P, and K.

Furthermore, the collection of root exudates of asparagus was made from the undisturbed root system (Tang and Young, 1982) and the phytotoxicity of collected root exudates was tested by asparagus seed bioassays. The results showed that the growth of both asparagus radicles and shoots was inhibited by root exu-

TABLE 6.1

Inhibition of root and shoot growth of each asparagus seedling in the presence of root exudates from donor plants in vermiculite culture

Donor Treatment	Inhibition of Root Growth[b] (%)	Inhibition of Shoot Growth[b] (%)
Control[a]	0a	0a
Mary Washington	59b	82b
California 309	42b	88b
California 711	65b	82b

Source. Adapted from Young (1984a).
[a]No donor plant was present.
[b]Inhibition in columns followed by the same letter are not significantly different ($P = 0.05$) by Duncan's multiple-range test.

TABLE 6.2

Effects of collected root exudates of asparagus on the inhibition of radicle and shoot growth of asparagus in paper bioassay

Added Volume (ml)	Inhibition of Radicle Growth (%)	Inhibition of Shoot Growth (%)
0.5	30[a]	74[b]
1	59[b]	100[b]
2	84[b]	100[b]

Source. Adapted from Young (1984a).
[a]Significantly different from the control, which is a pot without the presence of a donor plant, at $P - 0.05$ by Student's *t* test.
[b]Significantly different from the control, which is a pot without the presence of a donor plant, at $P = 0.001$ by Student's *t* test.

dates collected with that system (Table 6.2). Complete inhibition of shoot growth was achieved with larger amounts of root exudates. The inhibition of shoot growth by root exudates was greater than that of radicle growth. These results corresponded with those in the vermiculite-culture experiments of longer growth period. The characteristics of the collected root exudates of asparagus have been studied by paper chromatography, and they react chemically like phenolics after spraying with diazotized *p*-nitroaniline, followed by 10% sodium carbonate. The results given above indicated that asparagus can inhibit the growth of its own seedlings by the activity of root exudates.

AUTOTOXICITY FROM PLANT RESIDUES

There is ample evidence that crop residues that are allowed to remain in the soil have harmful effects on several succeeding crops (Chou and Lin, 1976; Chou et al., 1981; Kononova, 1966; Russell, 1973; Klein and Miller, 1980; Young and Bartholomew, 1981). Various studies support the presence of potentially allelo-pathic substances in many crops (Rice, 1974 and 1979). However, little is known about the effect of asparagus residues on the growth of its own species.

The vermiculite-nutrient culture system had been designed to examine the phytotoxicity of asparagus residues on the growth of asparagus seedlings (Young and Chou, 1985). The results indicated that root and shoot growth of asparagus seedlings was inhibited in the used vermiculite in which asparagus had been grown previously for 2 yr. All the asparagus residues inhibited either root or shoot growth of asparagus seedlings (Table 6.3). The highest degree of inhibi-tion was produced by either the root alone or root mixed with stem treatments. Asparagus seedlings in all residue treatments and in used vermiculite without any residue started to yellow and wilt after 21 days of growth in the unsterilized vermiculite cultures. Growth was normal in new vermiculite without any residue treatment and with residues in sterilized vermiculite. The results suggested that the death of plants in the residue treatments with unsterilized vermiculite was due to the interaction between residues and microorganisms.

To further confirm the autotoxicity in asparagus residues, the asparagus seed bioassay was used (Young and Chou, 1985). Asparagus radicle and shoot growth were inhibited by extracts of root, stem, and root litter (Table 6.4). The inhibi-

TABLE 6.3
Effect of asparagus residues on the shoot and root growth of asparagus seedlings in the unsterilized vermiculite culture

Treatment	Inhibition (%)	
	Root Growth	Shoot Growth
Control[a]	0a[c]	0a
Blank[b]	35.0b	35.5bc
Root litter	68.8bcb	60.0cd
Root	80.7cd	81.6de
Stem	77.1cd	76.5de
Root litter + stem	54.0bc	29.0b
Root + stem	91.2d	89.5e

Source. Adapted from Young and Chou (1985).
[a]New vermiculite without any residue in the treatment was the control.
[b]Used vermiculite for planting asparagus without any residue in the treatment was as the blank.
[c]Results within columns followed by the same letter are not significantly different ($P = 0.05$) by Duncan's multiple range test.

TABLE 6.4
Effect of shoot, root, and root litter extracts of asparagus on the growth of
asparagus seedlings

Extract	Amount of Extract Added (g Tissue/5 ml Solution)	Amount of Phenolics in the Solution (μg) (Folin–Denis Reagent)	Inhibition of Growth over Distilled Water (%)	
			Radicle Growth	Shoot Growth
Stem	0.224	2800	59.7a[a]	100.0a[a]
Root	0.224	1165	65.2a	91.2a
Root litter	0.224	269	31.4b	55.5b

Source. Adapted from Young and Chou (1985).
[a]Means in the same column followed by the same letter are not significantly different ($P = 0.05$) by Duncan's multiple range test.

tion of shoot growth was greater than that of radicle growth. The higher concentrations of extracts significantly retarded both radicle and shoot growth. Complete inhibition of shoot growth was affected by larger amounts of the stem extract. These results are similar to those obtained in root exudate bioassays (Young, 1984a) and growth-pouch tests of extracts from field-grown and tissue-cultured plants (Yang, 1982). These results indicate that asparagus is an autoinhibited species whose seedlings can receive inhibitors from root exudates and residues of its own species.

The results from the phenolic assays showed that the higher the phenolic content in the residue extracts, the higher the inhibition of growth from the extract (Young and Chou, 1985). Two phenolics assays indicated that mean values for the aqueous extracts of stem were greater than those for aqueous extracts of root and root litter. From field observations, Chen (1978) reported that the practice of removing asparagus residues before replanting can aid in improving the survival of asparagus seedlings. These results reveal that plant residues of asparagus are one potential source of physiologically active compounds in replanted fields.

AUTOTOXICITY FROM SOIL EXTRACTS

Many phenolic compounds in the environment are thought to play an important role in allelopathy (Chou and Muller, 1972; Rice, 1974; Swain, 1979; Wang et al., 1967; Whittaker and Feeny, 1971; Chou and Young, 1975). Phenolics are of direct importance in the interactions between plants and other organisms in a natural ecosystem. They include simple phenolic acids, flavonoids, polyphenols such as tannins, other classes of compounds common in plants, and many compounds of more restricted distribution (McKey et al., 1978). To extract minute

TABLE 6.5

Comparison of extracts from asparagus soil using acetone, methanol, and XAD-4 on the growth of asparagus seedlings

Extract	Amount of Extract Added (g Soil/5 ml Solution)	Amount of Phenolics in the Solution (μg) (Folin–Denis Reagent)	Inhibition of Growth over Distilled Water (%)	
			Radicle Growth	Shoot Growth
Acetone	4	10	53.0b[a]	83.7a[a]
Methanol	4	4	57.5b	90.7a
XAD-4	4	26	96.3a	100.0a

Source. Adapted from Young and Chou (1985).

[a]Means in the same column followed by the same letter are not significantly different ($P = 0.05$) by Duncan's multiple range test.

quantities of organic compounds in the soil complex by water is a difficult task because of the adsorption or protection by soil of these compounds. The XAD-4 resin method for extracting phenolics from soil surpassed either acetone or methanol (pH 9) methods (Table 6.5). A resin extraction to recover phenolics and chlorophenoxy herbicides from soils compared favorably to other methods (Young 1984b,c; Young and Tang, 1984; Tang and Young, 1982; Young and Song, 1983). The XAD-4 extracts from asparagus soil contained more inhibitory substances than acetone and methanol extracts (Young and Chou, 1985). Complete inhibition of shoot growth was obtained by small amounts of XAD-4 extracts (2 g soil/5 ml solution). Work has begun to isolate and characterize the substances in soil collected from the rhizosphere of asparagus. Some phenolic compounds were already identified, including 3,4-dihydroxybenzoic acid, 2,6-dihydroxybenzoic acid, 3,4-dihydroxyphenylacetic acid, 3,4-dimethoxyacetophenone and β-(m-hydroxyphenyl)propionic acid. A comparison of stem, root, and litter extracts of asparagus soil was conducted to detect phytotoxic substances. In seed bioassays the results indicated that asparagus shoot growth was more sensitive to the phytotoxic substances than root growth. These data also indicated that growth inhibition increased with increasing phenolic content of the extract.

DISCUSSION

There is ample evidence that plant exudates and residues influence the growth of other plants (Rice, 1974; Young, 1983). Autotoxicity could be significant in pure stands of long-term crops and some continuous cropping systems. The reasons for the reduction in yield of asparagus from old plantations and for asparagus replanting problems are poorly understood. The data from soil and vermiculite experiments show that the inhibition in establishing asparagus is due at least in part to phytotoxic substances produced by asparagus residues, which can inhibit the shoot and root growth of their own species. Autointoxication from root exudates and residues of asparagus can be a problem in pure stands and continuous cropping of asparagus.

Asparagus officinalis is indigenous to Europe (Thompson and Kelley, 1971). In a practical culture with short dormancy or without dormancy, as in tropical Taiwan, every crown is maintained with 3–5 mother stalks to provide nutrition for the growth of shoots during the harvest seasons. The profitable productive period in tropical Taiwan is much shorter than the 12–20 yr reported for temperate zones (Wang, 1974; Hung, 1978; Yang, 1982). Wang (1974) suggested that asparagus in Taiwan needs replanting even after 5 yr in some plantations. The reason for the shorter profitable productive period of asparagus in tropical plantations is not clear. The successive cropping of asparagus on the same soil was the most common situation where missing plants and replant problem have been reported (Hanna, 1938; Chen, 1978; Lin and Liou, 1978). The interval for replanting asparagus was recommended to be more than 2–3 yr in tropical Taiwan. Hanna (1938) reported that a 4-yr crop rotation of legumes and barley before replanting asparagus is still not sufficient to produce commercial yields. It is necessary to know the pattern of accumulation and life of the phytotoxic substances in the soil. However, very few studies relating to the stability of accumulated phytotoxins have been reported. Phytotoxins from wheat and oats were degraded in less than 8 wk after harvest; those in sorghum and corn degraded in 16 and 22 wk, respectively (Guenzi et al., 1967). Observations at Illinois suggested that a 1-yr rotation is sufficient to eliminate phytotoxins released by alfalfa (Klein and Miller, 1980). Most organic substances in the soil have similar mechanisms for adsorption, decomposition, polymerization, or protection by the soil components. For example, accumulated soil urease is more persistent than urease added to soils (Conrad, 1940; Zantua and Bremner, 1976 and 1977). Bremner and Mulvaney (1978) suggested that accumulated urease must be associated with and protected by soil components. Several workers have suggested that the urease in soils is protected by humus or clay colloids (Conrad, 1940; Burns et al., 1972a,b; McLaren, 1975). However, there is no published attempt to characterize the protection of phytotoxins by soil constituents. There is an obvious need for research to account for the remarkable stability of accumulated soil phytotoxins retained for several years and to explain the replant problems of asparagus.

The phenomenon of autointoxication of a plant throughout evolution and natural selection is obscure. From the principal guiding force of evolution, the organism–environment interaction produces genetic changes in populations through the operation of natural selection (Stebbins, 1966). Asparagus plants grow natively along the seacoast and riverside or in semidesert areas from the south of Europe to the south of Russia (Hung, 1976). Well-drained, sandy, and sandy loam soils are especially well suited to the production of good asparagus. Heavier-textured soils seldom produce profitable yields in tropical Taiwan.

Light-textured soils may allow leaching of the autotoxic substances and therefore produce good quality asparagus plants. Also, the better aeration in sandy-type soils, which allows higher levels of oxidation and degradation of autotoxic substances could be a necessary condition for good growth of asparagus.

The possibility exists that autointoxication of plants could be reduced by the utilization of biological, chemical, and physical methods. However, the methods to neutralize the accumulated phytotoxic substance remain unknown. In natural systems, there exist mechanisms for the detoxification and removal of phytotoxic substances in the soil. The study of these mechanisms undoubtedly could aid in solving the autotoxicity problem. Grodzinskiy and Golovko (1983) recommend promoting the metabolic process of the heterotrophic microflora to neutralize and destruct the phytotoxic substances. Soil microorganisms could be either the producer or decomposer of phytotoxins. They deserve much more attention. There is a great need to select for tolerance of autotoxic activity by crops. Yang (1982) indicated that breeding autotoxin-tolerant lines of asparagus is possible. However, much research is needed to solve the phytotoxicity problems in agriculture.

SUMMARY

To understand the replant problem in old asparagus fields autointoxication of asparagus from root exudates, residues, and soils was reviewed. The growth of asparagus seedlings was retarded by treatments of root exudates and residues in vermiculite cultures. The growth of asparagus seedlings was inhibited by the root exudates collected from a resin trapping system. Asparagus seed bioassay results obtained with root exudates where the inhibition of shoot growth by root exudates was greater than that of root growth corresponded to the results in vermiculite culture. The quantities of total phenolics from residue extracts corresponded to the autotoxicity in the bioassay. Soil extracts obtained by the XAD-4 method strongly inhibited the development of asparagus seedlings. These data suggested that asparagus is an autoinhibited species whose root exudates and residues inhibit its own growth. Reasons why old asparagus fields cannot be used for replanting in tropical Taiwan were explained and discussed. The strategy of autointoxication and the detoxification of phytotoxic substances in the soil were also discussed in the text.

REFERENCES

Bremner, J. M. and R. L. Mulvaney. 1978. Urease activity in soils. *In* R. G. Burns (ed.), *Soil Enzymes*. Academic, New York, pp. 149–196.

Burns, R. G., M. H. EL-Sayed, and A. D. McLaren. 1972a. *Soil Bio. Biochem.* 4:107.

Burns, R. G., M. H. EL-Sayed, and A. D. McLaren. 1972b. *Soil Sci. Soc. Amer. Proc.* 36:308.

Chen, W. Y. 1978. Study on improvement of asparagus decline and production problem. *In Proc. 2nd Symp. Asparagus Res. Taiwan*, National Chung Hsing University, Taiwan, pp. 17-25.

Chou, C. H. and H. J. Lin. 1976. *J. Chem. Ecol.* 2:353.

Chou, C. H. and C. H. Muller. 1972. *Am. Midl. Nat.* 88:324.

Chou, C. H. and C. C. Young. 1975. *J. Chem. Ecol.* 1:183.

Chou, C. H., Y. C. Chiang, and H. H. Cheng. 1981. *J. Chem. Ecol.* 7:741.

Conrad, J. P. 1940. *Soil Sci.* 50:119.

Dommergues, Y. R. 1978. Impact on soil management and plant growth. *In* Y. R. Dommergues, and S. V. Krupa (eds.), *Interactions Between Non-Pathogenic Soil Microorganisms and Plants*. Elsevier, Amsterdam, pp. 443-458.

Endo, R. M. and E. C. Burkholder. 1971. *Phytopathology.* 61:891.

Grodzinskiy, A. M. and E. A. Golovko. 1983. *Sov. Soil Sci.* (1):54.

Grogan, R. G. and K. A. Kimble. 1959. *Phytopathology.* 49:122.

Guenzi, W. D., T. M. McCalla, and F. A. Norstadt. 1967. *Agron. J.* 59:163.

Hale, M. G., L. D. Moore, and G. J. Griffin. 1978. Root exudates and exudation. *In* Y. R. Dommergues and S. V. Krupa (eds.), *Interaction Between Non-Pathogenic Soil Microorganisms and Plants*. Elsevier, Amsterdam, pp. 163-203.

Hanna, G. C. 1938. *Proc. Am. Soc. Hortic. Sci.* 36:560.

Hung, L. 1974. *J. Chin. Soc. Hortic. Sci.* 20:63.

Hung, L. 1976. *Asparagus: Cultivation Technique*. Taiwan University, Taipei, p. 1.

Hung, L. 1978. Asparagus breeding. *Proc. 2nd Symp. Asparagus Res. Taiwan*. National Chung Hsing University, Taiwan, pp. 133-154.

Klein, R. R. and D. A. Miller. 1980. *Commun. Soil Sci. Plant Anal.* 11(1):43.

Kononova, M. M. 1966. *Soil Organic Matter*. Pergamon, London. 360 pp.

Lacy, M. L. 1979. *Plant Dis. Rep.* 68:612.

Lin, C. H. and T. D. Liou. 1978. Investigation on the factors affecting the missing plants in asparagus fields. *Proc. 2nd Symp. Asparagus Res. Taiwan*. National Chung Hsing University, Taiwan, pp. 43-52.

McKey, D., P. G. Waterman, C. N. Mbi, J. S. Gartlan, and T. T. Struhsaker. 1978. *Science.* 202:61.

McLaren, A. D. 1975. *Chem. Scr.* 8:97.

Muller, C. H. 1966. *Bull. Torrey Bot. Club.* 93:332.

Rice, E. L. 1974. *Allelopathy*. Academic, New York.

Rice, E. L. 1979. *Bot. Rev.* 45:15.

Rovira, A. D. 1969. *Bot. Rev.* 35:35.

Russell, E. W. 1973. *Soil Conditions and Plant Growth*. Longman Group Limited, London, pp. 280-281.

Stebbins, G. L. 1966. *Processes of Organic Evolution*. Prentice-Hall, Englewood Cliffs, NJ.

Swain, T. 1979. Phenolics in the environment. *Biochemistry of Plant Phenolics*. Recent Advances in Phytochemistry 12. Plenum, New York, pp. 617-640.

Tang, C. S. and C. C. Young. 1982. *Plant Physiol.* 69:155.

Thompson, H. S. and W. C. Kelley. 1971. *Vegetable Crops*. McGraw-Hill, New York. 187 pp.

Wang, C. S. 1974. Experiment on the determination of economical productive period for asparagus. *Annu. Rep. Asparagus officinalis L. Exp. 1971 and 1972* Taipei D.A.I.S., pp. 14-15.

Wang, T. S. C., T. K. Yang, and T. T. Chuang. 1967. *Soil Sci.* 103:239.

Weibe, J. 1967. *Hortic. Res. Inst. Ont.* 1966:33.

Whittaker, R. H. and P. P. Feeny. 1971. *Science*. 171(3973):757.

Yang, H. J. 1982. *J. Am. Soc. Hortic. Sci*. 107:860.

Young, C. C. 1983. Phytotoxic effect in a diploid and a tetraploid pasture. *Food Fert. Technol. Cent. Tech. Bull*. 75:1.

Young, C. C. 1984a. *Plant Soil*. 16:377.

Young, C. C. 1984b. *Soil Bio. Biochem*. 16:377.

Young, C. C. 1984c. *Proc. Natl. Sci. Counc. B. ROC*. 8:119.

Young, C. C. and D. P. Bartholomew. 1981. *Crop Sci*. 21:770.

Young, C. C. and T. C. Chou. 1985. *Plant Soil*. 85:385.

Young, C. C. and S. C. Song. 1983. *J. Chin. Agric. Chem. Soc*. 21:224.

Young, C. C. and C. S. Tang. 1984. *Proc. Natl. Sci. Counc. B. ROC*. 8:26.

Zantua, M. I. and J. M. Bremner. 1976. *Soil Bio. Biochem*. 8:369.

Zantua, M. I. and J. M. Bremner. 1977. *Soil Bio. Biochem*. 9:135.

TECHNIQUES FOR STUDIES OF ALLELOCHEMICALS AND THEIR MODES OF ACTION

7

CONTINUOUS TRAPPING TECHNIQUES FOR THE STUDY OF ALLELOCHEMICALS FROM HIGHER PLANTS

CHUNG-SHIH TANG

Department of Agricultural Biochemistry
University of Hawaii, Honolulu, Hawaii

When a plant grows in a sterile, inorganic soil, the plant serves as the sole source of organic compounds to its rhizospheric environment. The root system, with its large surface area, continuously adds photosynthate to this medium. These compounds diffuse away from the rhizoplane, are adsorbed by mineral particles, and converted into new products through oxidation, reduction, isomerization, polymerization, and so on. Thus a dynamic and complex organic rhizosphere is created and maintained by the growing plant itself. In reality, however, other soil organisms play an important role in the modification of the rhizospheric chemicals, and further complicate the chemistry of this region.

Although knowledge of the organic chemistry of the rhizosphere is indispensable to the understanding of allelopathy and other related disciplines (e.g., soil microbiology, plant nutrition, and plant pathology), its importance so far has not been matched by research progress. The available information (Hale and Moore, 1979; Bowen and Rovira, 1976) has been mainly about sugars, amino acids, organic acids, and other ubiquitous primary metabolites in the root exudate. These compounds are readily soluble in water and serve as an important source of energy for microbial growth.

Allelochemicals reported to date are mainly secondary metabolites in plants (Whittaker and Feeny, 1971; Rice, 1984; Mandava, 1985). These terpenoids, alkaloids, phenolics, cyanohydrins, isothiocyanates, and polyacetylenes are widely recognized as an integral part of the plant's defense mechanisms (Harborne, 1982), and are thus logical candidates as allelochemicals. A distinctive difference in physical property between the primary and the secondary metabolites is their solubility. Except for long-chain fatty acids and their esters, primary metabolites are hydrophilic compounds. The secondary metabolites, however, are usually hydrophobic or partially hydrophobic. This characteristic makes it possible to selectively remove most of the secondary metabolites with a column packed with Amberlite XAD-4 hydrophobic resin. In this chapter, the principles and techniques of trapping these compounds from the undisturbed rhizosphere are described, preceded by a critical review of past efforts to collect root exudates for the chemical study of allelopathy.

ROOT EXUDATES

In 1832 DeCandolle ascribed the problem of "soil sickness" to the toxic exudates produced by crop plants. Since then, much effort has been devoted to the evaluation of root exudates as the source of allelochemicals. Although organs other than roots, as well as decomposing plant residues are also possible sources of allelochemicals, it is the root exudates that have inspired the imagination of many workers in the field. Plant root systems have extremely large surface areas and active metabolism to accomplish water and nutrient uptake. On the other hand, substantial quantities of organic matter are released from the roots. When sorghum plants were grown in soil and fed with $^{14}CO_2$, about 15% of the total ^{14}C recovered within 48 hours exposure was found in soil leachates (Lee and Gaskins, 1982). From the data of several authors, the rate of exudation from the roots of temperate cereals ranged from 50 to 150 mg/g dry weight of roots/day (Gardner et al., 1983). Thus plants constantly enrich the soil with substantial proportions of their photosynthate throughout their vegetative life. Secondary metabolites, though usually representing only a small fraction of the total organic exudates, may exert a disproportional impact on rhizospheric organisms or neighboring plants owing to their high biological activity. Also, unlike the aerial portions of a plant, roots exist in a relatively stationary environment. Compounds released into the soil solution may accumulate and form a gradient of concentration in the rhizosphere. Limited information is available regarding the quantitation of allelochemicals in root exudates (Fay and Duke, 1977; Tang and Takenaka, 1983), but it is necessary to bear in mind that an effective concentration must be reached before any allelopathic effects may be observed.

Many reports indicated that "root exudates" were collected for their experiments on allelopathy. Most of these claims should be carefully examined since there is no standard method of collection. Root exudates denote strictly chemical substances (except for water and CO_2) released from the roots under natural conditions. In an experimental process, any physical disturbance that could damage the fragile structure of the roots would trigger the release of intracellular contents (Ayers and Thornton, 1968). Results obtained without the assurance of this crucial requirement (i.e., that no disturbance occurred) are therefore difficult to evaluate in the context of root exudates.

Similar concerns have been raised regarding allelochemicals in general (Newman, 1978); in recent reviews (Rice, 1984; Mandava, 1985), a large proportion of compounds discussed are secondary plant metabolites with growth inhibitory activities. These compounds could very well be what had been suggested. It remains a haunting reality, however, that rigorous proof of the validity of many widely recognized "allelochemicals" is still lacking.

Methods for the study of root exudates are now readily available to plant nutritionists and soil microbiologists. This is because their main interest has been the primary metabolites, which are relatively abundant in the exudates and are therefore easier to determine both qualitatively and quantitatively. Sugars, amino acids, TCA cycle acids, and other primary metabolites have been identified under reliable and well-defined experimental conditions (Rovira and Davey, 1974; Hale et al., 1971, 1979).

The identification of allelochemicals in root exudates is far more challenging, and examples are found in the following reports:

1. Bonner and Galston (1944) attempted to isolate the inhibitory root exudates from 60 gal of nutrient solution that were previously recirculated in support of the growth of 240 guayule plants for 30 days. The experiment failed because of the high content of impurities. A second experiment used 20,000 young plants from the nursery; the roots were superficially washed, the plants were packed in crocks, and the roots were maintained in distilled water for 1 day. The leachate was collected and concentrated. From 460 g of solids collected, 1.6 g of t-cinnamic acid was obtained. A 50-ppm solution of this compound caused a 50% inhibition of growth in height of guayule seedlings. This painstaking pioneer work demonstrates the determination of the authors to use root exudates rather than the extracts as the source of allelochemicals. Unfortunately, in this case it is unlikely that physical injury to the roots could be avoided, and the identity of t-cinnamic acid as the allelochemical compound of guayule remains uncertain.

2. Eberhardt (1955) cultivated oat (*Avena sativa* L.) seedlings in distilled water and sand culture and identified scopoletin and scopoletin glucoside by paper chromatography. Although the samples were collected under undisturbed conditions, Martin (1957) discovered that the excretion of scopoletin from seedling roots of oat was 3.8 μg in nutrient solution but 121.9 μg in distilled water. It is probable, therefore, that the liberation of scopoletin in the experiments by Eberhardt chiefly resulted from the unfavorable conditions in distilled water (Borner, 1960). In view of this, Rovira (1969) stated that root exudate studies should not be conducted with distilled water.

Scopoletin from 3000 accessions of *Avena* sp. was determined in root exudates collected from sand culture by flushing and aspiration every 7 days with nutrient solution (Fay and Duke, 1977). The solution was concentrated and scopoletin was separated by TLC. Quantitative estimation was made by fluorimetry. Providing that the flushing and aspirating processes did not damage the roots, this work offers data on the quantitative aspects of an allelochemical in root exudates.

3. Plants were grown with their roots suspended in a chamber periodically sprayed with a mist of nutrient solution. This fog-box technique was first described by Went (1957). Clayton and Lamberton (1964) used the technique and detected the presence of amino acids and sugars in the root exudates from *Tagetes* species and *Albizzia lophantha* under this relatively undisturbed condition. They failed to detect thiophenes from *Tagetes*, probably because the total organic matter isolated from the root drip was very small. A collection from *A. lophantha* had no trace of the expected methane dithiol (formed from djenkolic acid). However, if the roots were handled gently or washed in running water, the odor of methane dithiol was easily detected. These results clearly demonstrated the vulnerability of roots. The authors pointed out that "It could be argued that the fog-box conditions are too artificial and give no information about events in the soil; on the other hand, unless care is exercised the amount isolated may be only a measure of the abrasion that has occurred in handling the plant."

4. A method widely used to support the belief that certain natural products could be root exudates or rhizosperic compounds has been the extraction of "rhizospheric soils." The soils close to the plants are carefully collected, the debris of roots removed, and the soils extracted for organic compounds. Phenolics were the dominating group of compounds identified in these samples. For example, quantities of vanillic, *p*-hydroxybenzoic, *p*-coumaric, salicylic, and syringic acid were found to be greater in the rhizospheric soil of *Zea mays* than the nonrhizospheric, that is, the soils collected from sites away from the roots (Pareek and Gaur, 1973). Recently, thiophenes have been detected in soils collected under *Tagetes patula* (Campbell et al., 1982).

Chemical evidence from soils supports the presence of rhizospheric allelochemicals; however, it is difficult to rule out the possibility that these compounds may actually be produced from the fine broken root tissues, are a result of plant residues, or are artifacts due to the drastic acid/base treatment of the soil sample (Young and Tang, 1984). Quantitative determination using soil samples is unreliable not only because the origin of the compounds is questionable but also because the recoveries are usually very poor.

To date, evidence of allelopathy through root exudation has been derived from the combined use of reliable sources of root exudates and bioassay (Rice, 1984). The stairstep technique (Bell and Koeppe, 1972), pots with donor and acceptor species separated by root dividers (Young and Bartholomew, 1981; Gilliland and Hayes, 1982), the U-tube technique containing donor and acceptor plants supported at opposite ends (Alsaadawi and Rice, 1982), recent methods using growth units made from PVC drain pipes and T-joints (Pope et al., 1985), and the root exudate recirculating system (Stevens and Tang, 1985) are a few examples. As for chemical evidence, a method for the collection of allelochemicals from an undisturbed root system is imperative. Ideally, the method should be simple, reliable, and capable of providing quantities of allelopathic root exudates for both chemical analysis and bioassay. A continuous root exudate trapping system (CRETS) using Amberlite XAD-4 hydrophobic resins was developed by Tang and Young (1982) and has since attracted some attention (Rice, 1984; Putnam, 1983; Putnam, 1985; Friedman and Waller, 1985; Curl and Truelove, 1986). The following is a description and discussion of this method, written in the hope that through its use and improvement the chemistry and biochemistry of the rhizosphere and of allelopathy can be better understood.

THE AMBERLITE XAD-4 RESIN

Many commercially available adsorbents are capable of trapping hydrophobic organics from water, but most are not suitable for collecting root exudates. For example, Tenax, certain types of Chromosorb and Porapak adsorbents, and silicate-based reverse phase (e.g., C-18) packing materials are all capable of adsorbing hydrophobic organics from an aqueous solution, but are usually too expensive for large-scale usage. Active charcoal granules are inexpensive but may cause irreversible adsorption of certain organics and catalytic destruction of the adsorbed compounds.

Amberlite XAD-2 and XAD-4 are products of the Rohm and Haas Company and are widely used in the study of trace organics in water (Junk et al., 1974; Smith et al., 1981). We found that the versatility and reasonable cost of the resins made them the best choice for the trapping system. Both XAD-2 and XAD-4 have similar adsorption properties, but XAD-4 has been used consistently in our studies since it has a higher adsorption capacity.

Amberlite XAD-4 is a styrene-divinyl benzene copolymer (Figure 7.1) which derives its hydrophobicity from the aromatic nature of its surface. The resin has macroreticular characteristics—each bead can be visualized as an agglomeration of a large number of microspheres. These microspheres are responsible for

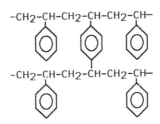

Figure 7.1   Chemical structure of Amberlite XAD-4 resin.

the high adsorptive capacity of the bead (725 m^2/g). Because of its highly cross-linked copolymer nature, XAD-4 has exceptional physical durability and can therefore be used repeatedly. The macroreticular structure results in a continuous solid phase as well as a continuous pore phase so that water or other solvents can readily penetrate the pores (Anonymous, 1978).

The XAD-4 functions much like active carbon in its ability to physically adsorb by binding to the hydrophobic end of an organic molecule. The adsorption occurs through "van der Waals" interactions, the binding ability of which is a function of the relative hydrophobicity of the solute, the solvents, and the adsorbent. Transport of solute molecules from the surrounding solution to the microsphere surfaces takes place readily via diffusion through the internal pores of the resin as well as by migration along the external surfaces of the resin particles (Gustafson et al., 1968).

Amberlite XAD-4 is most effective in adsorbing nonpolar solutes from polar solvents like water. Kauzmann (1959) showed that when organic molecules such as benzene, butanol, and diethyl ether are dissolved in water, the entropy changes of the solution are negative. This unfavorable entropy in solution of nonpolar molecules in water is produced by an orientation of water molecules in the so-called "iceberg formation" around the organic species (Frank and Evans, 1945). In the adsorption of organic matter on the surfaces of the copolymer or carbon, the "icebergs" are broken up with an accompanying large entropy gain. This entropy effect is the main driving force in many cases of physical adsorption from aqueous solution (Gustafson et al., 1968). Another mechanism involves the retention of ionic species of organic compounds through "salt-adsorption," in which a second molecule is attached to the ionic functional group of the first molecule through a salt bridge, while the hydrophobic end of the first molecule attaches to the XAD resin (Mohammed and Cantwell, 1978). This sandwich type of arrangement retains inorganic ions in between a pair of organic ions. However, the loss of mineral ions due to this mechanism is negligible in a regular nutrient solution.

A detailed description of the general properties and analytical techniques used with the resin is available (Junk et al., 1974; Smith et al., 1981). Its application in the analysis of drugs, steroids, amino acids, peptides, vitamins, antibiotics, and flavonoids was reviewed by Brusse et al. (1974). Partially hydrophobic compounds such as p-nitrophenyl glucuronide are more efficiently adsorbed in XAD-4 than XAD-2 (White and Schwartz, 1980). Table 7.1 shows the rate of recovery of model organic compounds from aqueous solution at ppb levels by XAD-4 (Nunez et al., 1984; Junk et al., 1974). In general, good recovery was achieved for hydrophobic secondary metabolites. For compounds such as benzoic acid, efficiency was low at neutral pH. This disadvantage was alleviated in CRETS because of the presence of mineral ions in the nutrient solution (Mohammed and Cantwell, 1978), the use of excess resins, and the recirculation of the solution. However, caution is needed in trapping ionic secondary compounds, especially in the case of quantitation.

Commercially available XAD-4 resins usually contain large amounts of impu-

TABLE 7.1
Recoveries of trace organics from water by Amberlite XAD-4 resins

Type of Compound	Name	Recovery (%)
Polynuclear aromatics	Naphthalene	80
	Biphenyl	87
Alkylbenzenes	Benzene	65
	p-Cymene	80
Halohydrocarbons	Chlorobenzene	85
	Benzylchloride	95
Alcohols	Cinnamyl alcohol	93
	n-Octanol	100
Aldehyde and ketone	Benzaldehyde	92
	2-Undecanone	100
Esters	Benzyl acetate	99
	Methyl salicylate[a]	96
Phenols	o-Cresol	84
	2-Naphthol	100
	Phenol[a]	41(40)
Acids	Octanoic[a]	22(108)
	Palmitic[a]	32(101)
	Benzoic[a]	7(107)

Source. Nunez and Gonzalez (1984). At 10–20 ppb levels.
[a]Junk et al. (1974). At 10–50 ppb levels. Figures in parentheses are percentage of recoveries after acidification.

rities. The following clean-up procedures have been suggested: (1) vacuum and gas-stream desorption under elevated temperatures (Junk et al., 1974); (2) successive washing using acetone, methanol, and methylene chloride or chloroform (Van Rossum and Webb, 1978); and (3) consecutive extraction by different solvents in a Soxhlet apparatus (Hunt and Pangaro, 1982). We have adopted a modified Soxhlet extraction procedure. The crude resin (ca. 1.5 l) was first cleaned with hot running tap water until the strong aroma was washed off, followed by Soxhlet extraction with acetone, acetonitrile, and methylene chloride, each for 24 hours. The cleaned resin was dried in a rotary evaporator under reduced pressure, and the drying process was repeated after saturating the resin with HPLC grade methanol. This last step was performed to completely remove the methylene chloride. The cleaned resin was then stored as a methanol slurry in a brown bottle. For regeneration of the used resins, it is necessary to first wash with a base and then mineral acid (0.1 N), followed by water to neutral pH before repeating the sequential Soxhlet extraction.

Whether the resin is sufficiently clean depends on the goal of a particular experiment. More rigorous cleaning and monitoring of contaminants are required if the root exudate samples are to be used for chemical identification. If one is interested only in the biological activities of the trapped compounds, impurities that do not interfere with the bioassay (as determined by the controls) may be tolerated.

Continuous recirculation of the nutrient solution allows a wide range of metabolites to be carried from the root systems of plants to the XAD-4 column. Only the hydrophilic compounds, including inorganic nutrients, sugars, and proteins, pass through the column. Hydrophobic and partially hydrophobic (e.g., phenolic acids) compounds are adsorbed and accumulated until the resin becomes saturated. Since the XAD-4 resin has a higher capacity than XAD-2, to which an adsorptive capacity of 123 mg phenol/g dry resin has been ascribed (Gustafon et al., 1968), and the rate of root exudation of these compounds is usually extremely low, it would be safe to leave the column in place for prolonged collection without exceeding its capacity. The length of time is determined by the rate of exudation and by the purpose of the experiment. In cases of quantitative collection, more frequent column change is recommended.

THE CONTINUOUS ROOT EXUDATE TRAPPING SYSTEM

The basic design of the continuous root exudate trapping system (CRETS) was published in 1982 (Figure 7.2). The following criteria were taken into consideration while designing the apparatus:

1. It should provide the plants with a healthy environment for growth.

2. It should be constructed with chemically inert material to minimize organic contaminants. In CRETS, glass and Teflon were used wherever possible. Rubber stoppers were wrapped with Teflon sealant tapes (Chemplast, Inc., NJ) to prevent direct contact of the nutrient solution with rubber. Silica sand and basaltic rocks were carefully washed and heat sterilized; nutrient solutions were pretreated by passing through an XAD-4 column. However, for practical reasons, the system was not operated under axenic conditions.

3. Since the allelochemicals are released at a very slow rate, the system must be capable of continuous operation to accumulate sufficient quantities for bioassay and chemical analysis.

4. Only nonionic resins can be used in the trapping system. Anion- and cation-exchange resins should not be used since they alter the pH and nutrient composition of the circulating solution.

5. Polymeric biomolecules such as polysaccharides and proteins accumulate in the circulating solution during collection, along with mucilage and sloughing-off tissue debris. They tend to mask the surface of resin particles and reduce their efficiency. A column with excess capacity is therefore required to ensure complete trapping.

6. The adsorption should be readily reversible.

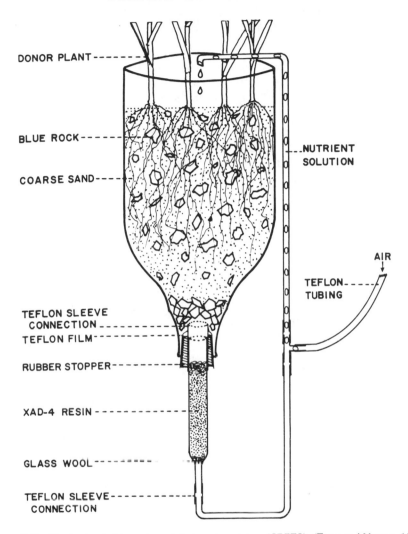

DONOR PLANT

BLUE ROCK

COARSE SAND

NUTRIENT SOLUTION

AIR

TEFLON TUBING

TEFLON SLEEVE CONNECTION

TEFLON FILM

RUBBER STOPPER

XAD-4 RESIN

GLASS WOOL

TEFLON SLEEVE CONNECTION

Figure 7.2   The hydrophobic root exudate trapping system (CRETS). (Tang and Young, 1982).

7. The materials used for constructing the CRETS should be readily available in a biology or chemistry laboratory. Cost is an important consideration, especially where a large-scale experimental set-up is required.

8. A major consideration is that the construction and operation of the system should be simple. Without simplicity, the experiment would be prone to failure, or even worse, would produce misleading results.

The CRETS technique basically fulfills the above requirements. Three major functions of CRETS have been demonstrated in this laboratory.

Determination of Allelopathic Activity

The first goal is to prove that the hydrophobic root exudates from a living, undisturbed donor plant are inhibitory to the seed germination and seedling growth of the acceptor plants. Bigalta limpograss, a tetraploid selection of *Hemarthria altissima* (Poir.) Stapf. and Hubb. inhibited the growth of a legume, *Desmodium intortum* (Mill.) Urb., in a mixed pasture in Hawaii. Greenhouse experiments established that the inhibition was due to allelopathy through root exudation of the limpograss (Young and Bartholomew, 1981). Using CRETS with established donor plants, root exudates were collected from the XAD-4 columns and separated into neutral, acidic, and basic fractions by adjusting the pH of the aqueous solution prior to extraction with methylene chloride. Both neutral and acidic fractions were inhibitory to the radicle growth of lettuce seedlings. At the lowest concentration tested for the acidic fraction, however, stimulation was observed. This is not surprising since growth stimulators may become inhibitory when exceeding certain concentrations.

Another example is the root exudates of guava (*Psidium guajava* L.) (Brown et al., 1983). In orchards treated with glyphosate to control weeds, rings devoid of vegetation were observed around each tree (Figure 7.3). It appeared that allelopathy by the guava root exudates or leachates from leaves by rain were inhibitory to the reestablishment of the weed population. Three to four-year old guava trees were transplanted and established in CRETS, and the hydrophobic root exudates collected were found inhibitory to both lettuce and bristly foxtail (*Setaria verticillata* L. Beuvois) (Table 7.2). No attempt was made to identify the allelochemicals, although tissue extracts of guava contain growth inhibitory terpenoids (Smith and Siwatibau, 1975).

Recently, the allelopathic effects of hydrophobic root exudates from an important weed, *Bidens pilosa* L., on several crop seedlings have been determined using a modified CRETS (Figure 7.4). In this system, the donor pot containing established *B. pilosa* and the acceptor pot with seedlings of the testing plants were irrigated by nutrient solution recirculating through both containers. Thus the inhibitory root exudates from the donor accumulated and the allelopathic effects on test crops were clearly shown when compared with the pot control, in which the donor pot did not contain *B. pilosa*. A resin control was also established with an XAD-4 column placed between the donor and the acceptor pots, hydrophobic root exudates were specifically removed by the column, and the inhibition of testing plants was alleviated to different degrees based on crop species. Table 7.3 shows the results of these three treatments on four test seedlings, including *Lactuca sativa* L., *Phaseolus vulgaris* L., *Zea mays* L., and *Sorghum bicolor* (L.) Moench.

Bidens pilosa significantly inhibited seedling growth of all crop species tested, with *L. sativa* being most sensitive. Larger and older *B. pilosa* plants caused greater inhibition. The hydrophobic root exudates eluted from the resin columns were also inhibitory to the test seedlings (Stevens and Tang, unpublished). The removal of hydrophobic root exudates (resin control) did not completely restore the growth rate to that of the pot control (without *B. pilosa*), suggesting that

Figure 7.3 Weed free "rings" around guava trees (Brown et al., 1983).

TABLE 7.2
Effects of guava root exudates on the radicle growth of lettuce and bristly foxtail[a]

Amount per Disc (μl)	Radicle Length (% of Pot Control)[b]	
	Lettuce	Bristly Foxtail
5	80	80
15	47	59
30	35	39
50	26	18

Source. Brown, et al. (1983).
[a]Root exudates were collected from 3 to 4-yr-old guava trees transplanted to the CRETS. Pot control did not contain guava.
[b]Means of 90 seedlings.

123

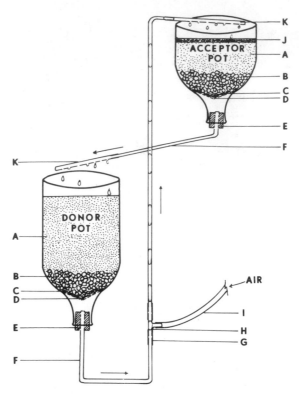

Figure 7.4 The root exudate recirculating system. A = silica sand, B = crushed basaltic rock, C = glass wool, D = perforated Teflon disk, E = robber stopper wrapped with Teflon sealant tape, F = glass tubing, G = Teflon sleeve connector, H = glass tube, I = Teflon tube to air pump, J = vermiculite, K = perforated Teflon tube. Arrows indicate direction of flow. A glass column containing XAD-4 resin was attached to the bottom of the donor pot for resin controls (Stevens and Tang, 1985).

TABLE 7.3
Effects of *Bidens pilosa* L. root exudates of dry weight of crop seedlings[a]

Treatment	Average Seedling Dry Weight (mg)[a]			
	L. sativa	*P. vulgaris*	*Z. mays*	*S. bicolors*
Pot control	5.8	1603	372	194
Resin control	4.9	1272	204	136
B. pilosa	2.2	878	175	113

Source. Stevens and Tang (1985).

[a]Values for all crop species except *L. sativa* represent an average of two replications of eight seedlings at 14 days after seeding. Values for *L. sativa* represent an average of eight seedlings at 11 days after seeding from replication 1 only.

hydrophilic allelochemicals also played a role in the inhibition. Thus using the XAD-4 resin, the recirculating bioassay system was able to differentiate the allelopathic effects caused by hydrophobic and hydrophilic root exudates.

The experiments described above do not involve sophisticated equipment, they are suitable for use by general biology or agricultural laboratories, especially if the goal of research is limited proving allelopathy through root exudation.

Identification of Allelopathic Root Exudates

The continuous root exudate trapping system is constructed with inert material and the root system is free of physical disturbance; therefore, compounds trapped on the column should be mainly the hydrophobic root exudates. Once adsorbed, the fixed compounds may increase in chemical stability (Berkane et al., 1977) and become less vulnerable to microbial decomposition due to high local concentrations. Microbial products remain as a source of contamination along with contaminants from other sources. These compounds may be recognized in the sample mixture by their presence in the CRETS control, which does not contain the donor plants. With the problems of contamination and stability under control, it becomes possible to accumulate trace root exudates for chemical and biological studies.

The most commonly used solvents for elution are methanol and acetone. They are water miscible and therefore provide good efficiency with the wet column. Diethyl ether, chloroform, methylene chloride, and ethylacetate are also used. The selection of solvents depends on the nature of the root exudates under consideration. For example, if the exudates contain mainly polar phenolics, water-miscible polar solvents should be used. If nonpolar, volatile terpenes are expected, methylene chloride would be a good choice because of its lower polarity and boiling point. Ether is an excellent solvent in terms of low boiling point and wide ranges of solubility. However, stored ether may contain peroxides which are explosion hazards and strong oxidizing agents. Elution of columns should be preceded with thorough washing using organic-free water to remove any water-soluble compounds such as salts or sugars. The ease with which these contaminants are removed is a major advantage for using XAD-4.

In the process of sample preparation, general precautions for working with trace organics must be observed, such as using high-purity solvents for extraction. Many bioactive secondary metabolites are unstable to heat, light, pH changes, and oxygen; it is always wise to assume that the unknown samples are susceptible to these factors.

Compounds adsorbed by XAD-4 are nonpolar and partially polar small organic molecules. High-performance liquid chromatography (HPLC) and gas chromatography (GC) are particularly suitable for the separation and identification of the individual components in this type of mixture. Paper and thin-layer chromatography are also useful although they have relatively less resolution power. When combined with specific color reactions and direct bioassay techniques (Tang and Young, 1982; DeFrank and Putnam, 1984), however, paper and thin-layer chromatography can serve as time-saving screening processes. These methods are simple and inexpensive, and should be utilized by laboratories lacking the support of modern instrumentation.

High-performance liquid chromatography has a wide range of capabilities; when a reversed-phase column such as C-18 is used, nonpolar compounds and partially polar compounds (e.g., glycosides) in the same mixture may be isolated in a single run with a gradient solvent system. The concentrated root exudates may be used directly for injection since samples collected from XAD-4 contain few interfering substances and usually do not need precleaning.

Compared to gas chromatography, HPLC is usually less sensitive and requires a larger sample size for detection. This requirement is not a problem since a relatively large sample is often required for bioassay. Initially, an analytical column may be used to obtain optimal conditions for the isolation. A semi-preparative column of the same stationary phase can then be installed using similar conditions for the purpose of collection. The collected peaks or fractions are suitable for either bioassay or further chemical characterization. Since only the bioactive components are of interest, a sensitive and reliable bioassay is essential to identify these individual fractions or peaks. We have used lettuce seed germination and radicle elongation for this purpose mainly because of its reproducibility and simplicity. If possible, however, the acceptor species should also be used.

The active fractions may be further purified by repeated HPLC and then subjected to various physical analyses for identification, including GC/MS, FTIR, NMR, and UV spectrometry. Comparing data obtained from the unknown sample with those of the authentic compounds is extremely helpful for positive identification. Laboratories with experience and a source of supply of natural products would always have an advantage in this regard.

Gas chromatography is the other indispensible instrument for working with trace organics. Compared to HPLC, GC is only suitable for the separation of volatile compounds. Chemical modification such as methylation, acetylation, or sylilation is needed to increase the volatility of samples containing nonvolatiles, such as certain phenolics, acids, and amines. Although the collection of separated fractions or peaks from the GC effluent for further analyses has been practiced from time to time, it is not as convenient as using HPLC. The strength of GC lies in its superb resolution power and its high sensitivity. A modern gas chromatograph equipped with fused silica capillary columns may achieve theoretical plates in the range of 100,000, with a detection limit in subnanograms.

Identification of the well-resolved peaks is possible using GC/MS. From the root exudates of *H. altissima*, 14 compounds were identified (Figure 7.5) of which 3-hydroxyhydrocinnamic, benzoic, phenylacetic, and hydrocinnamic acids were the major rhizospheric compounds with known growth regulatory activities. Most of the compounds identified were phenolic acids that required modification prior to GC/MS analysis. In this case, the root exudate mixture was first methylated with deuterated diazomethane. The CRETS technique has also been

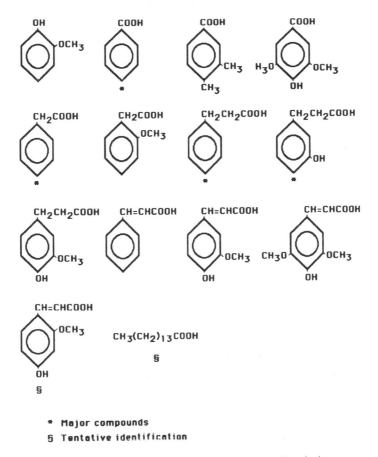

* Major compounds
§ Tentative identification

Figure 7.5 Compounds identified in the root exudates of biyalta limpograss.

used to collect rhizospheric metabolites of marigold (*Tagetes patula* L.) (Tang et al., unpublished). In this experiment, we did not pay special attention to preventing contamination, and a higher proportion of background appeared in the GC chromatogram as a consequence. Yet, using GC/MS equipped with a glass capillary column, benzofurans and thiophenes were detected among the unidentified volatile root exudates and contaminants.

Once positively identified, the allelochemicals may be obtained from commercial sources, chemical synthesis, or preparative liquid chromatography for further study.

Quantitative Determination of Allelochemicals in the Rhizosphere

The quantitation of individual allelochemicals in the rhizosphere under a given condition presents a great challenge to researchers who are interested in defining the microchemical environment of this important region. Since root exudation is a continuous process of a living plant, a gradient of these compounds with the highest concentration on the rhizoplane is expected. In this sense, the goal of quantitation extends beyond determining the concentration of allelochemicals in the soil solution. Furthermore, the rhizosphere is a highly dynamic region and the rate of exudation would respond to many factors that affect the physiology of the plant and its rhizospheric environment. All of these aspects would remain unexplored unless a method of quantitation is developed.

A recent attempt to determine the rate of exudation of a bioactive root exudate was made using CRETS (Tang and Takenaka, 1983). Benzyl isothiocyanate (BITC) is a hydrophobic secondary metabolite found in papaya (*Carica papaya* L.). With a broad spectrum of inhibitory activities against insects, plants, and microorganisms, BITC is considered as a potent defense compound in the papaya plant (Seo and Tang, 1982). Approximately 2 μg of BITC was released each day from the undisturbed root system of a 2-month old papaya tree. Table 7.4 shows a large standard deviation of BITC exudation rates, suggesting that further improvement is needed to perfect the methodology. A model system (Tang and Takenaka, unpublished) was established to optimize the recovery and reproducibility of the system. Replacing the roots of the living papaya plants, a dilute aqueous BITC solution was fed to the CRETS by a peristaltic pump. The rate of BITC input was adjusted to that of a 2-month old papaya, (i.e., 2 μg/ day). Recoveries were compared under various conditions. It was concluded that circulation of the nutrient solution by air-lift may reduce BITC recovery, probably because of oxidation. An improper method of concentrating the sample prior to GC analysis had the same negative effect owing to the volatility of BITC. For this reason, the use of micro-Snyder distillation apparatus instead of a rotary evaporator is recommended for solvent reduction (Smith et al., 1981). From this model study, we did not find any loss of BITC due to adsorption by sand. Addi-

TABLE 7.4
Rate of benzyl isothiocyanate release from roots of 2-month-old Higgins and Wiamanalo papaya trees

Month Collected (1982)	Benzyl Isothiocyanate [μg/tree/day, mean $\pm$ SD(N)[a]]	
	Higgins	Wiamanalo
January	2.30 $\pm$ 0.88 (20)	2.13 $\pm$ 0.83 (15)
February*	2.29 $\pm$ 0.90 (20)	2.01 $\pm$ 0.61 (18)
April**	3.09 $\pm$ 0.94 (8)	2.28 $\pm$ 1.15 (17)
June	2.37 $\pm$ 1.02 (16)	2.07 $\pm$ 1.51 (30)

Source. Tang and Takenaka (1983).
[a]The mean of Higgins significantly greater than that of Waimanalo at 0.1 (*) and 0.05 (**) levels according to the standard *t* test.

tion of an aqueous extract of soil did not affect BITC recovery, suggesting that microorganisms had little effect on the hydrophobic root exudates in CRETS.

CONCLUSIONS

For researchers interested in plant roots, it is tempting to state that "the problem of the root is the root of the problem." Understanding the chemical environment of rhizosphere, including our present effort in obtaining chemical proof of allelopathy by root exudates, is a part of the root problem. The major difficulty has been the lack of a reliable sample collection method. Nevertheless, knowledge concerning this chemistry will provide the answers to many fundamental questions. Using the continuous root exudate trapping system to collect the newly exuded and/or accumulated rhizospheric organics, followed by modern separation techniques to obtain individual compounds, bioassay to pinpoint active fractions, and instrumental methods for structural elucidation, it is possible to identify the allelochemicals. The challenge of quantifying these compounds is even greater, especially when dealing with plants grown under natural conditions, for example, in soil. The complexity of allelopathic chemistry is far from unique. Environmental chemists have devoted many years to the study of trace xenobiotics in natural waters and soils. They have consistently monitored toxic residues in the environment with standardized techniques. Their success is reassuring to the future of allelopathic chemistry.

In two recent texts on ecology (Lange et al., 1983; Cooley and Golley, 1984), allelopathy has received only limited treatment. Some basic issues in allelopathy have been rightfully questioned (Newman, 1978), suggesting that much chemical work is required to consolidate its very foundation. I believe that successful methodology and close working relationships among researchers of different disciplines will eventually lead to the accumulation of our knowledge in the chemistry of allelopathy. Only then can its role in natural and agricultural ecosystems be fully assessed.

REFERENCES

Alsaadawi, I. S. and E. L. Rice. 1982. *J. Chem. Ecol.* 8:993. Anonymous. 1978. *Amberlite XAD-4*. Technical Bull., Rohm and Haas Co., Philadelphia, PA.

Ayers, W. A. and R. H. Thornton. 1968. *Plant Soil* 28:193.

Bell, D. T. and D. E. Koeppe. 1972. *Agron. J.* 64:321.

Berkane, K., G. E. Caissie, and V. N. Mallet. 1977. *J. Chromatog.* 139:386.

Bonner, J. and A. W. Galston. 1944. *Bot. Gaz.* (Chicago). 106:185.

Borner, H. 1960. *Bot. Rev.* 26:393.

Bowen, G. D. and A. D. Rovira. 1976. *Annu. Rev. Phytopathol.* 14:121.

Brown, R. L., C. S. Tang, and R. K. Nishimoto. 1983. *Hortic. Sci.* 18:316.

Brusse, F. S., R. Furst, and W. P. Van Bennekom. 1974. *Pharmacol. Week.* 109:921.

Campbell, G., J.D.H. Lambert, T. Arnason, and G.H.N. Towers. 1982. *J. Chem. Ecol.* 8:961.

Clayton, M. R. and J. A. Lamberton. 1964. *Aust. J. Biol. Sci.* 17:855.

Cooley, J. H. and F. B. Golley, (eds.). 1984. *Trends in Ecological Research for the 1980s.* Plenum, New York. 344 pp.

Curl, E. A. and B. Truelove. 1986. *The Rhizosphere.* Springer-Verlag, New York. 288 pp.

DeFrank, J. and A. R. Putnam. 1984. Abstr., 1984 Int. Chem. Cong. Pacific Basin Soc., Honolulu, HI. Paper No. 04P43.

Eberhardt, F. 1955. *Ziets. Bot.* 43:405.

Fay, P. K. and W. B. Duke. 1977. *Weed Sci.* 25:224.

Frank, H. S. and M. W. Evans. 1945. *J. Chem. Physiol.* 13:507.

Friedman, J. and G. R. Waller. 1985. *TIBS*, February. pp. 47–50.

Gardner, W. K., D. A. Barber, and D. G. Parbery. 1983. *J. Plant Nutr.* 6:185.

Gilliland, T. J. and P. Hayes. 1982. *Rec. Agric. Res.* 30:39.

Gustafson, R. L., R. L. Albright, J. Heiser, J. L. Lirio, and O. T. Reid. 1968. *Ind. Eng. Chem. Prod. Res. Dev.* 7:107.

Hale, M. G. and L. D. Moore. 1979. *Adv. Agron* 31:93.

Hale, M. G., C. L. Foy, and F. J. Shay. 1971. *Adv. Agron.* 23:89.

Harborne, J. B. 1982. *Introduction to Ecological Biochemistry*, 2nd ed. Academic, New York. 278 pp.

Hunt, G. and N. Pangaro. 1982. *Anal. Chem.* 54:369.

Junk, G. A., J. J. Richard, M. D. Grieser, D. Witiak, J. L. Witiak, M. D. Arguello, R. Vick, H. J. Svec, J. S. Fritz, and G. V. Calder. 1974. *J. Chromatog.* 99:745.

Kauzmann, W. 1959. *Adv. Protein Chem.* 14:1.

Lange, O. L., P. S. Nobel, C. B. Osmond, and H. Ziegler (eds.). 1983. *Physiological Plant Ecology III. Responses to the Chemical and Biological Environment.* Encyclopedia of Plant Physiology New Series, Vol. 12C. Springer-Verlag, New York. 799 pp.

Lee, K. J. and M. H. Gaskins. 1982. *Plant Soil* 71:391.

Mandava, N. B. 1985. *In* A. C. Thompson (ed.), *The Chemistry of Allelopathy.* American Chemical Society, Washington, D.C., pp. 33–54.

Martin, P. 1957. *Z. Bot.* 45:475.

Mohammed, H. Y. and F. F. Cantwell. 1978. *Anal. Chem.* 50:491.

Newman, E. I. 1978. *In* J. B. Harborne (ed.), *Biochemical Aspects of Plant and Animal Coevolution.* Academic, New York, pp. 237–342.

Nunez, A. J. and L. F. Gonzalez. 1984. *J. Chromatogr.* 300:127.

Pareek, R. P. and A. C. Gaur. 1973. *Plant Soil.* 39:441.

Pope, D. F., A. C. Thompson, and A. W. Cole. 1985. *In* A. C. Thompson (ed.), *The Chemistry of Allelopathy.* American Chemical Society, Washington, D.C., pp. 219–234.

Putnam, A. R. 1983. *Chem. Eng.* News. 61:34.

Putnam, A. R. 1985. *In* S. O. Duke (ed.), *Weed Physiology. Vol. 1. Reproduction and Ecophysiology.* CRC Press. Boca Raton, FL, pp. 131–154.

Rice, E. L. 1984. *Allelopathy.* 2nd ed. Academic, Orlando, FL. 422 pp.

Rovira, A. D. 1969. *Bot. Rev.* 35:35.

Rovira, A. D. and C. B. Davey. 1974. *In* E. W. Carson (ed.), *The Plant Roots and its Environment.* University of Virginia Press. Charlottesville, pp. 155–240.

Seo, S. T. and C. S. Tang. 1982. *J. Econ. Entomol.* 75:1132.

Smith, S. R., J. Tanaka, D. J. Futoma, and T. E. Smith. 1981. CRC *Crit. Rev. Anal. Chem.* 10:375.

Smith, R. M. and S. Siwatibau. 1975. *Phytochemistry* 14:2013.

Stevens, G. and C. S. Tang. 1985. *Chem. Ecol.* 11:1411.

Tang, C. S. and T. Takenaka. 1983. *Chem. Ecol.* 9:1247.

Tang, C. S. and C. C. Young. 1982. *Plant Physiol.* 69:155.

Van Rossum, P. and R. G. Webb. 1978. *J. Chromatogr.* 150:381.

Went, F. W. 1957. *The Experimental Control of Plant Growth*. Chronica Botanica Co., Waltham, MA, pp. 79–81.

White, J. D. and D. P. Schwartz. 1980. *J. Chromatogr.* 196:303.

Whittaker, R. H., and P. P. Feeny. 1971. *Science.* 171(3973):757.

Young, C. C. and D. P. Bartholomew. 1981. *Crop Sci.* 21:770.

Young, C. C. and C. S. Tang. 1984. *Proc. Natl. Sci. Counc.* Part B: Life Science. Taiwan, ROC. 8:26.

8

BIOASSAYS IN THE
STUDY OF ALLELOPATHY

GERALD R. LEATHER

United States Department of Agriculture
Weed Physiology Laboratory
Frederick, Maryland

FRANK A. EINHELLIG

University of South Dakota
Vermillion, South Dakota

Bioassays are an integral procedure in all studies of allelopathy. They are necessary for evaluating the allelopathic potential of species and following the activity during extraction, purification, and identification of bioactive compounds. A bioassay, according to Webster, is defined as "the use of biological material to test the relative activity of a substance (as a drug) against a standard of known activity." We do not adhere strictly to this definition as we discuss the various techniques used in studies of allelopathy to identify the inhibitory activity of unknown compounds in extracts, leachates, and so on. Rather, we discuss the different types of bioassay, the organisms used, and the relative sensitivity of an assay compared to others used for the same purpose.

Nearly all the published reports on allelopathy describe some type of bioassay that was used to demonstrate allelopathic activity. Stowe (1979) pointed out that of 96 qualitatively different bioassays reported in the literature few if any are of value in relating the results to the demonstration of allelopathy under natural or field conditions. However, the assays that were compared all used altered seed germination as a criterion of allelopathic activity, whereas the sources, extraction methods, and concentrations used were the variables discussed (Stowe, 1979). One of the difficulties has been the lack of standardized bioassays, including incomplete information on the allelochemical source, method of extraction, fractionation concentrations, and the absence of comparisons with known compounds with demonstrated activity in the bioassay.

Bioassays are useful and necessary tools in the study of allelopathy, and should be employed to determine the allelopathic potential of organisms and at every step in the procedure of isolating and identifying allelochemicals (Figure 8.1).

In this chapter, we review the literature pertaining to the use of bioassays and discuss the general suitability of different assays for the determination of allelopathic activity among species. Unique bioassays that, to our knowledge, have not been used in studies of allelopathy are also discussed in terms of their applicability. In Table 8.1 we have categorized the bioassays and provided selected citations of the literature used in our discussion of this subject.

SEED GERMINATION

The most widely used bioassay to test for allelopathic activity is the inhibition (or sometimes stimulation) of seed germination. Since there are hundreds of reports describing this technique, we only reference a few that are pertinent. As noted

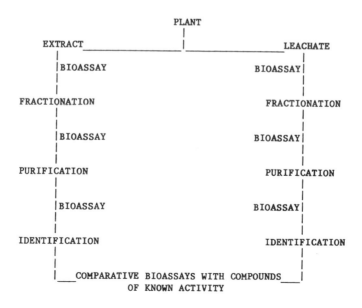

Figure 8.1   Utility of bioassays during fractionation and identification of allelochemicals.

TABLE 8.1
Bioassay methods used in the study of allelopathy

Bioassay	Reference
Seed germination	
1. Filter paper/petri dish	Gressel and Holm (1964), Wolf et al. (1977), Rasmussen and Einhellig (1979), Leather (1983a,b), Muir and Majak (1983), Nicollier et al. (1983), Nishimura et al. (1984)
2. Sand/petri dish	McPherson et al. (1971), Chou and Muller (1972), Pande et al. (1980)
3. Agar/petri dish	Tinnin and Muller (1972), Anaya and del Amo (1978)
4. Soil/petri dish	El-Din and El-Din (1981), Solomon (1983)
5. Felt/petri dish	Selleck (1972)
6. Sterile pumice/petri dish	Stewart (1975)
7. Sponge/petri dish	Muller et al. 1964), Muller and Muller (1964), Muller (1965), Chou and Muller (1972), Chou and Chung (1974), Chou and Lin (1976), Nagvi (1976)
8. Chromatographic	Chou and Muller (1972), Chou and Chung (1974), Thakur (1977)
9. Volatiles	
a. Sponge	Muller and Muller (1964), Muller et al. (1964), Halligan (1975)
b. Paper	French and Leather (1979), Lill et al. (1979)
Plant growth and development	Floyd and Rice (1967), Einhellig and Rasmussen (1973), Rasmussen and Einhellig (1975, 1977), Glass (1976), Cochran et al. (1977), Tang and Waiss (1978), Stevens and Merrill (1980), Einhellig et al. (1982), Blum and Dalton (1985), Leather and Einhellig (1985)
Aquatic	Wium-Anderson et al. (1982), Grace (1983), Yeo and Thurston (1984), Anderson (1985), Ashton et al. (1985), Einhellig et al. (1985a), Leather and Einhellig (1985), Saggese et al. (1985)
Microbial	Kapustka and Rice (1976), Murthy and Nagodra (1977), Averett et al. (1978), Moon and Martin (1981), Alsaadawi and Rice (1982), Mitscher et al. (1983)
Fungal spore	Singh (1974), Naim et al. (1976), Alfenas et al. (1982), Tyagi and Chauhan (1982)
Fern spore	Star (1980)
Algal plating	Chan et al. (1980)
Coleoptile elongation	Cutler and Cole (1983)
Tissue culture	Gressel (1984)
Stairstep/receiver-donor	Wilson and Rice (1968), Bell and Koeppe (1972), Nagvi and Muller (1975), Newman and Rovira (1975)

previously, there is little standardization governing seed germination bioassays. Even reports originating from the same laboratory differ in the conduct of the assay. One difficulty arises when comparisons are made of plant extracts prepared with different solvents and at differing concentrations. This was the basis upon which Stowe (1979) formed his argument against the use of seed germination bioassays for determining allelopathic activity. Another difficulty arises when evaluations are made with different species of seeds. Most crop seeds have been bred for coordinated germination and usually germinate to 100% over the incubation period. Thus only inhibition can be determined, whereas stimulation, which some allelochemicals may cause (Leather, 1983a), is undetected.

Seeds of many wild species are dormant and can be manipulated to germinate about 50% in the absence of further treatment (Leather, 1983a; Rizvi et al., 1980). In studies of allelopathy such seeds can be useful for detecting stimulation, but interpretation of the data requires some caution. With dormant seed, stimulation may be a result of dormancy release which would allow the seed to germinate under conditions unsuitable for growth and reproduction. Hence allelochemical stimulation under these circumstances does not necessarily confer an advantage to the target species.

In general, a seed-germination bioassay is conducted in petri dishes by placing the selected species of seed on substrata saturated with the test solution. The seed/test solution is incubated within an environmental chamber with a light/dark cycle and temperature regime optimum for germination of the selected seed. Germination is usually defined as the emergence of the radicle 2 mm beyond the seed coat and is scored over a period of time, usually 1–7 days depending on species. Most often, the results of seed germination bioassays have been expressed as germination percentages. To account for delays in germination and other factors presented by the unique actions of allelochemicals, several methods of quantification have been used. Pande et al. (1980) used a "speed of germination index" to evaluate delays in germination of treated seed. Germination was recorded daily for 5 days and the index was calculated as: $I = 2(5X + 4X + 3X + 2X + X)$, where X was the number of seeds germinated after each 24-hour period; $5X = $ 24-hour count, $4X = $ 48-hour count, and so on. Lehle and Putnam (1982) employed Richards' function to quantify the germination of cress (*Lepidium sativum* L.) seed. Three aspects of the cumulative germination results were used: germination onset, weighted mean rate, and final germination percentage.

We have found many factors that alter germination when testing allelochemicals, including temperature, light/dark cycle, oxygen availability, osmotic potential, and other interferences. For example, many wild species germinate best when incubated at alternating temperatures with specific maxima and minima. However, in some instances [e.g., phenolic acids on wild sunflower (*Helianthus annuus*) L.] the effect noted was a shift in the maximum and minimum temperature requirements owing to the properties of the solution rather than the chemical utilized. Inbibition rate is also affected by solution properties, temperature, and light. Another precaution that must be observed is the volume of solution per petri dish. This of course will vary with the size of dish and substrata used. A preliminary test should always be done to optimize the solution volume without inducing anaerobic conditions. Last, but not the least important, is the experimental design. Specifically, an ample number of seed and replications for statistical analysis must be used. We generally use a minimum of six replications of 50 seeds each, especially with small wild species of seed.

There are few reports (Solomon, 1983; Duke et al., 1983; McPherson et al., 1971; Chou and Chung, 1974; Leather, 1983a,b) of attempts to determine the osmotic potential of the plant extracts used in seed germination bioassays. Anderson and Loucks (1966) reported the influence of osmotic potential on seed germination and Bell (1974) demonstrated that osmotic potentials greater than 75 milliosmoles inhibited early radicle growth. We have reported that germination may be delayed at high osmotic potentials of 150 milliosmoles (Leather, 1983a; Leather and Einhellig, 1985). This is equivalent to extracting 4 gm of dried sunflower tissue in 100 ml of water. Thus consideration must be given to the plant/solvent ratio utilized and appropriate osmotic adjustment of the controls must be used.

In Table 8.1, we have subdivided germination bioassays into nine categories based mostly on the substrata used as the seed bed. It is difficult to compare the results of the bioassays and little experimental data exist to make such comparisons. McPherson et al. (1971) evaluated sand as a substratum in comparison with the sponge method described by Muller et al. (1964). They concluded that sand was superior, probably because it was better able to provide the necessary water for germination in the ranges of osmotic potential tested. We (unpublished) have observed that filter paper binds the components of extracts differentially, and the germination of seeds depends upon their proximity to the point of extract application. Thus uniform application of the extract to the entire filter paper disk is essential to avoid "chromatographic" separation of components.

Little work has been done using known allelochemicals in seed-germination bioassays. Wolf et al. (1977) reported the effect of phenolic acids on lettuce seed (*Lactuca sativia* L.) germination, Williams and Hoagland (1982) on the germination of nine weed and crop species, and Rasmussen and Einhellig (1979) determined the inhibiting threshold levels of ferulic, *p*-coumaric, *trans*-cinnamic, and vanillic acids on grain sorghum seed. Wink (1983) evaluated alkaloids isolated from *Lupinus albus* L. using lettuce seed germination bioassays. Others (Liebl and Worsham, 1983; Shilling et al., 1985) compared known compounds with extracts to follow activity during fractionation and identification of allelochemicals (see Figure 8.1). More often however, seed germination has been used only to identify the active fractions during chemical separation (Nicollier et al., 1983; Muir and Majak, 1983; Nishimura et al., 1984).

Volatile allelochemicals pose special problems not only in detection and measurement but also in the determination of activity in bioassays. Thus far seed germination has been the bioassay of choice. Muller et al. (1964) used the sponge/seed bioassay to test volatile substances produced by aromatic shrubs. Halligin (1975) used a variation of the sponge/seed bioassay to test the activity of volatile substances on seed germination. Heisey and Delwiche (1983) used the same method but measured radicle elongation. Regardless of the method used, it is a problem to determine active concentrations, whether the compounds emanate from ground litter (Lill et al., 1979) or other substrates. French and Leather (1979) found determination of effective quantities of known flavor compounds difficult because the compounds adsorbed to glass and plastic or diffused slowly in water.

Several investigators separated active fractions of extracts by paper chromatography and conducted subsequent seed-germination bioassays by placing the seed directly on the different zones of elution (Chou and Muller, 1972; Chou and Chung, 1974; Thakur, 1977). Those zones where germination is inhibited can be further evaluated.

As mentioned previously, germination bioassays use relatively small quantities of compound. Test solutions generally range from 3 to 10 ml, depending on the seed and substrata used for absorption. To our knowledge, only one reported germination bioassay used microquantities of allelochemicals. Gressel and Holm (1964) measured the germination of tomato (*Lycopersicum esculentum*) seed on small filter disks moistened with 0.15 ml of test solution and placed on glass microscope slide cover slips which in turn were placed on larger moist filter pads for humidification. The maximum number of seeds per replication was five and, given the variability of germination of wild seed, this assay would require nearly the same volumes as the general assay described previously to obtain sufficient replication for analysis. Few reports since that time describe the use of small quantities of allelochemicals in seed-germination bioassays.

There is no question that properly conducted seed-germination bioassays have great value in studies of allelopathy. They are simple, rapid, and require relatively small volumes of solution. Their sensitivity varies according to test species and allelochemical, and is less than that of other bioassay methods (Leather and Einhellig, 1985). Historically, seed germination was the bioassay of choice, since field observations supported the view that lack of a species in an area was due to inhibition of germination of its seed (Rice, 1984). However, other factors (such as germination without emergence of the seedling and altered dormancy of the seed by the allelochemicals) must be considered.

GROWTH AND DEVELOPMENT

Although we discuss plant growth and development bioassays separately from germination assays, many of the research reports on allelopathy use both. Numerous investigators have used radicle and hypocotyl/coleoptile elongation with germination percentages to demonstrate allelopathic activity (e.g., Bhowmik and Doll, 1982; Lovett and Jessop, 1982). Because of the tedious measurements of radicle and hypocotyl/coleoptile elongation required, many experimental designs lack sufficient replication for good statistical analyses. This problem can be alleviated by using dry weight to measure growth. When seeds are germinated in petri dishes, measurements of radicle elongation are often complicated by curling and other alterations in morphology.

In Dr. Alan Putnam's laboratory (personal communication), elongation of cress seed is measured rapidly by a modified Parker method (Parker, 1966). In this technique two square petri dishes completely filled with sand are arranged to join at one side. Pregerminated seeds are placed in a row in one dish along the common border with radicles extending to the second dish, which contains sand wet with the test solution. The plates are set at an angle to allow contact between the growing radicle and the sand containing the test solution. The position of the radicle tip is marked on the cover at the beginning of the incubation period and again after 24 hours. Radicle elongation is then easily determined by measuring the distance between marks. This method can also be used to determine shoot elongation. We further modified this technique by using thin-layer chromatography sandwich plates to maintain pregerminated seeds on a filter pad. The plates were placed at a 45° angle and the radicles grew into contact with a second filter paper containing an allelochemical. Growth of the radicle was determined by marking the plate after the radicles contacted the allelochemical and again 24 hours later (Leather and Einhellig, 1985). The radicles were less sensitive to allelochemicals than when incubated in petri dishes.

Seedling growth bioassays are often more sensitive than germination bioassays (Leather and Einhellig, 1985). For example, the threshold for inhibition of grain sorghum seedling growth by several cinnamic acids was approximately one-tenth the level found with radicle elongation and 1/25th the threshold for inhibition of germination (Einhellig et al., 1982). However, seedling growth tests usually require greater amounts of allelochemicals and this may limit their usefulness. Glass (1976) described a method for testing seedlings in hydroponic culture. This system permitted subsampling from a common source, or individual treatment with different allelochemicals. Floyd and Rice (1967) assayed for allelopathic activity in a hydroponic system using 25 ml of solution for each seedling. Einhellig et al. (1970) evaluated the effects of scopoletin on tobacco seedlings growing in opaque vials containing about 40 ml of solution. We have since made extensive use of grain sorghum seedlings growing in vials as an assay. Several sizes of vials (40–140 ml) have been employed, depending on the length of the assay and the plant parameters monitored. This assay is well suited for interaction studies and action mechanism research (Einhellig et al., 1985a; Einhellig et al., 1982). After a 7–10 day growth period, the effects on seedling dry weights provide a sensitive indicator of overall allelochemical interference. If desired, seedlings can be divided into shoot and root segments, thus allowing calculation of shoot/root ratios and comparison of effects on these two aspects of growth. Although this assay is quite sensitive, its drawbacks are the work and time required and the amount of allelochemical necessary.

When amounts of available allelochemicals are limited, as during fractionation of plant extracts, agar culture can be used, provided the incubation times are short (Cochran et al., 1977; Tang and Waiss, 1978; Stevens and Merrill, 1980). Pregerminated seed can be placed on the surface of agar containing the desired concentration of allelochemical or extract within a test tube. Increments of radicle and shoot growth can be determined nondestructively over a period of time and the seedlings are easily removed for further analysis.

We hesitate to term the stairstep method of isolating allelopathic interactions from other forms of plant–plant interference a bioassay. It does, however, utilize seedling growth and has been successfully employed by several investigators (Wilson and Rice, 1968; Bell and Koeppe, 1972). Others using the receiver-donor concept in sand culture have also measured modified plant growth from root exudates and/or leached tissue (Newman and Rovira, 1975; Nagvi, 1976; Leather, 1983a,b).

Tissue elongation, fresh and dry weights, seedling height, and root/shoot ratios have been measured in growth and development assays. Blum and Dalton (1985) reported the effects of ferulic acid on the rate of leaf expansion of cucumber (*Cucumis sativus* L.) seedlings growing in nutrient culture. Based on data from a leaf area meter, they developed a nondestructive model using length and width measurements to determine leaf expansion by the equation: seedling leaf area $= 1.457 + 0.00769(L \times W)$; leaf area is in square centimeters and length (L) and width (W) are in millimeters.

AQUATIC PLANTS

The use of aquatic organisms as bioassays for allelopathy has a relatively short history. Typically, they have been employed in evaluating suspected allelopathy of aquatic plants such as *Typha latifolia* (Grace, 1983), dwarf spikerush (*Eleocharis coloradoensis*) (Yeo and Thurston, 1984; Ashton et al., 1985), and *Charales* (Wium-Anderson et al., 1982). Recently Anderson (1985) described a bioassay for determining the allelopathic potential of aquatic plants using explants of *Hydrilla verticillata* and vegetative propagules of *Potamogeton nodosus* and *P. pectinatus*. Treatments were made in 500-ml Erlenmyer flasks with 250 ml of medium. He suggested that bioassays should be done with the target organism when possible.

Parks and Rice (1969) evaluated the effects of a number of phenolic acids and several suspected allelopathic weeds on selected blue-green algal species. Bioactivity was assessed in terms of chlorophyll content. This use of these nitrogen-fixing soil algae was never developed further as a bioassay. Chan et al. (1980) described an algal plating technique using pennate diatoms on an agar medium to evaluate potential allelochemicals extracted from microalgal cultures. However, subsequent reports using this technique were not found.

We described a bioassay for allelochemicals using *Lemna* species (duckweed) which is rapid, sensitive, and requires only microvolumes of test solution (Einhellig et al., 1985b; Leather and Einhellig, 1985). Although *Lemna* species have been used in plant growth studies for many years (Gorham, 1941; Lin and Mathes, 1973), the large amounts of chemicals needed appeared to prohibit these techniques for assays of suspected allelochemicals isolated by HPLC and other recently employed separation procedures. We found that duckweed cultured on 1.5 ml of mineral solution for 7 days in 24-well tissue culture cluster plates maintained a linear growth rate. We can measure growth rate (frond production), growth (dry weight), reproduction (frond number), and, depending upon the *Lemna* species, chlorophyll and anthocyanin production. Each cluster plate contains four treatments with six replications in a randomized block design. We use 5 μl of test solution for each well or a total of 30 μl per treatment. All treatments are expressed in terms of the final concentration within each well. Anthocyanin production by *L. obscura* has thus far been the most sensitive attribute determined. Salicylic acid inhibited anthocyanin production at levels to 0.5 μM (Leather and Einhellig, 1985). This bioassay has also proven useful for following bioactivity during fractionation and identification of allelochemicals (Saggese et al., 1985).

MICROBIAL

Bioassays using microorganisms have been employed to evaluate allelochemical effects (Murthy and Nagodra, 1977; Averett et al., 1978) and to identify antimicrobial agents (Mitscher et al., 1983; Culter 1985). Moon and Martin (1981) saturated paper disks with material elaborated by marine blue-green algae, placed these on agar impregnated with yeast, and evaluated their action according to the extent a zone of inhibition developed. Some investigators (Kapustka and Rice, 1976; Kaminsky, 1981; Alasaadawi and Rice, 1982) have used soil microorganisms to determine the primary and/or secondary events in allelopathy. However, to our knowledge, microorganisms are not used presently for routine bioassays of allelochemicals. We suggest that increased emphasis be placed in this area since modification of allelochemicals by microorganisms plays an important role in the soil environment (Kaminsky, 1981).

OTHER BIOASSAYS

Several bioassays have been used for specific interactions on target species (Table 8.1), including fungal-spore germination (Singh, 1974; Alfenas et al., 1982) and fern-spore germination (Star, 1980). Other bioassays that are useful to evaluate herbicides, plant growth regulators, and plant hormone response have found limited use as allelochemical bioassays. For example, coleoptile growths (Cutler and Cole, 1983) and tissue cultures (Gressel, 1984) have been used. Cutler (1985) reported that the use of etiolated wheat coleoptiles was an extremely useful tool for screening fungal metabolities with plant growth regulator properties.

TABLE 8.2
Relative threshold levels of inhibition by an allelochemical in bioassays

Bioassay	Measured Parameter	Threshold (μM)[a]
		Ferulic Acid
Sorghum seed	Germination	2500
Sorghum radicle	Elongation	1000
Sorghum seedling	Growth (weight)	200
		Salicylic Acid
Sorghum seedling	Growth (weight)	250
Lemna (*L. minor*)	Frond (number and weight)	100
Lemna (*L. minor*)	Chlorophyll	50
Lemna (*L. obscura*)	Anthocyanin	0.5

[a]Varies with allelochemical and environmental conditions.

CONCLUSIONS

Historically, a variety of bioassays have been used to study allelopathy (Table 8.1). The selection of a bioassay has, in many cases, depended on the availability of species (especially seed), with little regard for standardization. Therefore, many reports of allelopathy are questionable because the bioassays were not suitable indicators, and as suggested by Stowe (1979), do not correlate well with the plant-to-plant interference observed in the field. There is no perfect assay that will meet all the requirements for detecting bioactivity of allelochemicals and it would be prudent to use several for each case of suspected allelopathic interaction. The bioassay selected may depend on the suspected mode of action of the allelochemical. For example, seed germination would not be the bioassay of choice for a suspected photosynthesis inhibitor.

We compared several bioassays to determine threshold levels of inhibition for some known allelochemicals (phenolic acids) (Leather and Einhellig, 1985) and found that the most sensitive was the one in which anthocyanin production in *L. obscura* was inhibited by 0.5 μM salicylic acid. The least sensitive was either germination or radicle elongation (Table 8.2). The most widely used bioassays (seed germination) are often the least sensitive.

The bioassays of the future will depend upon available technology and may employ, for example, monoclonal antibodies or tissue cultures. However, we must standardize those currently in use and show that bioassay results can be interpreted in relation to field observations.

REFERENCES

Alfenas, A. C., M. Hubbes, and L. Couto. 1982. *Can. J. Bot.* 60:2535.

Alsaadawi, I. S. and E. L. Rice. 1982. *J. Chem. Ecol.* 8:1011.

Anaya, A. L. and S. del Amo. 1978. *J. Chem. Ecol.* 4:289.

Anderson, L. W. J. 1985. Use of bioassays for allelochemicals in aquatic plants. *In* A. C. Thompson, (ed.), *The Chemistry of Allelopathy.* American Chemical Society, Washington, D.C., pp. 351–370.

Anderson, R. C. and O. L. Loucks. 1966. *Science.* 152:771.

Ashton, F. M., J. M. Ditomaso, and L. W. J. Anderson. 1985. Spikerush (*Eleocharis* spp.): A source of allelopathics for the control of undesirable aquatic plants. *In* A. C. Thompson, (ed.), *The Chemistry of Allelopathy.* American Chemical Society, Washington, D.C., pp. 401–414.

Averett, J., W. Banks, and D. Boehme. 1978. *Biochem. Syst. Ecol.* 6:1.

Bell, D. T. 1974. *Trans. Ill. State Acad. Sci.* 67:312.

Bell, D. T. and D. E. Koeppe. 1972. *Agron. J.* 64:321.

Bhowmik, P. C. and J. D. Doll. 1982. *Agron. J.* 74:601.

Blum, U. and B. R. Dalton. 1985. *J. Chem. Ecol.* 11:279.

Chan, A. T., R. J. Andersen, M. J. LeBlanc, and P. J. Harrison. 1980. *Marine Biol.* 59:7.

Chou, C. H. and Y. T. Chung. 1974. *Bot. Bull. Acad. Sin.*, New Ser. 15:14.

Chou, C. H. and H. J. Lin. 1976. *J. Chem. Ecol.* 2:353.

Chou, C. H. and C. H. Muller. 1972. *Am. Midl. Nat.* 88:324.

Cochran, V. L., L. F. Elliott, and R. I. Papendick. 1977. *Soil Sci. Soc. Am. J.* 41:903.

Cutler, H. G. 1985. Secondary metabolites from plants and their allelochemic effects. *In* P. A. Hedin, (ed.), *Bioregulators for Pest Control.* American Chemical Society, Washington, D.C., pp. 455–468.

Cutler, H. G. and R. J. Cole. 1983. *J. Nat. Prod.* 6(5):609.

Duke, S. O., R. D. Williams, and A. H. Markhart, III. 1983. *Ann. Bot.* 52:923.

Einhellig, F. A. and J. A. Rasmussen. 1973. *Am. Midl. Nat.* 90:79.

Einhellig, F. A., E. L. Rice, P. G. Risser, and S. H. Wender. 1970. *Bull. Torrey Bot. Club.* 97:22.

Einhellig, F. A., M. K. Schon, and J. A. Rasmussen. 1982. *J. Plant Growth Regul.* 1:251.

Einhellig, F. A., M. Stille Muth, and M. K. Schon. 1985a. Effects of allelochemicals on plant–water relationships. *In* A. C. Thompson, (ed.), *The Chemistry of Allelopathy.* American Chemical Society, Washington, D.C., pp. 170–195.

Einhellig, F. A., G. R. Leather, and L. L. Hobbs. 1985b. *J. Chem. Ecol.* 11:65.

El-Din, S. M. S. Badr and H. Gamal El-Din. 1981. *Egypt J. Microbiol.* 16:25.

Floyd, G. L. and E. Rice. 1967. *Bull. Torrey Bot. Club.* 94:125.

French, R. C. and G. R. Leather. 1979. *J. Agric. Food Chem.* 27:829.

Glass, A. D. M. 1976. *Can. J. Bot.* 54:2440.

Gorham, P. R. 1941. *Am. J. Bot.* 28·98.

Grace, J. B. 1983. *Oecologia.* 59:366.

Gressel, J. B. 1984. *Adv. Cell Cult.* 3:93.

Gressel, J. B. and L. G. Holm. 1964. *Weed Res.* 4:44.

Halligan, J. P. 1975. *Ecology.* 56:999.

Heisey, R. M. and C. C. Delwiche. 1983. *Bot. Gaz.* 144:382.

Kaminsky, R. 1981. *Ecol. Monogr.* 51:365.

Kapustka, L. A. and E. L. Rice. 1976. *Soil Biol. Biochem.* 8:497.

Leather, G. R. 1983a. *Weed Sci.* 31:37.

Leather, G. R. 1983b. *J. Chem. Ecol.* 9:983.

Leather, G. R. and F. A. Einhellig. 1985. Mechanisms of allelopathic action in bioassay. *In* A. C. Thompson, (ed.), *The Chemistry of Allelopathy.* American Chemical Society, Washington, D.C., pp. 197–205.

Lehle, F. R. and A. R. Putnam. 1982. *Plant Physiol.* 69:1212.

Liebl, R. A. and A. D. Worsham. 1983. *J. Chem. Ecol.* 9:1027.

Lill, R. E., J. A. McWha, and A. L. J. Cole. 1979. *Ann. Bot.* 43:81.

Lin, A. and M. C. Mathes. 1973. *Am. J. Bot.* 60:34.

Lovett, J. V. and R. S. Jessop. 1982. *Aust. J. Agric. Res.* 33:909.

McPherson, J. K., C-H. Chou, and C. H. Muller. 1971. *Phytochemistry.* 10:2925.

Mitscher, L. A., G. S. Raghav Rao, I. Khanna, T. Veysoglu, and S. Drake. 1983. *Phytochemistry.* 22:573.

Moon, R. E. and D. F. Martin. 1981. *Microbios Lett.* 18:103.

Muir, A. D. and Walter Majak. 1983. *Can. J. Plant Sci.* 63:989.

Muller, C. H. 1965. *Bot. Gaz.* 126:195.

Muller, C. H., W. H. Muller, and B. L. Haines. 1964. *Science.* 143:471.

Muller, W. H. and C. H. Muller. 1964. *Bull. Torrey Bot. Club.* 91:327.

Murthy, M. S. and T. Nagodra. 1977. *J. Appl. Ecol.* 14:279.

Naim, M. S., A. F. Afifi, and A. A. El-Gindy. 1976. *J. Indian Phytopathol.* 29:412.

Nagvi, H. H. 1976. *Pak. J. Bot.* 8:63.

Nagvi, H. H. and C. H. Muller. 1975. *Pak. J. Bot.* 7:139.

Newman, E. I. and A. D. Rovira. 1975. *J. Ecol.* 63:727.

Nicollier, G. F., D. F. Pope, and A. C. Thompson. 1983. *J. Agric. Food Chem.* 31:744.

Nishimura, H., T. Nakamura, and J. Mizutani. 1984. *Phytochemistry.* 23:2777.

Pande, P. C., P. K. Dublish, and D. K. Jain. 1980. *Bangladesh J. Bot.* 9:67.

Parker, C. 1966. *Weeds.* 14:117.

Parks, J. M. and E. L. Rice. 1969. *Bull. Torrey Bot. Club.* 96:345.

Rasmussen, J. A. and F. A. Einhellig. 1975. *Am. Midl. Nat.* 94:478.

Rasmussen, J. A. and F. A. Einhellig. 1977. *J. Chem. Ecol.* 3:197.

Rasmussen, J. A. and F. A. Einhellig. 1979. *Plant Sci. Lett.* 14:69.

Rice, E. L. 1984. *Allelopathy,* 2nd ed. Academic, New York.

Rizvi, S. J. H., D. Mukerji, and S. N. Mathur. 1980. *Indian J. Exp. Biol.* 18:777.

Saggese, E. J., T. A. Foglia, G. Leather, M. P. Thompson, D. D. Bills, and P. D. Hoagland. 1985. Fractionation of allelochemicals from oilseed sunflowers and Jerusalem artichokes. *In* A. C. Thompson, (ed.), *The Chemistry of Allelopathy.* American Chemical Society, Washington, D.C., pp. 99–112.

Selleck, G. W. 1972. *Weed Sci.* 20:189.

Shilling, D. G., R. A. Liebl, and A. D. Worsham. 1985. Rye (*Secale cereale* L.) and wheat (*Triticum aestivum* L.) mulch: the suppression of certain broadleaved weeds and the isolation and identification of phytotoxins. *In* A. C. Thompson, (ed.), *The Chemistry of Allelopathy.* American Chemical Society, Washington, D.C., pp. 243–272.

Singh, R. S. 1974. *Indian Phytopathol.* 27:553.

Solomon, B. P. 1983. *Am. Midl. Nat.* 110:412.

Star, Aura E. 1980. *Bull. Torrey Bot. Club.* 107:146.

Stevens, K. L. and G. B. Merrill. 1980. *J. Agric. Food Chem*. 28:644.

Stewart, R. E. 1975. *J. Chem. Ecol*. 1:161.

Stowe, L. G. 1979. *J. Ecol*. 67:1065.

Tang, C. S. and A. C. Waiss, Jr. 1978. *J. Chem. Ecol*. 4:225.

Thakur, M. L. 1977. *J. Exp. Bot*. 28:795.

Tinnin, R. O. and C. H. Muller. 1972. *Bull. Torrey Bot. Club*. 99:287.

Tyagi, V. K. and S. K. Chauhan. 1982. *Plant Soil*. 65:249.

Webster's New Collegiate Dictionary. 1980. G and C Merriam Co., Springfield, MA.

Williams, R. D. and R. E. Hoagland. 1982. *Weed Sci*. 30:206.

Wilson, R. E. and E. L. Rice. 1968. *Bull. Torrey Bot. Club*. 95:432.

Wink, Michael. 1983. *Planta*. 158:365.

Wium-Anderson, S., U. Anthoni, C. Christophersen, and G. Houen. 1982. *OIKOS*. 39:187.

Wolf, F. T., M. L. Martinez, and R. H. Tilford. 1977. *J. Tennessee Acad. Sci*. 52:104.

Yeo, R. R. and J. R. Thurston. 1984. *J. Aquat. Plant Management*. 22:52.

9

ISOLATING, CHARACTERIZING, AND SCREENING MYCOTOXINS FOR HERBICIDAL ACTIVITY

HORACE G. CUTLER

Plant Physiology Unit
United States Department of Agriculture
Agricultural Research Service
Richard B. Russell Research Center
Athens, Georgia

'Twas brillig, and the slithy toves
Did gyre and gimble in the wabe;
All mimsy were the borogoves,
And the mome raths outgrabe.
"JABBERWOCKY" LEWIS CARROLL
(1960) * (CA. 1855)

The first reading of Lewis Carroll's somewhat nonsensical verse gives one the impression that either the author had taken leave of his senses or there is some complex riddle that compellingly invites a solution. Those who take the bait discover that Lewis Carroll did not exist. He was, in fact, disowned by his creator, the Reverend Charles L. Dodgson, who was a minister and mathematician,

---

*A possible translation of *Jabberwocky* is: It was brilliant and the slimy (wet) toffs/Did gyrate and gambol (frolic) in the waves;/All mime (active or, farsical) were the brothers gov (erness, or governor = father)/And their mummy was outraged.

loved children, and enjoyed puzzles and pranks. Thus what appears to have been an enigma is the playing with language and sounds, as children are wont to do, the shunting of bits of words (mathematical equations), and a mirror (it's all done with mirrors) of the social mores of the times.

What has Lewis Carroll's enigmatic verse to do with the isolation of mycotoxins that possess herbicidal activity? Simply, the path to the successful separations of natural products from fungal sources is strewn with false clues, trapdoors, and paradoxes. At any given point during the isolation procedure anomalies are guaranteed, but in the end all the pieces of the puzzle fit nicely together and logic prevails. Although certain rules may be followed to ensure a successful outcome, and we shall look at a general protocol for isolating natural products from fungi, a good deal of art and intuitive thinking is involved. However, all novices must start somewhere and are subject to the same pitfalls and errors. In this chapter I discuss some of my own experiences as well as those of my colleagues, to whom I am grateful for sharing both their knowledge and significant discoveries. For ease of presentation the protocol is divided into six parts.

SELECTION OF FUNGUS

In choosing an organism for study there is an almost overwhelming temptation to go to established culture collections and either select an organism at random or pick one that is a plant pathogen. The reasons for doing so are based on at least two facts. First, the organisms have been accessed, fully identified, and cataloged. Second, plant pathogenic fungi probably exert their influence on plants, though they may be plant-species specific, because they produce biologically active natural products. Unfortunately, we have found that some fungi lose their ability to produce certain metabolites after as few as three transfers on culture medium. For example, the *Fusaria* tend to be notoriously unstable. A case in point was the isolation of moniliformin (Figure 9.1) from *Fusarium moniliforme*, where the organism no longer produced the desired metabolite after three to four transfers (Cole et al., 1973). However, the initial isolate had been lyophilized and, fortunately, had remained stable. Often the novice investigator is puzzled by the sudden cessation of production of a highly active natural product that has been detected in a bioassay, and becomes even more so when the resident mycologist says that the useless organism is precisely the same genus and species as the original isolate, and that there are no discernible morphological changes. This leads to the following conclusions. Not all fungi of the same genus and species are biochemically equal, and it is wise to make several lyophilized vials for safekeeping when accessing interesting fungi. The latter is not always true, for some *Fusaria* undergo changes on freeze drying. Generally, we have been disappointed by organisms held in collections, with the exception of certain lyophilized accessions, and our most rewarding experiences have been with organisms collected in the wild. Such was the case with an *Aspergillus ustus*

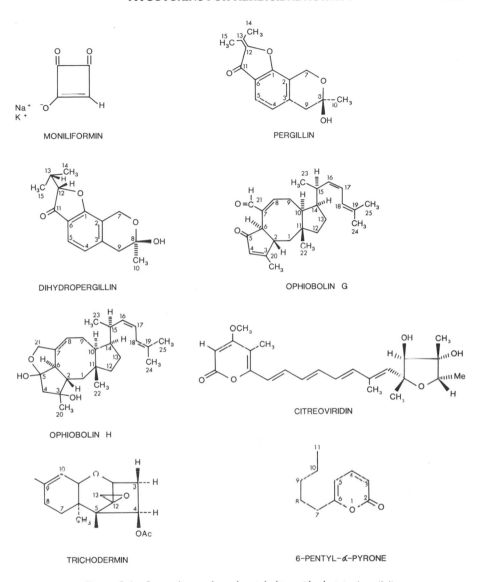

Figure 9.1 Some diverse fungal metabolites with phytotoxic activity.

encountered, by chance, as a result of a trip to buy corn (*Zea mays* L.) for use in secondary plant bioassays. A stray packet of edible pod peas (*Pisum sativum* var. *macrocarpon* cv. Oregon Sugarpod) had been delivered to the store and, being curious about the organoleptic properties of edible pea pods, we planted them in a greenhouse in early spring. As the temperature increased with approaching summer, the plants became less vigorous and chlorotic. Pods became infected with *A. ustus*, and we cultured it out of curiosity. The result was the discovery of a new class of fungal metabolites comprising pergillin (Cutler et al., 1980) and dihydropergillin (Cutler et al., 1981), and the isolation and characterization of two new ophiobolins (Figure 9.1). These were ophiobolin G and ophiobolin H (Cutler et al., 1984), both of which differed structurally from all previously isolated ophiobolins. Another incident involved a colleague who had purchased a truckload of pecan shells to mulch azaleas. While looking out of his window, he noticed that squirrels were making off with some of the shells and on closer examination the shells were found to contain pecan fragments that were in many cases yellow, owing to the presence of the fungus *Penicillium charlesii*. The fungus was cultured *in vitro* and produced citreoviridin (Figure 9.1), which was shown to have specific phytotoxic properties against corn plants and caused growth inhibition for 2 months following treatment, but it was not active against 6-week-old tobacco seedlings (*Nicotiana tabacum* L.) (Cole et al., 1981). Thus even the strain of organism is an important factor in selecting for novel fungal metabolite production.

Many of the organisms that produce phytotoxic and inhibitory metabolites are not obligate plant parasites. Some of the soil-borne organisms, supported on plant debris and organic matter, produce such compounds. A small sampling includes *Trichoderma viride*, which produces the 12,13-epoxytrichothecene trichodermin (Godtfredsen and Vangedal, 1965; Cutler and Lefiles, 1978); *T. viride*, which produces 6-pentyl-α-pyrone (Collins and Halim, 1972) (Figure 9.1); and *Diplodia macrospora*, the agent that synthesizes chaetoglobosin K (Cutler et al., 1980) (Figure 9.2). Other nonplant pathogenic organisms also produce phytotoxic materials, such as *Aspergillus candidus*, which produces terphenyllin (Figure 9.2) and hydroxyterphenyllin (Cutler et al., 1978) and *A. niger*, which produces orlandin (Cutler et al., 1979). The list of new metabolites from fungi grows almost daily, as does the isolation of identical metabolites from totally different genera and species. However, a word of caution is given here to those who would espouse the chemotaxanomic approach to fungal classifications; for example, the isolation of cladosporin (Figure 9.2) from *Cladosporium cladosporioides* (Scott et al., 1971) *Aspergillus flavus* (Grove, 1972), *Aspergillus* sp. (Ellestad et al., 1973), *Eurotium* spp. (Anke et al., 1978), and *Aspergillus repens* DeBary (Springer et al., 1981).

The second important element in isolating novel natural fungal products is the choice of a suitable substrate on which the organism will be grown in mass culture. After accession and before use in the laboratory, the organism may be stored in a number of ways after freeze-dried cultures have been made. Generally, potato–dextrose agar, dilute V-8 agar, or any other of a number of substrate slants may be made to support the organism, which may be kept at ca. 5°C until such time as it is transferred to fermentation flasks. Transfer may be accomplished by adding sterile distilled water to the slants, scraping the surface of the culture with the tip of the transfer pipette, and adding a few drops of the suspension to the culture flasks.

The most obvious medium of choice is a liquid. It offers many attractive features that include relative ease of handling and simple solvent extraction, which

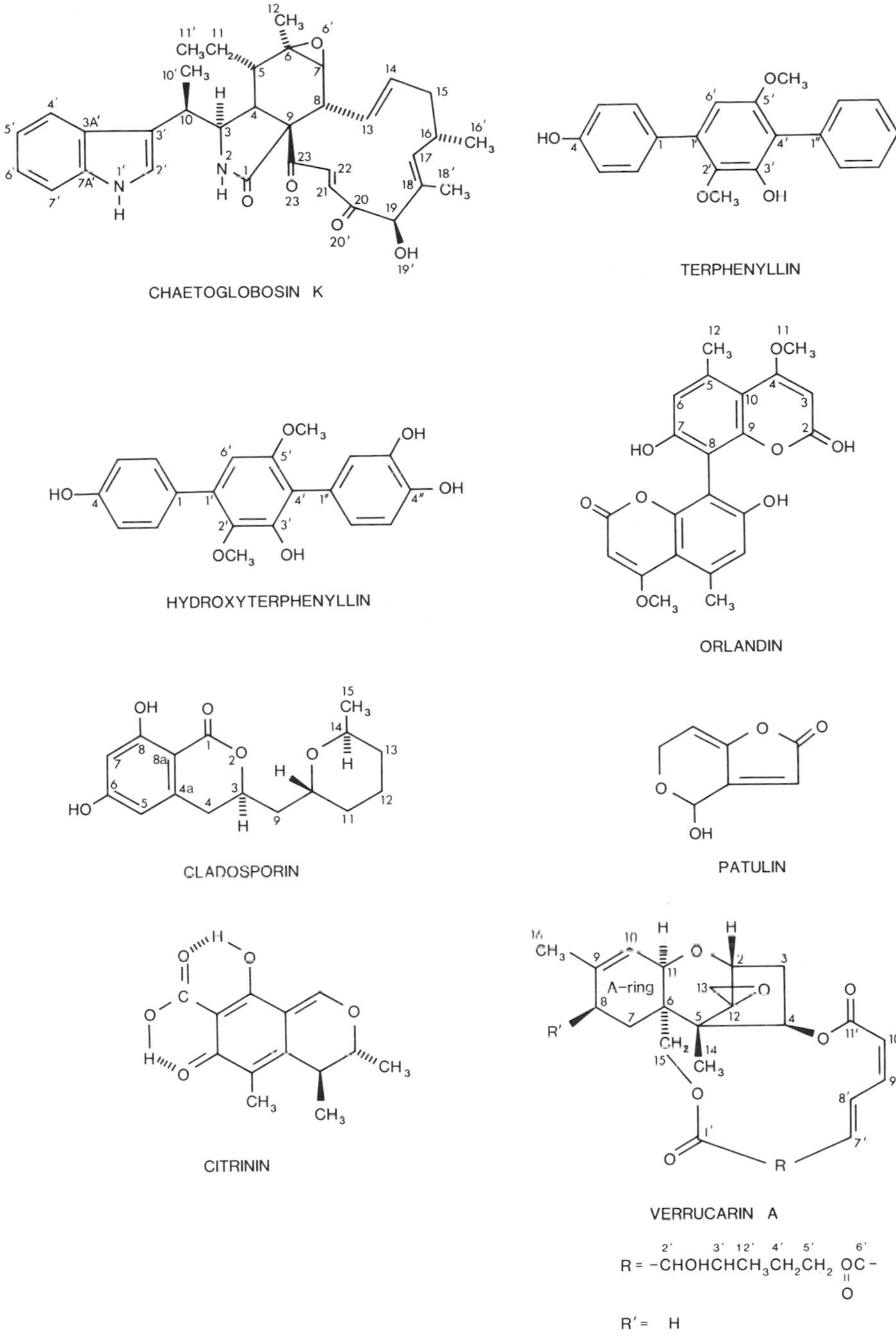

CHAETOGLOBOSIN K

TERPHENYLLIN

HYDROXYTERPHENYLLIN

ORLANDIN

CLADOSPORIN

PATULIN

CITRININ

VERRUCARIN A

R = -CHOHCHCH₃CH₂CH₂ OC -

R' = H

Figure 9.2   Additional fungal metabolites with herbicidal activities.

151

can allow for exhaustive extraction of the metabolite from the substrate. It is as alluring as the song of the Lorelei was to the Rhine fishermen, with almost as disasterous results.* And it has been our experience, with relatively few exceptions (patulin and citrinin) (Figure 9.2), that liquid culture is not the most effective substrate for successfully producing novel metabolites.

We have used solid media as a general substrate for mass culturing of organisms. The most routine of these consists of 100 g of shredded wheat, 200 ml of Difco mycological broth (pH 4.8), 2% yeast extract, and 20% sucrose (Kirksey and Cole, 1974) in 2.8-l Fernbach flasks. The number of flasks is generally kept at 60 and metabolites may be obtained in quantities that range from approximately 10 mg to 1 g. Other solids have been used, including black-eyed peas (*Vigna sinensis*) and rice (*Oryza sativa*), with moderate success. The isolation of moniliformin (Figure 9.1) from *Fusarium moniliforme* involved the use of cracked corn as the fungal substrate (Cole et al., 1973), but the difficulty with corn is the large quantity of oil isolated during the extraction process. The problem seems to be exacerbated if nonpolar solvents are used. The complete separation of metabolites from corn oil is theoretically simple, especially if the compounds are relatively polar, but in practice the resolution of the matter is far from easy.

Our initial examination of any organism consists of seeding three Fernbach flasks (2.8 l) containing shredded wheat medium and incubating them at room temperature (22–27°C, depending on the time of the year) for 14–30 days. The actual time depends on how vigorously the organism grows and how well sporulation occurs. A general rule is that organisms are ready for harvest if the flask, when viewed from the underside, contains ample mycelium on the underside of the shredded wheat. Once a metabolite has been successfully isolated for the first time, the correct parameters can be established for temperature and time at which the organism should be incubated for optimal production. Admittedly, the procedure seems to be haphazard, but at least two statements must be made in favor of this approach. One is that we have been remarkably successful using this method. The other is that constructing and operating a critically controlled incubation facility and initially introducing large numbers of inoculated flasks so that aliquots may be drawn at frequent intervals and bioassays run for every new fungal accession is extremely costly and, presently, labor intensive. If this could be accomplished, the ideal culture facility would also have provisions for culturing each organism on several solid and liquid media concurrently. The study of metabolite production, culture medium, and optimum production conditions is a science in itself. Perhaps this will be accomplished robotically in the future.

---

*Lorelei was a legendary siren whose song lured boatmen on the Rhine river to their destruction on a reef.

SELECTION OF PRIMARY BIOASSAY

Although collecting fungi and choosing a culture medium may seem to be riddled with serendipity, the next step, selecting a primary bioassay, must be performed with deliberation. This is a critical step in the isolation of biologically active natural products and if it is not done properly the chances of success are very slim. The bioassay is the detection system that enables the investigator to track the quarry when all else fails. Therefore, it must be unerring in its ability to give reproducible results, yet broad enough to detect a wide spectrum of biological activity. A common misconception is that the primary bioassay should be highly specific. For illustration, if the search is for a compound that will control a specific weed, common dandelion (*Taraxacum officinale*), then only dandelion should be the subject for bioassay. Our experiences have taught us that it is better to start with an assay that detects biological activity, even if the final outcome is a new bacteriocide, fungicide, phytotoxin, herbicide, insecticide, plant growth regulator, or other "biologically active" compound, then to tailor the metabolite, by chemical synthesis, to the function desired. A new metabolite, which possess general biological activity, thus becomes a template from which to design useful molecules for further application. Such a broad bioassay does exist. The wheat coleoptile bioassay is most versatile and is capable of detecting bacteriocides, fungicides, plant growth promoters and inhibitors, mycotoxins, phytotoxins, and other "biologically active" compounds (Cutler, 1984). These categories may appear to be quite compartmentalized; in fact, many compounds have multiple functions. The assay is quite simple to perform and results are uniform from test to test and can be obtained in less than 24 hours. The bioassay that we use is a hybrid of two methods, a coleoptile straight-growth assay (Hancock et al., 1964) and an oat mesocotyl assay, which will be described shortly (Nitsch and Nitsch, 1956).

Wheat seed (*Triticum aestivum* L., cv. Wakeland) is sown in moist coarse sand in plastic trays and covered with a layer of sand equal to twice the width of the seed. The tray is sealed with a sheet of aluminum foil and placed in the dark at $22 \pm 1°C$ for 4 days and the seeds are left to germinate and grow. Seedlings are removed from the trays under a green safelight (540 nm), the roots and caryopses are removed, and apices are fed into a Van der Weij guillotine so that the first 2 mm may be removed and discarded. The next 4 mm of each seedling are cut and retained for bioassay. Ten of these segments are placed in a test tube containing 2 ml of phosphate–citrate buffer at pH 5.6 supplemented with 2% sucrose (Nitsch and Nitsch, 1956), with an aliquot of the crude extract, or pure metabolite, to be tested for activity. Generally, the fraction to be tested is introduced to the test tube in a suitable solvent and dried under a stream of nitrogen, though pure metabolite may be added in a small amount of acetone ($7.5 \mu l / 1$ ml of buffer solution) (Cutler, 1968) before adding the buffer solution. The test tubes are placed in a roller-tube apparatus and rotated at 0.25 rpm for approximately 18 hours. Then sections are removed, blotted on paper towels, placed on a glass sheet, and put into a photographic enlarger to produce a magnified ($\times 3$) image. Last, coleoptile lengths are measured (Figure 9.3). Controls consist of sections floated in buffer–sucrose solution. Data are statistically analyzed (Kurtz et al., 1965).

Test solutions of pure metabolites are routinely analyzed from 10^{-3} to 10^{-6} M, but when the end point of activity has not been reached at 10^{-6} M the range is extended to 10^{-9} M. The most active class of mycologically derived herbicides to date has been the macrocyclic 12,13-epoxytrichothecenes, specifically, verrucarin A (10^{-8} M) (Figure 9.2) with verrucarin J and trichoverrin B (Figure 9.4) exhibiting significant activity ($P < 0.01$) at 10^{-7} M (Cutler and Jarvis, 1985). During separations on column chromatography 25-ml fractions are collected and 25-μl aliquots are bioassayed.

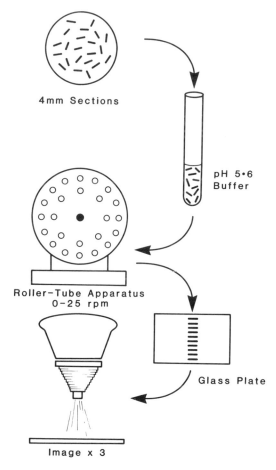

Figure 9.3 Schematic of the etiolated wheat (*Triticum aestivum* L., cv. Wakeland) bioassay system. All manipulations are performed at 540 nm.

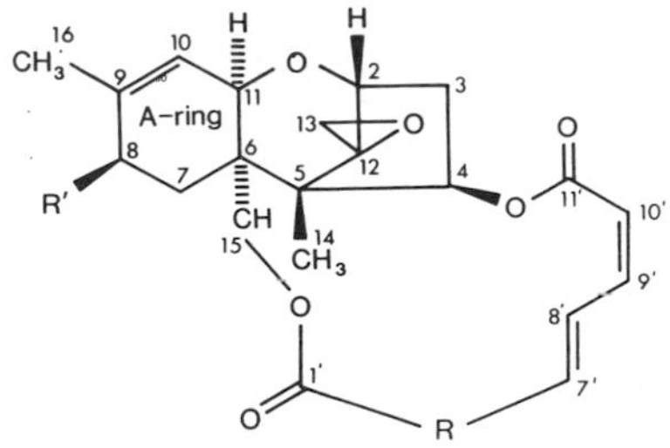

VERRUCARIN J

R = –CH=CCH$_3$CH$_2$CH$_2$OC–
     (E)   O

R′ = H

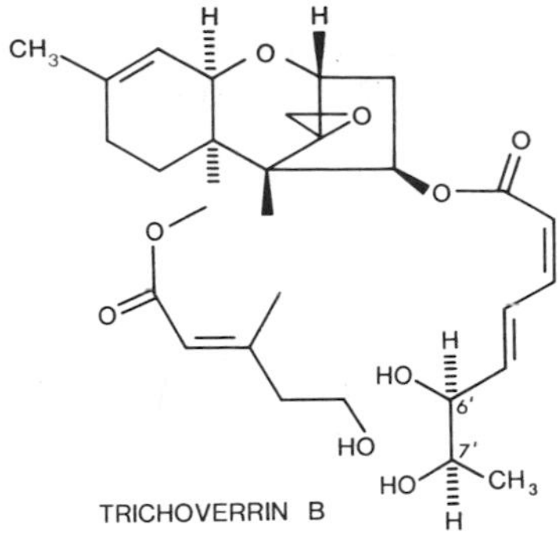

TRICHOVERRIN B

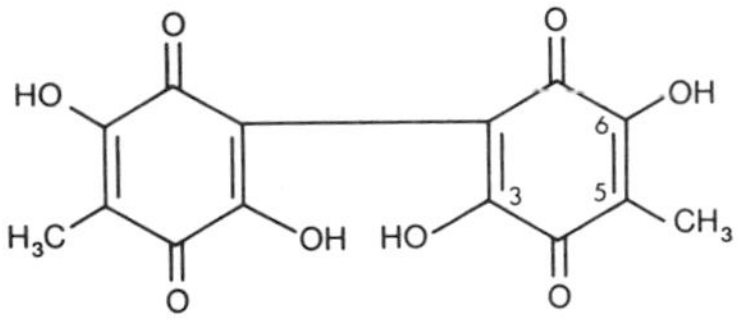

OOSPOREIN

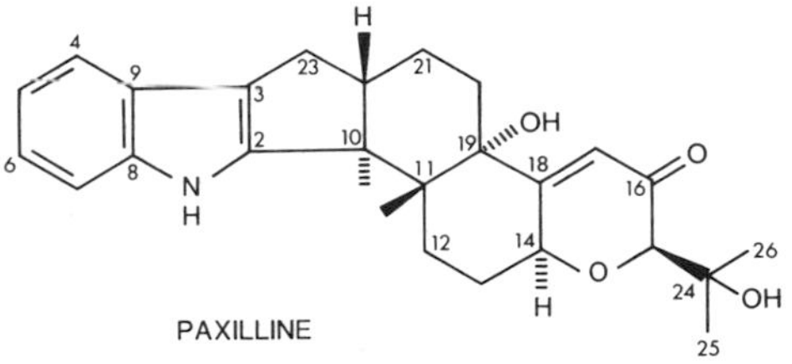

PAXILLINE

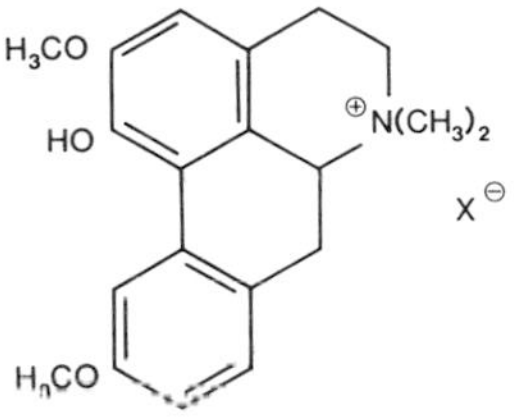

LAURIFOLINE

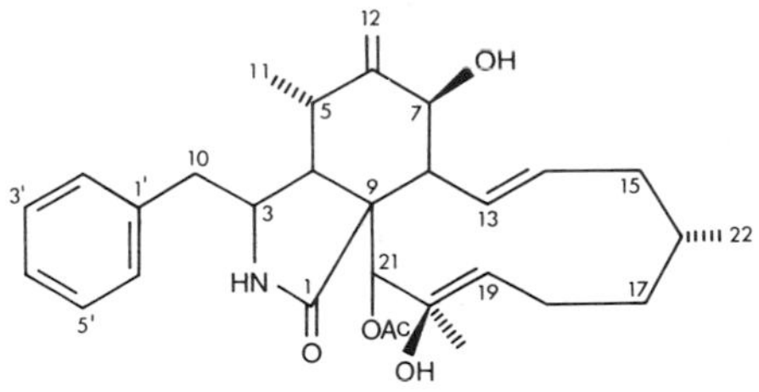

CYTOCHALASIN H

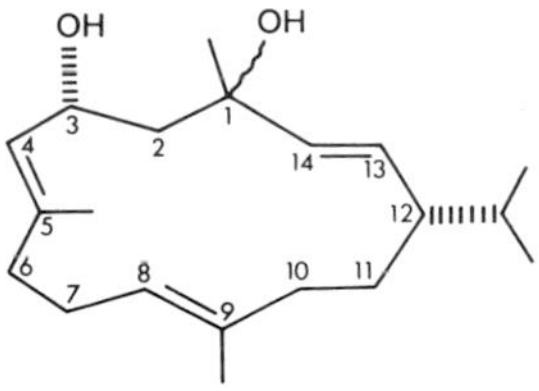

α & β–4,8,13–DUVATRIENE–1,3–DIOLS

Figure 9.4   Additional fungal metabolites with herbicidal activities.

155

Another bioassay that we used routinely was the oat mesocotyl or first internode (Nitsch and Nitsch, 1956). Oat seed (*Avena sativa* L., cv. MO-0205) is grown in moist sand in the same way that wheat is prepared and seedlings are harvested after 4 days. The coleoptiles are completely removed from the stems after the roots and caryopses have been cut away, and the decapitated sections are fed into the Van der Weij guillotine. The first 2 mm are discarded and the next 4-mm segments are saved for bioassays. Ten of these are placed in a test tube with buffer-sucrose solution and the extract, or metabolite, to be tested. The procedures for incubation, measurement, and data analyses are the same as for the wheat coleoptile bioassay. However, the assay takes about twice as long to prepare. In addition, the tissue preparations for the oat mesocotyl assay require more manual dexterity and the possibility for error is greatly increased because the removal of the coleoptile, using scissors, has to be exact.

Although the physical procedures may approximate each other, the responses obtained with an individual metabolite differ. An example is oosporein (Figure 9.4), a fairly common fungal metabolite that has been isolated from diverse fungal sources. Among these are the fungi imperfecti (Kogl and Van Wessem, 1944; Itahashi et al., 1955; Smith and Thompson, 1960; Vining et al., 1962), the ascomycetes (Lloyd et al., 1955), and the basidiomycetes (Divekar et al., 1959). More recently (Cole et al., 1974) oosporein was reported to be toxic to plants and animals, a fact missed by earlier studies. Oat mesocotyls were significantly inhibited ($P < 0.01$) 90, 66, and 27% at 10^{-3}, 10^{-4}, and 10^{-5} M relative to controls; wheat coleoptiles were inhibited 100, 44, and 14% at 10^{-3}, 10^{-4}, and 10^{-5} M. Some questions, still unanswered, pertain to the relative activities. Why were oat first internodes inhibited only 90% whereas wheat coleoptiles were inhibited 100% at 10^{-3} M? Why were oat first internodes inhibited 20 and 10% more than wheat coleoptiles at 10^{-4} and 10^{-5} M? Does the apparently increased specific activity reflect what may happen in other plant species? We do not know the answers, but we do know that either bioassay would have compelled us to carry out further tests on higher plants. In the end, we chose the easier of the two bioassays for routine testing.

Other bioassays are available and they may include fungi, bacteria, the bioluminescent properties of bacteria, pollen, and so on. Cole et al. (1986) discuss bioassays for mycotoxins and describe a number of reliable microsystems for assaying mycotoxins and phytotoxins. Whatever the outcome, the bioassay must be simple, rapid, sensitive, reproducible from test to test, and of low maintenance cost.

ISOLATION AND SEPARATORY PROCEDURES

The art of separation is something of a mystery and although there may appear to be many rules governing this discipline, it is quite surprising to listen to the stories that colleagues tell of their experiences, experiences that they would hesitate to document or relate at a scientific meeting. The first step is to obtain crude extracts from the mass fungal culture. It has been our experience, after using chloroform (which is chosen by some investigators because of its nonflammable properties), chloroform:methanol, ethyl acetate, and ethanol:water solvents that acetone is a superior extracting solvent because it is midway in the eluotropic series. That is, it will dissolve a wide range of polar and nonpolar compounds. Moreover, the mycelial mass contains myriad organic compounds that act as co-solvents, thereby extending the useful range of this solvent. At harvest time, approximately 300 ml of acetone is added to each flask and the contents are macerated with a Super Dispax homogenizer to produce a pulp. This is filtered through Whatman No. 1 filter paper on a Buchner funnel and the clarified filtrate is reduced in volume to an aqueous phase, under vacuum, at 50°C. The aqueous phase is extracted twice with ethyl acetate, each volume of ethyl acetate being equal to that of the aqueous phase, and the ethyl acetate is dehydrated over anhydrous sodium sulfate and reduced to a small volume under vacuum at 50°C. The resulting liquid, if it has been shown to have activity in the etiolated wheat coleoptile bioassay, is placed on a silica gel (70–230 mesh) chromatography column 9.0 cm in diameter, packed 10 cm high. The crude extract is eluted stepwise through the silica gel with, generally, 1.0 l each of benzene, ethyl ether (tertiary butyl methyl ether may be substituted and is safer to use), ethyl acetate, acetone, and acetonitrile or methanol. Each solvent is allowed to drain to the top of the column before the next is added. After each bulk fraction exits the column it is reduced in volume, under vacuum, and an aliquot of each is bioassayed in the coleoptile test.

The crude ethyl acetate extract that is put on the column is generally 75–100 ml and is a viscous fluid. Many of the active sites on the silica gel become quickly saturated and, in many instances, this appears to be helpful in separations, especially those in which hydroxylated compounds are being isolated. Of course, at this stage the chemical characteristics of the molecule are totally unknown. The only piece of solid information is that the crude extract and the subsequent class eluates contain a biologically active metabolite.

Biological activity is usually found in the ethyl ether, ethyl acetate, and acetone fractions. Activity is very rarely limited to one fraction. But in either case this is the point at which the first thin-layer chromatography plates are run. Plates of silica gel 60 F254 (E.M. Laboratories, Inc.)* are developed in one of several solvents. These may include toluene:ethyl acetate:formic acid (5:4:1, v/v/v), CH_2Cl_2:acetone (9:1, v/v); acetone:CH_2Cl_2 (9:1, v/v), or ethyl acetate:benzene (55:45, v/v), though each researcher has a favorite system. Developed plates may be viewed at 254 and 366 nm and the absorbing or fluorescent spots noted. Chromogenic responses may be obtained with a suitable spray reagent, such as anisaldehyde, phosphomolybdic acid, or ethanol:concentrated sulphuric acid, and heating (Stahl, 1965). The result may be quite deceiving, for

---

*The mention of a firm name does not imply endorsement by the U.S. Department of Agriculture of the products. No discrimination or preference is intended. The use of firm names and products is for identification only.

while there are discreet spots it often happens that the active metabolite is not the most intense spot on the plate, and the active material is masked by another spot, or is so weak as to be ignored.

Test tubes containing biologically active material are now pooled together, reduced in volume, and placed on the top of a silica gel 60 (70–230 mesh) chromatography column (3.5 × 45 cm) that has been slurry packed in benzene. After allowing a 600-ml head of benzene to percolate, a gradient of benzene (1.0–1.5 l) to acetone (1.0–1.5 l) is used as the developing solvent, 25-ml fractions are collected, and the tubes are bioassayed for activity. Modifications of the procedure may include a benzene to ethyl acetate gradient, or a benzene to benzene:ethyl acetate gradient. The packing may also be replaced by Florisil for certain compounds (Cutler et al., 1978), but usually this decision is arrived at only after experiencing the loss of a compound on a column. Another way of stating this is that sometimes, after exhaustive elution with a series of solvents, none of the aliquots from the various fractions possess biological activity after passage through silica gel. There are variations on this theme—the active sites on a 9 × 10-cm column become quickly deactivated and biologically active material will pass through in different solvents, but the sites on a longer column do not become totally inactivated, because the extract has been greatly purified and the biologically active material is adsorbed to the active sites. Hence it is possible to lose some, none, or all of the unknown metabolite. And once attached to the active sites it has been our experience that nothing will make it available. But assuming that biological activity is detected at this stage of separation, TLC plates should be made of the active fractions.

If impurities still exist, and this will most likely be the case, the tubes containing biologically active material may be combined and again run through the same type of column, freshly packed, using either the same solvent system or one with slightly modified proportions. If a spot is visible on the TLC plate following further purification, certain conjectures may be made. For example, if a spot correlates exactly with tubes that exhibit biological activity and if that spot is initially weak in the early tubes, becomes more intense in the middle, and fades out in final set of tubes, a bell-shaped concentration curve is evident. Exactly this pattern is made by metabolites eluting from a chromatography column and it may be assumed, until otherwise proved, that the spot observed on the plate is the active metabolite. Also, the position of the presumptive metabolite on the plate, in a specific solvent, indicates the nature of its polarity and a suitable elution solvent for column chromatography can be chosen. Sometimes a series of compounds run closely together on a silica gel TLC plate and no amount of solvent manipulation can separate them. At this time it is wise to change the solid phase to a reverse phase. This may be C_{18} silica gel. Thin-layer C_{18} plates (J.T. Baker Chemical Co.) may be developed in a suitable solvent, acetonitrile:water (75:25, v/v) is ideal, and the results noted. Unfortunately, acetonitrile:water (50:50, v/v) cannot be used with C_{18} TLC plates because the phase tears away from the glass plate. Having obtained the information from the TLC plate, a column of C_{18} may now be packed in water or acetonitrile:water (50:50, v/v).

The packing may be taken from a PrepPAK 500/C_{18} (Waters Associates) cartridge. Column dimensions may range from 1×15 cm to 3.5×35 cm, but the amount of C_{18} packing used is much less relative to that used to pack silica gel columns. Furthermore, columns may be reused after cleaning with acetone:water (50:50, v/v). The partially purified metabolite may be added to the top of the C_{18} packing in a minimum volume of neat acetonitrile and allowed to percolate into the top of the packing. The eluting solvent is now added (acetonitrile:water, 50:50, v/v) and fractions are collected. If the room is kept dark and a long-wave UV lamp is used, various fluorescent and absorbing bands can be seen and fractions can be collected as they exit the column. For exceptional cases, a quartz column may be packed with C_{18} and a short-wave UV lamp used. Removal of the acetonitrile from the water may be easily accomplished by placing the test tubes containing the fractions in a freezer overnight. Either the water freezes or there is an interface formed between the acetonitrile and water phases. In any case, the acetonitrile portion, which contains the organic materials can be gently removed with a pipette and reduced in volume under a stream of nitrogen at 65°C. Compounds tend to be of high purity when separated by this method.

Intermediate phases that have properties between silica gel 60 and C_{18} may also be used. Silanized Silica gel 60-RP-2 (E.M. Laboratories, Inc.) may also be used like C_{18} using acetonitrile:water, but it has slightly different properties (Cutler et al., 1982). Other phases, including ion exchange, are ideal for certain metabolities—each scientist tends to pick a medium that he can manipulate with consummate skill. Recently we attempted to separate two metabolities that were very closely associated, on TLC, by a series of open-column methods, high-pressure liquid chromatography (HPLC), ion exchange, radial plate chromatography, and preparative TLC, all to no avail. Finally, argentation chromatography solved the problem quite easily. An aqueous solution of silver nitrate, 1% of the dry weight of the silica gel 60 (70–230 mesh), was added to silica gel 60, stirred, and dried overnight in a forced-air oven at 105°C. The dehydrated material was packed into a chromatography column (2.0×17 cm) in a slurry with CH_2Cl_2: acetone (9:1, v/v), and 5-ml fractions were collected using the identical solvent to develop the column (Cutler and Cox, unpublished). Separation of the two metabolites was easily effected.

Radial plate chromatography is an excellent method for resolving separation problems (Figure 9.5). It combines the features of static TLC, can handle relatively large amounts of material, and separation is rapid and can be carried out in an inert atmosphere. The solvent flow rate can be adjusted to control development and a quartz lid allows for the use of UV lamps to detect fluorescent and adsorbing bands. The separated band may be reduced in volume and reapplied to the system for further separation or two chromatotrons may be hooked together in tandem to effect separation. Many phases are available for preparing radial plates.

Certain separation problems lend themselves readily to HPLC systems that are usually carried out on conventional or preparative columns. Although this may appear to speed up the process, we find that it is necessary to know some-

Figure 9.5   The Chromatotron—radial, spinning-plate chromatography. (Reproduced with permission, Harrison Research, 840 Moana Court, Palo Alto, CA 94306.)

thing about the properties of the metabolite before using HPLC. Also, open glass columns are extremely useful when watching the separation of colored bands, especially if these bands are associated with biological activity.

Another useful device is droplet countercurrent distribution (DCC) (Figure 9.6). A colleague uses this method with great dexterity and has recently separated a number of fungal ophiobolins with excellent results (Gueldner, personal communication). The method is nondestructive, but the initial stages of purification are carried out by open-column chromatography. Again, while there may appear to be certain "rules" that should be followed in separating metabolites from their crude matrix, some extraordinary mistakes have led to interesting isolations. Two examples are offered for instruction and amusement. Some years ago, I received a telephone call from a colleague who had separated paxilline (Figure 9.4), for the first time, from a crude liquid with relative ease. The tremorgen had been extracted by homogenizing the mycelium, grown on shredded wheat, in hot chloroform (Cole and Kirksey, 1974). The crude extract had been filtered, dehydrated over anhydrous sodium sulfate, and reduced to a small volume under vacuum. The crude material was added to a silica gel column (3.5 × 45 cm) slurry packed in hexane, which was then sequentially eluted with 1.0 l each of hexane, benzene, ethyl ether, and chloroform. As the ether started to percolate through the column, it was observed that a layer of powder, approximately 5 cm thick, had formed on top of the silica and was a slightly different color and texture from the silica gel. It had initially been planned to continue the elution range past chloroform to ethyl acetate, acetone, and methanol. On an impulse, the investigator stopped the solvent flow toward the end of the ether phase, siphoned off the excess solvent from the top of the column, carefully

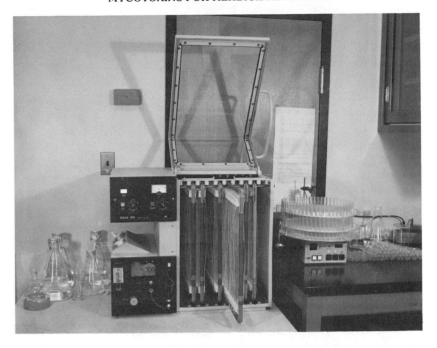

Figure 9.6 Droplet countercurrent distribution apparatus.

scraped off the powder, and recrystallized pure paxilline from acetone. Being somewhat embarrassed at reporting this method, the investigator proceeded to work out another method for isolating paxilline (Cole and Kirksey, 1974) from an advantageous position because an authentic sample now existed. Had the investigator proceeded with the original elution series, history might have recorded a different outcome. The second case involved the isolation of laurifoline (Cole and Kirksey, personal communication) (Figure 9.4) from southern prickly ash (Toothache Tree) (*Zanthoxylum clavaherculis*). The bark was formerly chewed by country folk because of its local anaesthetic properties. The gums became quite numb and the active substance made a toothache bearable. However, the isolation in this narrative was in conjunction with a toxicosis in cattle that had chewed the bark. The scientist had weighty financial matters on his mind the day of the separation and had packed what was thought to be a reverse-phase column in water:methanol (50:50, v/v). The crude material had been placed on top of the packing, a gradient of 1.5 l (water:methanol, 50:50, v/v) to 1.5 l of pure methanol was running when the investigator noted some odd behavior. The material passed rapidly through the column and, at this point, it was obvious that something was very wrong. Upon examination it was discovered that the bottle used to fill the column was not C_{18} reverse phase but was in fact silica gel. At this point most investigators would have quickly disposed of the column contents, said nothing, and started over. Undaunted by the ribbing from his colleagues, who took great care to point out his lack of professionalism, he continued the procedure. The eluting fluid turned crystal clear and, some fractions later, pure laurifoline was detected in the test tubes in large quantities.

The moral of these anecdotes is that rules of separation are to be cautiously broken once in a while. Even water has a place in the preparation of some silica gel columns. Above all, be prepared to experiment with phases and if a lucky accident happens, accept it humbly, gracefully, and gratefully.

ANALYTICAL PROCEDURES

Ultraviolet spectrometry is a useful method for obtaining initial information about a freshly isolated metabolite. The ether fraction often yields either a powdery or crystalline substance in large quantity that ranges from white to pale yellow in color. Frequently, this is one of the fungal sterols and UV spectrophotometry can quickly resolve whether it is or not. When analyzed in ethanol, most of these related metabolites exhibit the characteristic "four fingers" at approximately 262, 270, 280, and 290 nm. Ergosterol, for example, has λ_{max}^{EtOH} 260, 271, 281, and 293 nm. But not all crystalline or powdery material in the ethyl ether fraction is a sterol or sterol mixture. Cytochalasin H (Figure 9.4) collects as an off-white powder in ether fractions from silica gel chromatography (Wells et al., 1976) and after recrystallization in the same solvent exhibits λ_{max}^{EtOH} 218 and $\pi \rightarrow \pi^*$ transition owing to the aromatic ring. The first thought upon seeing the cytochalasin H precipitating in the ether fraction is that it must be a sterol because of the large quantity. Thus UV can save time and some embarrassment. Crystals obtained from ether fractions can be time-consuming, false leads for novices.

Absorption values obtained from UV spectrophotometry may also be useful in attempting to solve the structures of metabolites even when mass spectral and other data are not available. Relative energies may be obtained and together with wavelength values they may be indexed against known metabolites using a suitable handbook, for example, the *Handbook of Toxic Fungal Metabolites* (Cole and Cox, 1981). Even if a match is not found the rules of UV spectrophotometry still hold. The numbers of exocyclic bonds, bonds in conjunction, aromaticity, α,β-unsaturation, and chromophores still affect the wavelength values.

Infrared (IR) spectrophotometry is extremely useful in determining functional groups in the molecule. The characteristic absorptions for hydroxyl groups in a molecule means that derivatization is possible. A carbonyl function presages the possibility of making a crystalline 2,4-dinitrophenylosazone. A basic nitrogen implies the synthesis of a picrate salt. The goal in each case is the synthesis of enough crystalline material to allow for single crystal X-ray analysis. This route is especially necessary if there is less than 500–600 mg of liquid. If there is more, the problem of structural analysis for a novel metabolite can be solved by NMR techniques. Fourier transform IR allows the investigator to obtain spectra on very small samples (<1 mg) and again functional groups are revealed. The accession of IR data may also be useful for comparison of known spectra with the one obtained. Several sources are available, including the Sadtler series.*

Mass spectrometry (MS) is a very powerful tool for the identification of a natural product. The parent ion peak is indicative of the molecular weight of the metabolite and gives the researcher a firm idea of the size of the molecule being studied. The fragment ions may complement the data derived from IR analyses so that if, for example, the first fragment lost from the parent ion peak is $M^+ - 18$, a loss of H_2O is reasonable. If the next fragment is $M^+ - 36$ (or $M^+ - 18, -18$), a loss of $2H_2O$ is suggested. Thus that which appears to be a single hydroxyl in IR is, in fact, two hydroxyl groups. Depending on whether one, or both, is a primary, secondary, or tertiary hydroxyl, the possibility for derivatization may or may not exist. Depending on the results of such synthetic work the exact nature of the molecule may be deduced. Other chemical games may be played with mass spectral data and concomitant syntheses which will yield valuable information about a molecule under examination. An interesting case was α,β-4,8,13-duvatriene-1,3-diol ($C_{20}H_{34}O_2$), a natural product from the trichomes of *Nicotiana tabacum* L., cv. Hick's (Figure 9.4). The only knowledge that we had about this molecule was the IR and MS data from which we deduced the following. First, the IR data implied the presence of a hydroxyl group, CH_2, CH_3 (gem dimethyl), and at least one trans double bond. The MS data yielded, upon computer analysis, the molecular formula $C_{20}H_{34}O_2$. Using the ring/double bond rule, total rings and/or double bonds $= x - 1/2y + 1/2z + 1$, where C is x, H is y, N is z, and O is n, the total rings/bonds $= 20 - 1/2(34) + 1 = (20 - 17) + 1 = 4$. However, we knew that at least one double bond existed from the IR. Hence possible theoretical structures existed with one double bond and three rings, two double bonds and two rings, and so on. When the molecule was saturated with hydrogen, in the presence of Adam's catalyst, and subjected to MS, the mass changed and the molecular formula was $C_{20}H_{40}O_2$ and, concomitantly, the trans double-bond absorption in the IR was conspicuously absent. Thus from the molecular formula given above, the possible number of rings was one. Therefore, the molecule had one ring and three double bonds. The main reason for choosing this molecule for illustration (it is not a fungal metabolite, but a higher plant product) is because one hydroxyl group is tertiary. Tertiary hydroxyl groups may disappear without a trace in electron impact MS, giving a false molecular formula. Always analyze compounds by electron impact, chemical ionization, and other MS techniques.

The science of nuclear magnetic resonance spectroscopy (NMR) has advanced considerably in the past 10 yr. Proton and ^{13}C NMR are exceedingly useful in the identification of fungal natural products. At worst, data complement those obtained from IR and MS. At best, the techniques offered by two-dimensional NMR can solve the questions posed by some very complex structures and

---

*Sadtler Spectra. Sadtler Research Laboratories, 3316 Spring Garden Street, Philadelphia, PA, 19104. Several volumes are available for UV, IR, FTIR, ¹HNMR, and ¹³CNMR.

NMR becomes most useful when crystals are not available for X-ray studies. Presently it is difficult to solve structures with <500–700 mg, unless the spectra of related compounds are known, but the range of detection improves daily and it is only a question of time before structures will be elucidated on smaller quantities.

The most elegant procedure for the natural product researcher is single crystal X-ray diffraction analysis. With a computer-generated drawing the beauty of the molecule becomes readily apparent. The spatial relationship of the various atoms is made plain without having to resort to molecular models. Within the past decade we have moved from the necessity of synthesizing heavy atom derivatives to using the pure crystalline metabolite. Unfortunately, only a few metabolites are suitably crystalline in their native state. Other metabolites may crystallize but the product may be too small, too fine, or twinned in such a way as to be unsuitable. This was the case with ophiobolin H (Cutler et al., 1984), a compound isolated from *Aspergillus ustus* which produced phototoxicity in corn (*Zea mays* L.) but not in bean (*Phaseolus vulgaris* L.) or tobacco (*Nicotiana tabacum* L.). The crystals were twinned and quite unsatisfactory for X-ray crystallography. After 2 yr of difficult work ophiobilin G (Cutler et al., 1984) was isolated using the etiolated wheat coleoptile as the detection system. During the first crystallization small but workable crystals formed. During the second crystallization crystals measuring approximately $4 \times 2 \times 1$ mm grew following seeding with the smaller species. After a few days, the crystals appeared to go through some transformation due to solvent evaporation from the matrix of the crystal. The X-ray crystallographer, being resourceful, received freshly prepared crystals in the mail and immediately coated them with epoxy. Having defined the structure of ophiobolin G by X-ray crystallography, the heretofore intractable solution to the structure of ophiobolin H was determined by NMR using data obtained for ophiobolin G.

The art of crystallization is a difficult topic to discuss. There has to be an almost intuitive feeling for the nature of the compound being handled. The way in which a compound dissolves in a specific solvent, the amount of heat necessary to dissolve the material in solvent, the amount of solvent to be added, and the evaporation rate, are critical factors. Even the type of vessel in which the crystallization is carried out may vary from investigator to investigator. Our most successful crytallizations took place in a 7°C room measuring 8 ft (l) $\times$ 5 ft (w) $\times$ 7 ft (h) which, at one end near the ceiling, had a fan blowing across cooling coils into the room. Test tubes containing purified metabolites in solvent were left on a table for several weeks in the dark. Crystallizations took place easily and probably minute particles of dust, blown in by the fan, acted as a nucleus for the crystals to form. We have never achieved that level of success in a refrigerator or freezer.

At times, simple lessons are forgotten. Such was the case with chaetoglobosin K (Cutler et al., 1980), a fungal metabolite isolated from *Diplodia macrospora*. After several months of hard work a small volume of yellow liquid was the ultimate product of purification. It had high specific activity in the etiolated wheat coleoptile bioassay and the UV, IR, NMR, and MS properties were known. All the data, including the TLC properties in several solvents, pointed to a pure product, but it defied crystallization. As a last resort, we attempted to produce a very high purity product using HPLC, when it seemed remotely possible that we had been wrong, that in fact it might not be pure. The centrifuge tube containing the material was sitting in a test-tube rack in front of a window when the syringe was introduced to extract an aliquot for injection into the HPLC. A few seconds later, light refracted from what appeared to be water droplets. They were, on closer examination, perfect parallepiped crystals. The needle had not been scrutinized and had, on one edge of the tip, a burr that accidently scratched the side of the glass tube. This is a technique that all students learn in elementary chemistry which we, in our relative sophistication, had forgotten.

The crystallization of a novel metabolite has an added bonus, especially if the endeavor has been difficult. It means that in successive isolations there is crystalline material that may be used for seeding. The usual procedure involves taking the metabolite, which may be an apparently uncrystallizable liquid in solvent and is the final product of purification. The solvent may be evaporated under a stream of nitrogen on a water bath at 65°C until the remaining liquid is fluid, but not viscous. This is placed in a refrigerator and allowed to equilibrate at ca. 5°C, then two–three crystals of the parent, and the fines associated with them, are dropped into the solution and the seeded liquid is left undisturbed until crystals grow. The process may be instantaneous or may take several days.

SECONDARY BIOASSAYS

At this point the investigator has a wealth of knowledge about the metabolite. The primary bioassay, which was used as a monitoring agent for the isolation of the metabolite, has also yielded the specific activity. The physical properties, including the melting point and structure of the compound, have been demonstrated. It is now time to consider further biological testing, especially if plant growth regulatory activity is a goal. The most limiting factor is not the assortment of plant species to be tested, but rather the amount of metabolite available. Responses elicited in plants may be diverse and may range from chlorosis and necrosis to mild formative effects, stunting, and phytotoxicity. Other effects may be more subtle, such as inhibition of flowering, and these may be considered indirect herbicide responses, because if an annual does not produce seed in a temperate climate, there can be no progeny.

Secondary bioassays, carried out in greenhouses, consist of three plant genera in our regimen. These are bean, corn, and tobacco. Bean is chosen because it is a nitrogen fixer. Bean plants (*Phaseolus vulgaris* L., cv. Black Valentine) are grown, six to a pot, in 6-in. clay pots and when they are 10 days old the four sturdiest plants are retained. Each pot is then treated with 1 ml of metabolite formulated at 10^{-2}, 10^{-3}, and 10^{-4} M in 10% acetone and 90% water containing 0.1% Tween 20. The solution is applied in aerosol and effects are noted for 2 wk, even though plants may be kept through flowering and fruiting.

Corn plants (*Zea mays* L., cv. Norfolk Market White) are used as experimental test plants because of their monocotyledonous characteristics. Plants are grown six to the 6-in. pot and are thinned to four when they are 1 wk old. At that time, each plant is treated by pipetting 100 μl of a 10^{-2}, 10^{-3}, and 10^{-4} M solution of the metabolite into the innermost bundle sheath of each plant. The solutions are prepared as described (*vide supra*). Effects are generally noted daily for the first 2 wk and at weekly intervals thereafter.

Tobacco (*Nicotiana tabacum* L., cv Hick's) is also included as a test species in greenhouse assays because of its unique habits. When grown under conditions in which the days are short and night temperatures are below 10°C the plants flower, even when small, and the effects of fungal metabolites on this phenomenon can be easily studied. In addition, the phytotoxic effects of metabolites against axillary shoot growth in individual plants closely resemble those obtained in field-grown tobacco. During the culturing of tobacco it is necessary to top the plants before they flower to prevent the transport of the nutrients from the rest of the plant to that organ. Compensatory growth then takes place in the lower leaves, which are harvested, cured, and marketed, but axillary shoots grow because of the destruction of the apical meristem at deflowering, and they must be selectively destroyed without the slightest damage to the main leaves.

Tobacco seedlings are individually grown in 6-in. clay pots and are each treated with 1 ml of metabolite in solution, in aerosol, at 10^{-2}, 10^{-3}, and 10^{-4} M, when 6 wk old. At that time they are approximately 5 cm in diameter. The effects noted include chlorosis, phytotoxicity, and influences on flowering. All experiments are at least triplicated. The results of the secondary bioassays are important not only for the direct implication of phytotoxicity and growth regulatory activity, but also for the indirect. If, for example, corn plants are killed by a metabolite but tobacco is not, the possibilities then exist for developing that compound for use as a grass herbicide in tobacco or some other dicotyledonous crop. The converse may also hold true. We are constantly surprised at the target specificity of fungal metabolites and it is almost impossible to predict that specificity based on chemical structure. Perhaps in the future, when more is known about the mode and site of action of metabolites, the relationship between activity and structure will be easily determined. At present, the investigator is constantly confounded in attempting a prediction.

The inclusion of microbial bioassays may seem irrelevant in the search for new herbicides of fungal origin, but recent experiences indicate that some metabolites (*vide infra*, cladosporin) which have antimicrobial properties also have phytotoxic properties (Springer et al., 1981). Disk assays, while they are indicative of antimicrobial activity, are useful for at least two reasons. They require only small amounts of metabolite and when there is an indication of biological activity the range of activity may be extended, for use on higher plants, by synthetic modification. Also, such activity suggests that further effort should be expended to obtain more metabolite.

Bioassays are routinely run in our laboratories by challenging microorganisms with fungal metabolites. These include *Bacillus subtilis* (gram positive), *Escherichia coli* (gram negative), and, if compounds are active, *Mycobacterium thermosphactum* (gram positive), *E. cloacae* (gram negative), and *Citrobacter freundii* (gram negative). Organisms are densely streaked on DST-Oxoid medium and 4-mm disks impregnated with the metabolite dissolved in acetone or dimethylsulfoxide and dried are immediately placed on the surface of the agar. Treated plates are incubated at 37°C overnight and zones of inhibition are measured and recorded. The identical procedure is used with fungi, specifically *Curvularia lunata* and *Aspergillus flavus*, except that the culture medium is potato-dextrose agar, plates are incubated at room temperature, and observations are made at 3–5 days, but not more than 7. *A. flavus* is included in the assay system because it produces the aflatoxins, which may contaminate food sources, and in working with biologically active natural products all avenues of investigation should be left open for exploration.

SYNTHESIS

The possibilities are good that some synthesis has been done earlier in the exercise. Most probably this has taken place prior to the introduction to mass spectrometry, where the available hydroxyl groups on the parent may have been peracetylated so that the acetate fragments could be noted. The same derivative may have been used to ascertain the NMR shifts. If these materials were made in sufficient quantity, they will have been tested in both the primary and secondary bioassays, sometimes with an extraordinarily successful outcome. Such was the case with cladosporin, a fungal metabolite originally described for its antifungal properties (Scott et al., 1971). Both cladosporin and cladosporin diacetate were equally as active in the etiolated wheat coleoptile bioassay, even though it has generally been observed that acetylation of hydroxyl groups decreases activity (Springer et al., 1981). When cladosporin was applied to bean, corn, and tobacco plants, no effects were noted. But when cladosporin diacetate was applied there was a response within 24 hours in corn. Leaves became chlorotic with 10^{-2} and 10^{-3} M treatments and within 2 wk plants were quite stunted. Neither bean nor tobacco were affected. Hence very simple syntheses may have major implications.

The syntheses of useful natural products and their derivatives fall into three major categories. Total synthesis by the organism, partial production by the organism followed by additional laboratory synthesis, and total laboratory synthesis using the natural product as a template. The last of these categories has been epitomized with the moniliformin molecule (Figure 9.1) to make 3-hydroxycyclobut-3-ene-1,2-dione experimental herbicides. The metabolite was isolated from *Fusarium moniliforme*, found as a secondary invader of corn which was toxic to chicks (Cole et al., 1973). Because of its relative structural simplicity and because it inhibited the growth of etiolated wheat seedlings, corn, and tobacco plants in a series of bioassays, it became a candidate for total synthesis. The fungus produced either the sodium or potassium salt, though which one could not be predicted in advance of the extraction. Chemists synthesized the free hydroxyl compound, which does not seem to be as stable as the salt upon storage at room temperature, but which retained water-soluble properties. From this, simple stable derivatives were made including alkyl, substituted phenyl, CH_2CH_2SEt, $CH_2CH:CH_2$, and 1-octyloxycyclobutene-3,4-dione to control, by postemergence application, *Setaria italica* (L.) Beauv., *Lolium perenne* L., *Sinapis alba* L., and *Stellaria media* (L.) Villars. The free hydroxyl compound controlled axillary shoot growth in field-grown tobacco at 266 and 26.6 g/ha compared with the standard, maleic hydrazide at 3.39 kg/ha, but the effects were not as long lasting. At the end of 3 wk, the moniliformin treated axillary shoots had started to grow; maleic hydrazide treated shoots were held in check

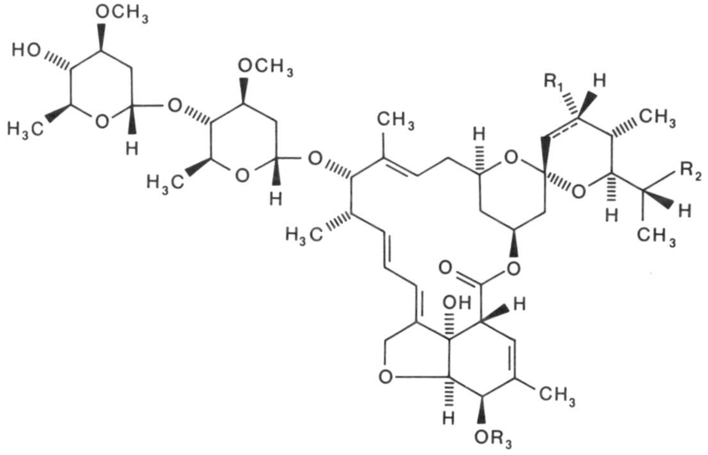

Avermectin	R_1	R_2	R_3
A_{1a}		C_2H_5	CH_3
A_{1b}		CH_3	CH_3
A_{2a}	OH	C_2H_5	CH_3
A_{2b}	OH	CH_3	CH_3
B_{1a}		C_2H_5	H
B_{1b}		CH_3	H
B_{2a}	OH	C_2H_5	H
B_{2b}	OH	CH_3	H

Where R_1 is absent, the double bond (===) is present.
Both sugars are α-L-oleandrose.

Figure 9.7 Skeletal structure and analogs of the avermectins, several of which show promising pesticidal activities. (Reprinted with permission from ACS Symposium Series #276, "Bioregulators for Pest Control," page 415, copyright 1985, American Chemical Society.

for 6 wk. The yields of axillary shoots expressed as ratios of maleic hydrazide: moniliformin:control were 1:2:7 at 21 days (Cutler et al., 1976).

The complexity of certain natural products may appear to act as a deterrent to their use as agricultural chemicals and arguments made along these lines are reasonable. But the avermectins (Figure 9.7), lactone metabolites from the actinomycete *Streptomyces avermitilis* are extraordinarily complex, are subjected to further synthesis after isolation and purification, and have high specific activity (Whitehead, 1985). Their potential promise for agriculture and medicine is extremely high. No doubt other natural products, including fungal products, will be equally as rewarding for those who are intrepid enough to logically exploit them.

DEDICATION

To all my colleagues who earn their daily bread by isolating and identifying natural products. And to those who will follow us.

REFERENCES

Anke, H., H. Zahner, and W. A. Konig. 1978. *Arch. Microbiol.* 116:253.

Carroll, L. 1960. Jabberwocky. *The Annotated Alice.* Bramwell House, New York.

Cole, R. J. and R. H. Cox. 1981. *Handbook of Toxic Fungal Metabolites.* Academic, New York.

Cole, R. J. and J. W. Kirksey. 1974. *Can. J. Microbiol.* 20:1159.

Cole, R. J., J. W. Kirksey, H. G. Cutler, B. L. Doupnik, and J. C. Peckham. 1973. *Science.* 179:1324.

Cole, R. J., J. W. Kirksey, H. G. Cutler, and E. E. Davis. 1974. *J. Agric. Food Chem.* 22:517.

Cole, R. J., J. W. Dorner, R. H. Cox, R. A. Hill, H. G. Cutler, and J. M. Wells. 1981. *Appl. Environ. Microbiol.* 42:677.

Cole, R. J., H. G. Cutler, and J. W. Dorner. 1986. *In* R. J. Cole (ed.), *Modern Methods in the Analysis and Structure Elucidation of Mycotoxins*, Academic, New York.

Collins, R. P. and A. F. Halim. 1972. *J. Agric. Food Chem.* 20:437.

Cutler, H. G. 1968. *Plant Cell Physiol.* 9:593.

Cutler, H. G. 1984. *Proc. 11th Annu. Meet. Plant Growth Regul. Soc. Am.* 1.

Cutler, H. G. and B. Jarvis. 1985. *Environ. Exp. Bot.* 25:115.

Cutler, H. G. and J. H. LeFiles. 1978. *Plant Cell Physiol.* 19:177.

Cutler, H. G., R. J. Cole, and J. M. Wells. 1976. *Proc. 6th Int. Tobacco Sci. Congr.* 124.

Cutler, H. G., J. H. LeFiles, F. G. Crumley, and R. H. Cox. 1978. *J. Agric. Food Chem.* 26:632.

Cutler, H. G., F. G. Crumley, R. H. Cox, O. Hernandez, R. J. Cole, and J. W. Dorner. 1979. *J. Agric. Food Chem.* 27:592.

Cutler, H. G., F. G. Crumley, R. H. Cox, R. J. Cole, J. W. Dorner, J. P. Springer, F. M. Latterell, J. E. Thean, and A. E. Rossi. 1980. *J. Agric. Food Chem.* 28:139.

Cutler, H. G., F. G. Crumley, J. P. Springer, R. H. Cox, R. J. Cole, J. W. Dorner, and J. E. Thean. 1980. *J. Agric. Food Chem.* 28:989.

Cutler, H. G., F. G. Crumley, J. P. Springer, and R. H. Cox. 1981. *J. Agric. Food Chem.* 29:981.

Cutler, H. G., F. G. Crumley, R. H. Cox, E. E. Davis, J. L. Harper, R. J. Cole, and D. R. Sumner. 1982. *J. Agric. Food Chem.* 30:658.

Cutler, H. G., F. G. Crumley, R. H. Cox, J. P. Springer, R. F. Arrendale, and P. D. Cole. 1984. *J. Agric. Food Chem*. 32:778.

Divekar, P. V., R. H. Haskins, and L. C. Vining. 1959. *Can. J. Chem*. 37:2097.

Ellestad, G. A., P. Mirando, and M. P. Kunstmann. 1973. *J. Org. Chem*. 38:4204.

Godtfredsen, W. O. and S. Vangedal. 1965. *Acta Chem. Scand*. 19:1088.

Grove, J. F. 1972. *Chem. Soc. Perkin Trans*. 1:2400.

Hancock, C. R., H. W. Barlow, and H. J. Lacey. 1964. *J. Exp. Bot*. 115:166.

Itahashi, M., Y. Murakami, and H. Nishikawa. 1955. *Tohoka J. Agric. Res*. 5:281.

Kirksey, J. W. and R. J. Cole. 1974. *J. Mycopathol. Mycol. Appl*. 54:291.

Kogl, F. and G. C. Van Wessem. 1944. *Rec. Trav. Chim. Pays-Bas*. 63:5.

Kurtz, T. E., R. F. Link, J. W. Tukey, and D. L. Wallace. 1965. *Technometrics*. 7:95.

Lloyd, G., A. Robertson, G. B. Sankey, and W. B. Whalley. 1955. *J. Chem. Soc*. 2163.

Nitsch, J. P. and C. Nitsch. 1956. *Plant Physiol*. 31:94.

Scott, P. M., W. VanWalbeek, and W. M. MacLean. 1971. *J. Antibiot*. 24:747.

Smith, J. and R. H. Thomson. 1960. *Tetrahedron*. 10:148.

Springer, J. P., H. G. Cutler, F. G. Crumley, R. H. Cox, E. E. Davis, and J. E. Thean. 1981. *J. Agric. Food Chem*. 29:853.

Stahl, E. 1965. *Thin-layer Chromatography*. Academic, New York, pp. 47(4c), 486, and 498.

Vining, L. C., W. J. Kelleher, and A. E. Schwarting. 1962. *Can. J. Microbiol*. 8:931.

Wells, J. M., H. G. Cutler, and R. J. Cole. 1976. *Can. J. Microbiol*. 22:1137.

Whitehead, D. 1985. *In* P. A. Hedin, H. G. Cutler, B. D. Hammock, J. J. Menn, D. E. Moreland, and J. R. Plimmer (eds.), *Bioregulators for Pest Control*, Chapter 29. American Chemical Society, Washington, D.C.

10

MECHANISMS AND MODES OF ACTION OF ALLELOCHEMICALS

FRANK A. EINHELLIG

*Department of Biology, University of South Dakota,
Vermillion, South Dakota*

The elucidation of the methods whereby allelopathic chemicals alter plant growth and development has been a difficult and ongoing challenge. A clear insight into the precise physiological perturbations caused by these substances has not been obtained, and it must be emphasized that much additional information is needed. One reason for this frustration is the limited amount of work that has focused on this question. Typically, putative allelochemicals have been evaluated by their impact on seed germination or some aspect of plant growth without regard to the sequence of cellular events causing the growth reductions. Clarification of the action of allelochemicals has also been hampered because there are so many different compounds involved.

Most of the chemicals implicated in allelopathy are either secondary compounds that have their origin in the shikimic acid and acetate pathways or structures that arise with components from each of these sources. Rice (1984) classified allelopathic agents into 14 chemical categories and a miscellaneous group. They include a variety of phenolic acids, flavonoids, quinones, terpenoids, steroids, purines, long-chain fatty acids and acetylenes, organic acids, unsaturated lactones, and others. The number and diversity of identified allelochemicals is rapidly growing, yet most of the substances currently known have not even been tested in ways that allow evaluation of their mode of action. Undoubtedly, it is too simplistic to assume that these diverse substances have a common mode of action.

I also suspect it is a rare exception when a single substance is responsible for allelopathy. Isolation of several compounds and different classes of compounds from a particular allelopathic situation has been the consistent pattern. Experiments indicate that either additive or synergistic inhibition may occur from combinations of terpenoids, benzoic acids, organic acids, derivatives of cinnamic acid, p-hydroxybenzaldehyde with coumarin, and the three-way concert action of a flavonoid, coumarin, and phenolic acid (Einhellig, 1986). How such combinations act at the cellular level has not been determined.

The consideration of mechanisms and modes of action in this chapter focuses primarily on the effects of inhibitors on higher plants. Literature cited illustrates major points, since space limitations preclude an inclusive review. More explicit references concerning certain mechanisms have been given by Horsley (1977), Rice (1984), Balke (1985), Einhellig (1985), and Einhellig et al. (1985). I do not attempt to discuss the synonymy between the mode of action of allelochemicals and herbicides, but clearly many parallels exist. As a group herbicides impair the same plant processes as those altered by allelochemicals and compilations by Moreland (1980) and Duke (1985) can be consulted for comparisons.

Much of the information available for evaluating the action of allelochemicals deals with the effects of phenolic compounds, particularly cinnamic and benzoic acids, compounds with the coumarin skeleton, flavonoids, and various other polyphenols. These substances are widely distributed in nature, they have been frequently implicated in allelopathy, and more research has been directed toward assessment of their action than other allelochemicals. However, this does not mean they should automatically be assumed to be the most important compounds or those with the greatest biological activity.

TRANSFER AND TRANSFORMATION OF ALLELOCHEMICALS

Many chemical constituents are volatilized or lost by root exudation and leaching from shoots of actively growing plants, and through residue decomposition. Nonvascular plants and microorganisms also are donors. Some organic compounds that escape into the environment are inhibitors of the growth or development of receiving species. Alternatively, other allelochemicals arise after transformation by microbial metabolism. Most allelochemicals cycle through the soil in terrestrial communities, but in some situations leachate components and volatiles may transfer directly between plants. Root contacts may also provide an avenue for direct transfers.

The biological activity of primary allelochemicals may be multiplied through degradation products. For example, peach roots release amygdalin, a cyanogenic glycoside that eventually degrades to hydrogen cyanide and benzaldehyde. Phlorizin deposition in apple orchards yields phloroglucinol, p-hydrocinnamic acid, and subsequently p-hydroxybenzoic acid. The tissues of many plants of the walnut family (Juglandaceae) contain hydrojuglone, which is oxidized to a more toxic form (juglone) when exposed to air. These few examples demonstrate the common principle of allelochemical transformations, yet illustrate that these changes cannot be assumed to mean detoxification.

Although some allelochemicals are quickly altered, others, like the alkaloid caffeine released from coffee (*Coffea arabica* L.) litter, have considerable anti-microbial activity that prolongs their retention. Factors of the physical environment also influence retention. Rietveld et al. (1983) proposed that soil accumulation of juglone in walnut (*Juglans nigra* L.) plantation sites with poorly drained soil was greater than well-drained areas because the former had less aerobic microorganism activity. Increased retention of allelochemicals in water-logged soils has also been shown for organic and phenolic acids (Wang et al. 1968, 1971; Shindo and Kawatsuka, 1977). Conversely, organic matter breakdown under well-aerated conditions may provide an environment that enhances the growth of organisms that produce new and highly toxic allelochemicals. Decomposing crop residues in Nebraska supported a bloom of *Penicillium urticae* Bainier, a fungus that produces patulin, which is a potent inhibitor limiting wheat growth (McCalla and Nordstadt, 1974).

The unique features of each allelochemical, plus the edaphic and climatic conditions, prohibit generalizations concerning their fate in the soil matrix. Fixing of organic compounds by soil minerals and the humus complex often occurs, yet this may also provide a pool for future release to the soil solution or direct transfer to roots. Although events in the soil are shrouded with mystery, the soil environment is a major source whereby receiver plant contact occurs. Even volatile allelochemicals like some of the terpenes may be adsorbed on soil particles (Muller, 1966). These may solubilize in the cutin of the root epidermis as contact occurs with particle surfaces. The soil air space may contain volatiles from decomposing residue. The soil solution will be the source of allelochemicals, and even some of those are relatively insoluble in water. Water washes of leaves often contain sterols and other water-insoluble lipids, plus long-chain fatty acids, and it has been suggested that micelle formation between these may enhance their solubility in the aqueous medium. In short, it is generally accepted that lower plants, germinating seeds, and roots of seedlings will take up allelochemicals from their soil microenvironment.

Herbicides have frequently been characterized according to (1) whether they cause their damage on contact, (2) whether they have foliar or root absorption, and (3) the extent of their translocation. Although little definitive work of this type has been done with the spectrum of allelochemicals, many of the low molecular compounds can be expected to have entry into cells. We found tobacco and sunflower roots removed scopoletin from a nutrient medium with substantial quantities translocated to the leaves, and a colleague followed this uptake by radioisotope labeling (Einhellig et al., 1970). Incorporation and metabolism of ^{14}C-labeled cinnamic, caffeic, and ferulic acids and coumarin by yeast cells and germinating seeds have been demonstrated (Van Sumere et al., 1971). Labeled arbutin, hydroquinone, and salicylic acid transfer into roots (Glass and Bohm,

1971; Harper and Balke, 1981), and in recent work with Dr. Gerald Leather we found rapid translocation of ^{14}C-labeled salicylic throughout the plant. Metabolism of these phenolic inhibitors probably involves reducing toxification by binding or conjugating with sugar components, and recently an inducible UDPG glycosyltransferase active in salicylic acid metabolism was reported (Balke et al., 1984).

ALLELOCHEMICAL EFFECTS ON PLANT PROCESSES

A frequent approach to assessing the mode of action of specific allelochemicals has been through monitoring their effects on major plant functions. The bioassay species and conditions have not been uniform in these studies. However, they still lead to the conclusion that allelochemicals interfere with many of the primary metabolic processes and growth regulatory systems of higher plants. The action of ferulic acid illustrates this broad spectrum (Table 10.1). Thus separation of the primary action of an allelopathic substance from its secondary effects has been difficult, and it is possible that such a separation is not even valid.

Before discussing the physiological alterations caused by allelochemicals, it should be recognized that their biological activity is concentration dependent with a response threshold. Mild growth reductions above the threshold often show no visible signs of injury, and in some instances growth is stimulated below the threshold concentration. The inhibition threshold for a specific substance is not a constant, but is intimately related to the sensitivity of the receiving species, the plant process, and environmental conditions. For example, the threshold for reducing the growth of grain sorghum [*Sorghum bicolor* (L.) Moench.] seedlings by several cinnamic acids was 1/25th that required to inhibit germination, and initiation of ferulic acid inhibition of these seedlings under high temperatures occurred at about one half the concentration required at moderate temperatures (Einhellig et al., 1982; Einhellig and Eckrich, 1984). Nevertheless, for many phenolics the threshold for inhibition of postgermination growth and plant functions is often in the 100–1000 μM range (Table 10.1).

Regulation of Growth

Cell Division. Cell division and elongation are essential phases of development and one might assume that most inhibitors of growth would modify these events. Efforts to characterize the action of allelochemicals support this inference. Root mitosis was slowed or prevented by parasorbic acid, coumarin, and scopoletin (Cornman, 1946; Avers and Goodman, 1956). Jankay and Muller (1976) found that umbelliferone increased radial expansion in cucumber (*Cucumis sativus* L.) root cells, whereas it decreased elongation. Volatiles from *Salvia leucophylla* L., primarily cineole and camphor, reduced cell division and elongation in radicles and hypocotyles of germinating cucumber, but these cells had a greater diameter than controls (Muller, 1965). The root cells had irregular

TABLE 10.1.
Effects of ferulic acid on plant metabolism

Plant Function or Condition	Test Species	Concentration (μM)	Action	Reference
Protein synthesis	Lettuce	500	Inhibited	Cameron and Julian (1980)
Lipid synthesis	Cell culture	100	Increased	Danks et al. (1975a)
IAA-Induced growth	Pea	500	Increased	Tomaszewski and Thimann (1966)
ABA Inhibition	Amaranthus caudatus	10	Released	Ray et al. (1980)
Respiration rate	Yeast	500	Increased	Van Sumere et al. (1971)
Photosynthesis	Soybean	1000	Decreased	Patterson (1981)
Stomatal conductance	Sorghum	500	Decreased	Einhellig et al. (1985)
Chlorophyll content	Soybean	500	Decreased	Einhellig and Rasmussen (1979)
Phosphorus content	Sorghum	250	Decreased	Kobza (1980)
Ion uptake	Barley	500	Decreased PO_4^{3-}	Glass (1973)
	Oats	500	Decreased K^+	Harper and Balke (1981)
Membrane potential (mv)	Barley	250	Decreased	Glass and Dunlop (1974)
Water potential	Sorghum	250	Lowered	Einhellig et al. (1985)

nuclei and large internal globules, indicating food reserves were not properly utilized.

Organic Synthesis.   Several phenolic allelochemicals modify biosynthesis of major plant constituents or the distribution of carbon in cellular pools. Van Sumere et al. (1971) found that coumarin and a group of cinnamic and benzoic acids and aldehydes inhibited uptake and incorporation of phenylalanine-^{14}C by yeast (*Saccharomyces cerevisiae*) cells. Additional tests with coumarin and ferulic acid showed these events were suppressed in lettuce (*Lactuca sativa* L.) seeds and barley (*Hordeum vulgare* L.) embryos. Cameron and Julian (1980) indicated that inhibition of protein synthesis by cinnamic and ferulic acids was a reliable indicator of their action on lettuce growth. In contrast, introduction of approximately 50 μM catechin or chlorogenic acids in *Prunus avium* L. callus culture caused proliferation of tissue which correlated with an elevated rate of protein synthesis, suggesting that several enzymes were stimulated by this level of polyphenols (Feucht and Schmid, 1980). Apparently the influence of phenolic compounds on metabolic pools is not always predictable. Danks et al. (1975a) found that incubation of cells with glucose-UL-^{14}C plus ferulic (100 μM) or cinnamic (10 μM) reduced carbon flow into proteins, but the effects of the two compounds on other pools differed. Ferulic acid promoted incorporation of the label into soluble lipids, and decreased organic acids and soluble amino-acid fractions. Cinnamic acid elevated soluble amino acids and did not promote lipids.

Interactions with Hormones.   How allelochemicals influence the delicate balance of hormones involved with regulating plant growth has been an intriguing question. Evidence from the last 20 yr affirms that one mechanism of allelopathic action for several phenolic substances is an alteration of the level of indoleacetic acid (IAA). Either inhibition or stimulation of IAA has been reported by scopoletin and chlorogenic, cinnamic, and benzoic acids. Assays with oat (*Avena sativa* L.) coleoptile and pea (*Pisum sativum* L.) sections incubated with phenolics and IAA illustrated that monophenols stimulated the decarboxylation of IAA, while polyphenols synergized IAA-induced growth by counteracting IAA destruction (Tomaszewski and Thimann, 1966). Many investigations have corroborated that the phenolic acids can be divided into two groups, suppressors of IAA destruction (i.e., chlorogenic, caffeic, ferulic, and protocatechuic) and those that stimulate IAA oxidase (*p*-coumaric, *p*-hydroxybenzoic, vanillic, syringic, phloretic, etc.) (Lee et al., 1982). Lee (1980) also reported that *p*-coumaric acid, ferulic acid, and 4-methylumbelliferone inhibited the formation of bound IAA and consequently caused an accumulation of free IAA independent of enzymatic oxidation.

Some regulatory polyphenols may reduce growth by binding gibberellic acid (GA), whereas others promote growth by binding abscisic acid (ABA). A variety of tannins inhibited GA-induced growth in dwarf pea bioassays, and reduced amylase and acid phosphatase synthesis in endosperm half-seeds of barley (Corcoran et al., 1972; Jacobson and Corcoran, 1977). The tannins apparently blocked amylase synthesis, not secretion. Conversely, Ray et al. (1980) reported that many phenolics released ABA inhibition. Coumarin and ferulic, gallic, tannic, and cinnamic acids removed ABA inhibition of hypocotyl growth at 10 μM, and rutin, morin, quercetin, and chalcone were effective at 1 μM.

Additional work is needed to clarify how allelopathic inhibition of growth is related to hormonal activity, but certainly it is more complex than simply an interaction with auxin. From the information presently available it can be inferred that in some cases, and to some degree, allelochemical effects are mediated by these interactions.

Effects on Enzymes.   In addition to the known activity of phenolic compounds on IAAO and amylase, allelochemicals alter either the synthesis or function of many enzymes. Jain and Srivastava (1981) reported that nitrate reductase activity in corn (*Zea mays* L.) was increased by 10 μM salicylic acid and inhibited above 1000 μM. The effects were thought to be on enzyme synthesis. Phenylalanine ammonia–lyase was inhibited by most of the caffeic and gallic acids tested by Sato et al. (1982), indicating that they could alter phenylpropanoid metabolism in higher plants. In various situations allelochemicals have inhibited proteinase and pectolyic enzymes, catalase, peroxidase, phosphorylase, sucrase, cellulase, succinic dehydrogenase, and others (Rice, 1984), but no case has been established that such effects are the primary actions causing growth reductions.

Respiratory Metabolism

Muller et al. (1969) found that volatile monoterpenes (cineole, dipentene, etc.) from *Salvia leucophylla* were potent inhibitors of O_2 uptake by mitochondrial suspensions, although camphene, α-pinene, and β-pinene did not alter respiration of excised roots. Inhibition was localized in that part of the Kreb cycle following succinate, and later work showed a decrease in oxidative phosphorylation (Lorber and Muller, 1980). Weaver and Klarich (1977) reported that volatile substances from *Artemisia tridentata* Nutt. suppressed seedling growth and the respiration rate of juvenile plant material, but respiration of mature leaves was elevated. Monoterpenes were the presumptive toxins and laboratory tests with wheat (*Triticum aestivum* L.) demonstrated that certain of these compounds raised and others lowered respiration. *Artemisia* also contains esculin and several sesquiterpenoid lactones, including arbusculin-A, achillin, and viscidulin-C, which can elevate respiration (McCahon et al., 1973).

A wide range of phenolic compounds interfere with mitochondrial functions, including quinones, flavonoids, and phenolic acids. Koeppe (1972) reported that juglone-induced reduction of respiration *in vivo* resulted from inhibition of the coupled intermediates of oxidative phosphorylation, slowing electron flow to O_2. Stenlid (1970) found that the extent of flavonoid interference with ATP production depended upon the hydroxylation pattern. Stimulation of O_2 uptake by yeast cells has been observed after 1000 μM treatments with coumarin, various aldehydes, cinnamic acids, and benzoic acids (Van Sumere et al., 1971). Quinones inhibited O_2 uptake in this study, but at reduced levels some of the quinones were stimulatory. Further work showed that p-coumaric acid, cinnamic acid, salicylaldehyde, and 2-methylnaphthoquinone depressed the ADP/O ratio in yeast mitochondrial suspensions, indicating uncoupling of oxidative phosphorylation. Inhibition of seed germination by caffeic and ferulic acids could not be explained by uncoupling. Inconsistencies were also found by Demos et al. (1975) when comparing the effects of a spectrum of phenolics on respiration. Tannic, gentisic, and p-coumaric acids inhibited mung bean (*Phaseolus aureus* L.) hypocotyl growth, reduced respiration in isolated mitochondria, released respiratory control, and prevented substrate-supported Ca^{2+} and PO_4^{3-} transport. However, ferulic, caffeic p-hydroxybenzoic, and syringic acids reduced hypocotyl growth without affecting any of the mitochondrial processes.

All allelochemicals must not be categorized as uncouplers or as agents that otherwise perturb respiration. However, the literature on adverse affects on respiration at the level of enzymes, isolated mitochondria, excised tissue, and the whole plant argues for respiratory metabolism stress as one mode of action of some allelochemicals. This is supported by the fact that water extracts from allelopathic plants and residues may either decrease or increase respiration of test plants (Patrick and Koch, 1958; Lodhi and Nickell, 1973).

Photosynthesis and Related Processes

Effects on Photosynthesis. The increase in dry matter of higher plants is linked to carbon fixation, so any loss in efficiency of photosynthesis might be detrimental to growth. This logic fostered our investigation of the effects of scopoletin on the net photosynthetic rate of several species grown in nutrient culture (Einhellig et al., 1970). A one-time amendment of either 500 or 1000 μM scopoletin in the growth medium of tobacco (*Nicotiana tobacum* L.) reduced net photosynthesis by the second day, rates continued to decline for two more days, then a slow recovery occurred. Dark respiration was unaltered. The inhibition of expansion of leaf area paralleled reductions in photosynthesis. Even though the photosynthetic rate returned to normal, at the end of the 11-day experiment plants were stunted and a calculation of CO_2 fixed per illumination hour per plant showed 1000 μM scopoletin-treated tobacco fixed only 51% as much CO_2 as the controls. Scopoletin also depressed the photosynthetic rate of sunflower (*Helianthus annuus* L.) and pigweed (*Amaranthus retroflexus* L.). Hence the data indicate the same impact on C_3 and C_4 species.

A subsequent study demonstrated that p-coumaric and caffeic acids inhibited grain sorghum photosynthesis at treatments that reduced growth (Kadlec, 1973). Patterson (1981) reported that 1000 μM p-coumaric, caffeic, ferulic, gallic, t-cinnamic, and vanillic acids severely reduced the photosynthesis of soybeans [*Glycine max* (L) Merr.], whereas 100 μM had no effect. The 1000 μM-treated soybeans also had lower dry-matter production, leaf expansion, height, leaf productivity, and net assimilation rate. Recent manometric tests in my laboratory showed that an array of coumarins and cinnamic and benzoic acids suppressed photosynthesis of *Lemna minor* L. at concentrations corresponding to their individual thresholds for growth inhibition (Einhellig et al., 1985). Inhibition was not found in similar tests with two flavonoids, catechin and rutin.

Photosynthesis may be altered by a variety of mechanisms, both direct effects at the chloroplast level and indirect actions, such as stomatal closure. Arntzen et al. (1974) found that kaempferol, a flavonol implicated in several allelopathic situations, inhibited coupled electron transport and both cyclic and noncyclic photophosphorylation in pea chloroplasts. Kaempferol (25 μM) blocked energy transfer in the sequence prior to the terminal step in phosphorylation. In contrast, a dihydrochalcone glucoside, phlorizin, inhibited the activity of the chloroplast membrane ATPase. Moreland and Novitsky (1986) reported that several cinnamic and benzoic acids, coumarins, and flavonoids inhibited CO_2-dependent oxygen evolution in isolated chloroplasts. However, this occurred at fairly high concentrations. In thylakoid assays the compounds primarily affected photophosphorylation, but they neither acted like uncouplers nor classical energy transport inhibitors. At present, the link between chloroplast effects and growth inhibition is difficult to evaluate.

Stomatal Response.   We observed that plants grown with scopoletin in the nutrient solution often showed some loss of leaf turgor accompanying reductions in photosynthesis, and we hypothesized that stomatal closure may have limited CO_2 availability. Subsequent work demonstrated that 500 and 1000 μM scopoletin and chlorogenic acid closed stomates in tobacco and sunflower for several days following treatment (Einhellig and Kuan, 1971). There was a good correlation between the effects of scopoletin on stomatal aperture and photosynthetic rate. Likewise, Kadlec (1973) and Patterson (1981) found that phenolic acid-induced reductions in photosynthesis of grain sorghum and soybeans were accompanied by reduced stomatal conductance. A precise determination of cause and effect from these observations is obscure, because interference with photosynthesis would alter light-induced stomatal opening, just as stomatal closure would hinder CO_2 diffusion.

The relationship between allelochemical interference with stomatal function and growth reduction has not always been clear. Tobacco stomates were closed by 1000 μM tannic acid, but they were not affected at the threshold for growth reduction (100 μM) (Einhellig, 1971). Grain sorghum grown with 1000 μM vanillic and *p*-hydroxybenzoic acids had slightly higher leaf resistances, but Einhellig and Rasmussen (1978) concluded that this was not the primary factor responsible for growth inhibition. However, the response of test plants to allelopathic conditions supports the conjecture that stomatal interference is one mechanism of action for some allelochemicals. The introduction of water extracts from velvetleaf (*Abutilon theophrasti* Medic), *Kochia* [*Kochia scoparia* (L.) Schrad.], Jerusalem artichoke (*Helianthus tuberosum* L.), and cocklebur (*Xanthium pensylvanicum* Wallr.) into the nutrient solution for growing grain sorghum and soybeans caused stomatal closure, which corresponded with growth reductions (Colton and Einhellig, 1980; Einhellig and Schon, 1982; Einhellig et al., 1985). The growth of these test plants in soil containing dried residue from these weeds was also reduced, and plants in some treatments had a higher leaf resistance than did controls.

Chlorophyll Content.   A chlorotic appearance has occasionally been reported as one symptom of allelopathic interference. We found that 6 days after treatment with ferulic, *p*-coumaric, and vanillic acids soybean plants weighed less than controls and had less leaf chlorophyll on a dry-weight basis (Einhellig and Rasmussen, 1979). In contrast, grain sorghum seedlings did not have less chlorophyll even though growth was inhibited. Chlorophyll loss could contribute to a lower rate of photosynthesis. However, the different effects on chlorophyll content of soybean and grain sorghum indicates the need for caution in suggesting what primary event alters growth.

Allelochemicals from the trichomes of *Parthenium hysterophorus* L. caused chlorophyll loss in the leaves of beans (*Phaseolus vulgaris* L.) used as the assay species (Kanchan and Jayachandra, 1980). The investigators surmised that this reduction in chlorophyll was primarily due to enhanced degradation. Alternatively, allelochemicals may inhibit the synthesis of the Mg-porphyrin. As pointed out by Rice (1984), allelopathic interference in nodulating legumes often causes both a reduction in hemoglobin content of the nodules and chlorotic leaves. Hemoglobin and chlorophyll are both porphyrin containing compounds.

Nutrient Uptake and Associated Processes

Mineral Accumulation.   Relatively few studies have tested the effects of specific allelochemicals on the mineral content of intact plants. McClure et al. (1978) reported that ferulic acid inhibited the absorption, but not translocation, of PO_4^{3-} by soybeans. Work in my laboratory demonstrated that ferulic acid levels near the threshold for growth inhibition, 250 and 500 μM, interfered with mineral nutrition of grain sorghum (Kobza, 1980). Analysis of tissue 3 and 6 days after treatment showed that ferulic acid treatments reduced PO_4^{3-} concentration in both roots and shoots, and roots had significantly lower K^+ and Mg^{2+}. Danks (1975b) found that the effects of ferulic acid on ion uptake of a cell-suspension culture varied with the age of the culture. At the end of a 14-day growth cycle, 100 μM ferulic acid-treated cultures had taken up less PO_4^{3-} and more K^+, Mg^{2+}, and Ca^{2+} than the controls. Chlorogenic acid also alters nutrient balance. Pigweed plants grown in soil to which chlorogenic acid was added had an elevated level of N and a reduced amount of PO_4^{3-} (Hall et al., 1983). Growth reductions and mineral changes in these seedlings were both overcome by nutrient supplements.

Alterations of the mineral content of plants subjected to nonspecific allelopathic conditions has been shown in more than a dozen investigations (Balke, 1985). In some cases the experimental design placed donor and receiver plants together, whereas in others the receiving plant was exposed to residues, leachates, or exudates from the suspected allelopathic source. The work of Chambers and Holm (1965) showed that a single bean plant absorbed much more labeled (^{32}P) PO_4^{3-} than when grown in association with other beans, redroot pigweed, or green foxtail (*Setaria viridis* L.), and the evidence suggested allelopathy. Nitrogen and K^+ levels of corn (*Zea mays* L.) were decreased when grown with quackgrass (*Agropyron repens* L.) and an increase in fertilizer did not overcome this effect (Buchholtz, 1971). One of the autotoxicity effects of berseem clover (*Trifolium alexandrinum* L.) was a reduction in phosphorus content (Katznelson, 1972). Mineral imbalance has also been reported in woody species. Sweetgum (*Liquidambar styraciflua* L.) subjected to leachates from the rhizosphere or residue of fescue (*Festuca arundinacea* Shreb.) had less PO_4^{3-} content and more K^+ than controls, but N was not different (Walters and Gilmore, 1976). Fisher et al. (1978) found similar effects on sugar maple (*Acer saccharum* Marsh.) grown with mulches from goldenrod (*Solidago canadensis* L.) and aster (*Aster nova-angliae* L.).

It is difficult to generalize about changes in mineral content incurred from allelopathic interference. Phosphorus reductions have been the most consistent finding, although sometimes this has not been the case. The data of Bhowmik and Doll (1984) on the effects of five annual weed residues on corn and soybean illustrate these inconsistencies. Even though the residues reduced dry-matter production, inhibition of N, PO_4^{3-}, and K^+ uptake varied according to the residue source and the test crop. They suggested residue-induced growth suppression in these experiments was probably not related to nutrient imbalance. Certainly at this point it is not clear how these two factors are coupled in allelopathy interference. The data is also mixed concerning the effectiveness of fertilizer enhancement in overcoming allelochemical-induced growth suppression.

Ion Uptake by Excised Roots.   Monitoring mineral absorption of excised roots has given direct evidence that most of the benzoic and cinnamic acids, hydroquinone, juglone, naringenin, genistein, kaempferol, and phloretin are interfering agents. These investigations are typified by the work of Glass (1973, 1974), which showed that benzoic and cinnamic acids inhibited the uptake of PO_4^{3-} and K^+ by barley roots. The tests were of short duration, using a treatment concentration of 500 μM with most of the PO_4^{3-} studies; 250 μM levels were utilized for the K^+ uptake work. The extent of inhibition of ion uptake varied among the dozen compounds tested and it correlated well with their lipid solubility. Cinnamic acid, the most inhibitory, caused almost complete suppression of ion uptake. Harper and Balke (1981) found that as the pH of the test medium was reduced from 6.5 to 4.5, salicylic acid caused greater inhibition of K^+ uptake by oat roots, more salicylic acid was absorbed, and a lower treatment concentration resulted in inhibition. Thus under the right conditions phenolic acids may be effective inhibitors of mineral uptake.

Effects on Membranes.   Allelochemical interference with mineral absorption is probably the result of effects on cell membranes. Transfer of minerals across membranes would be inhibited by any action that changes permeability or alters the energetics associated with the process. Glass and Dunlop (1974) found that membranes of barley root cells were rapidly depolarized by the addition of 500 μM salicylic acid to the buffer medium (pH 7.2). The potential difference changed from -150 mV to near zero in 12 minutes. Other benzoic and cinnamic acids tested (250 μM) also caused depolarization with the extent of disruption correlated with their lipid solubility. Similarly, micromolar levels of benzoic and butyric acids depolarized oat coleoptile cell membranes, whereas hyperpolarization occurred with high (10 μM) concentrations of butyric acid (Bates and Goldsmith, 1983).

Plausible explanations for the decreases observed in membrane potential include inhibition of ATPases, which produce the potential difference and reduction in available energy for their function. In tests to evaluate these factors, Balke (1985) found salicylic acid decreased the ATP content of excised oat roots, but had little effect on ATPase activity. The reduction in tissue ATP was pH-dependent, similar to the action of salicylic acid on both depolarization and absorption of potassium ions. However, Balke (1985) summarized several experiments that showed jugone and several flavonoids inhibited plasma membrane ATPase activity. Certain allelochemicals may modify the supply of ATP, as previously noted. Some of the terpenoids, quinones, and phenolic acids uncouple oxidative phosphorylation or stimulate oxygen consumption in other nonproductive ways. Thus either interference with ATP production or its use in transport processes may contribute to changes in membrane functions.

Glass and Dunlop (1974) summarized their work on the effects of phenolic acids on barley root cells by suggesting that these compounds caused an increase in membrane permeability to both cations and anions, allowing a nonspecific efflux of ions. Balke (1985) confirmed that salicylic acid caused K^+ leakage from oat tissue. Whether phenolic acid-induced cell leakage is explained by the drop in membrane potential or by structural alterations as these compounds solubilize in the membrane remains unclear, but such changes infer other physiological effects.

Effects on Water Relationships.   A logical outcome of some of the observed effects on ion uptake and membrane functions would be changes in plant–water balance, and this appears to be the case. Our studies have shown that p-coumaric and ferulic acids lowered the leaf water potential of grain sorghum and soybean seedlings within 1 day following treatment, and this persisted during the 6-day experiments (Einhellig et al., 1985). Sorghum grown with 250 μM p-coumaric or ferulic acids in the nutrient medium had midday leaf water potentials of approximately -10 bars, compared to -5 bars for controls. This difference resulted from reductions in both osmotic and turgor pressure. Since the threshold for causing partial stomatal closure was 500 μM, water potential was the more sensitive indicator of phenolic acid stress. Patterson (1981) reported that the depressed growth of soybeans treated with 1000 μM caffeic, ferulic, or gallic acids was accompanied by reductions in water potential, although several other phenolic acids did not alter water balance.

Compounds from allelopathic plants that reduce the growth of test seedlings may also create part of their impact through disruption of water relationships. We found that water stress was one action of sunflower, velvetleaf, Jerusalem artichoke, and cocklebur (Colton and Einhellig, 1980; Schon and Einhellig, 1982; Einhellig et al., 1985). Rice (1984) summarized observations indicating that allelopathic compounds may interfere with xylem flow by clogging the vessels. Certainly an interdependence between water balance and allelochemical effects seems likely, since we have observed that even a mild water stress can be very detrimental in conjunction with allelochemical stress (Einhellig, 1986).

INDIRECT EFFECTS ON HIGHER PLANTS

Optimum growth and vigor of higher plants often depends upon functions of associated organisms whose populations may be suppressed by allelochemicals. Mycorrhizal fungi improve nutrient and water-absorbing properties and confer protection against root disease for many plants in natural communities. Brown and Mikola (1974) found that allelopathic inhibition of these symbionts resulted in less PO_4^{3-} uptake and inhibited the growth of pine and spruce seedlings. Subsequent work has shown that the suppression of fungal and root colonization is a contributing factor to delays in conifer reforestation on disturbed sites (Rose et al., 1983). There also are many situations where flourishing of certain higher plants is hindered owing to allelochemical interference with nitrification or the activity of nitrogen-fixing organisms, *Rhizobium* spp., free-living fixers, and blue-green algae (Rice, 1984). It is logical that the actions of allelochemicals on microorganisms and fungi may be analogous to some of the perturbations found in higher plant cells. However, specific mechanisms may involve synthesis systems and enzyme functions unique to these organisms.

Allelopathic conditions may foster ancillary detriments to plant growth. It has long been apparent that plants subjected to allelochemicals are more susceptible to disease (Patrick et al., 1964). The reasons for this reduced resistance must be varied and complex. Although not strictly an allelochemical effect, the residues of some allelopathic plants have concert actions that reinforce allelopathy. Harvey and Linscott (1978) reported that the decomposition of quackgrass rhizomes in the soil raised ethylene to levels that could inhibit crop growth.

EVALUATION AND INTERPRETATIONS

Every effort was made to highlight information from treatment situations where some parallel might be drawn with effects on growth, but little is known about the concentration of an allelochemical at presumed active sites. The prior discussion dealt almost exclusively with benzoic and cinnamic acids, coumarins, and flavonoids, with a few references to terpenoids. Even within one of these groups there are differences in the effects of specific compounds, but some similarities are also evident. These similar effects provide help in trying to sort primary from secondary modes of action.

Considerable speculation is still involved in any attempt to draw together the various inferences and interrelationships about allelochemical functions. Figure 10.1 suggests a possible sequence of events for the action of phenolic acids and closely related substances. Many probable interactions and the details of effects on each altered function have been omitted. In fact, currently there is almost no evidence to pinpoint specific biochemical and biophysical events responsible for these effects on functions. I am suggesting that perturbations of membranes and interactions with phytohormones may be two primary modes of action of these

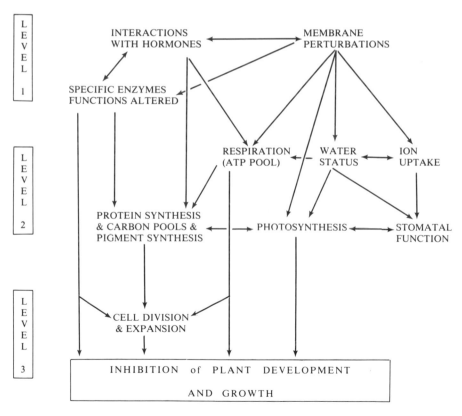

Figure 10.1 Hypothetical action sequence for the effects of phenolic allelochemicals on higher plants. Each arrow suggests a negative impact on plant processes. Modified with permission from Einhellig et al. (1985), copyright, American Chemical Society.

compounds, and these two need not be mutually exclusive. Interference with specific protein functions are included at this level because several of these effects cannot be separated from membrane and hormone activity.

Phenolic acids solubilize in membranes according to their lipid solubility, altering permeability and transmembrane transport. It is easy to relate dysfunction of the plasma membrane and tonoplast to the failure of cells to maintain proper mineral nutrition and a favorable water balance. Likewise, efficiency in the energy systems of respiration and photosynthesis demands precise membrane organization, charge separation, and the work of membrane-associated proteins. If phenolics partition into these membranes, perhaps binding to lipids, carbohydrates, or proteins, several mechanisms for interference are possible. These include various opportunities for disruption of electron flow and uncoupling of electron transport from phosphorylation. Allelochemical-induced membrane dysfunction may also arise if the synthesis of membrane constitutents is blocked, a possibility suggested because phenolics alter protein synthesis. Certainly several other biochemical and biophysical mechanisms for membrane perturbations are possible.

The actions of several·allelochemicals on hormone-induced growth and development responses might occur from alterations in hormone synthesis or reception sites, hormone inactivation, or combinations of these. Hypothetical mechanisms for the action of IAA, GA, cytokinins, and ABA all include potential actions at membranes. Hence it is possible that an allelochemical that perturbs membrane functions would interfere with these responses. Allelochemicals might also act by inducing the release of a secondary agent, like ethylene, but this has not been investigated.

All the functions listed in Figure 10.1 will probably not apply in a specific case of allelochemical inhibition, and factors that are modified will not have equal impact. For example, effects on photosynthesis would not apply to blocking seed germination, and hormone actions would not apply to inhibition of microorganisms. The greatest breadth of secondary effects probably occurs in the inhibition of seedling growth, where I suspect it is the collective disruption of multiple plant functions that translates into disorganization and impairment of growth. Perhaps this gives some rationale for the fact that seedling growth is typically more sensitive to allelochemical interference than seed germination.

The interaction model presented may be useful for evaluating nonphenolic allelochemicals. Because of their lipid solubility, terpenes will absorb into membranes and loss of membrane integrity may be a starting point for their actions. One can only speculate that several other classes of allelochemicals must have primary effects on membrane functions.

CONCLUSIONS

A refined understanding of the mode of action of allelochemicals is clearly not at hand, yet I have hypothesized that several phenolic compounds have their primary effects on membrane functions and interactions with hormones. The evidence for this proposal is most substantial for benzoic and cinnamic acids, but coumarins, flavonoids, and other polyphenols may also fit the scheme. Undoubtedly there are many subtle differences among these groups, and even within compounds of a particular class, as to the precise mechanisms and metabolic perturbations whereby they elicit changes in plant systems. Impacts on mineral nutrition, water balance, photosynthesis, respiration, protein synthesis, hormone-induced growth, and other processes are the actions which can be rationalized as directly affecting growth. Probably it is the combined detriment from alterations of several of these that reduces the efficiency of cells. The relative importance of any one of these actions depends on the specific allelochemical, associated environmental stresses, and what facet of plant growth is inhibited (i.e., germination, radicle elongation, or seedling growth).

Very little information is available on the mode of action of allelochemical alkaloids and cyanohydrins, organic acids and aldehydes, long-chain fatty acids and polyacetylenes, steroids, unsaturated lactones, steroids, and several others. One can only project that the structural variety among these substances dictates potential differences in the way they interfere with biochemical and biophysical events. An explanation is also needed for allelochemical-sensitivity differences, both among species and when comparing the phases of growth of higher plants. These distinctions probably relate to which plant processes are altered, uniquenesses in metabolism, detoxification mechanisms, and the extent of allelochemical accumulation at an active site. The many unanswered questions demonstrate the challenges involved with interpreting how allelochemicals alter plant growth.

REFERENCES

Arntzen, C. J., S. V. Falkenthal, and S. Bobick. 1974. *Plant Physiol.* 53:304.

Avers, C. J. and R. H. Goodwin. 1956. *Am. J. Bot.* 43:612.

Balke, N. E. 1985. Effects of allelochemicals on mineral uptake and associated physiological processes. *In* A. C. Thompson, (ed.). *The Chemistry of Allelopathy.* American Chemical Society, Washington, D.C., pp. 161–178.

Balke, N. E., C. C. Lee, and M. P. Davis. 1984. *J. Cell Biochem.* 8b:255.

Bates, G. W. and M.H.M. Goldsmith. 1983. *Planta.* 159:231.

Bhowmik, P. C. and J. D. Doll. 1984. *Agron. J.* 76:383.

Brown, R. T. and P. Mikola. 1974. *Acta For. Fen.* 141:1.

Buchholtz, K. P. 1971. The influence of allelopathy on mineral nutrition. *In* U. S. Natl. Comm. for IBP, (ed.). *Biochemical Interactions Among Plants.* National Academy of Science, Washington, D.C., pp. 86–89.

Cameron, H. J. and G. R. Julian. 1980. *J. Chem. Ecol.* 6:989.

Chambers, E. E. and L. G. Holm. 1965. *Weeds.* 13:312.

Colton, C. E. and F. A. Einhellig. 1980. *Am. J. Bot.* 67:1407.

Corcoran, M. R., T. A. Geissman, and B. O. Phinney. 1972. *Plant Physiol.* 49:323.

Cornman, I. 1946. *Am. J. Bot.* 33:217.

Danks, M. L., J. S. Fletcher, and E. L. Rice. 1975a. *Am. J. Bot.* 62:311.

Danks, M. L., J. S. Fletcher, and E. L. Rice. 1975b. *Am. J. Bot.* 62:749.

Demos, E. K., M. Woolwine, R. H. Wilson, and C. McMillan. 1975. *Am. J. Bot.* 62:97.

Duke, S. O. (ed.). 1985. *Weed Physiology Vol. II: Herbicide Physiology.* CRC Press, Inc., Boca Raton, FL.

Einhellig, F. A. 1971. *Proc. South Dakota Acad. Sci.* 50:205.

Einhellig, F. A. 1985. Allelopathy—A natural protection, allelochemicals. *In* N. B. Mandava, (ed.). *Handbook of Natural Pesticides: Methods*, Vol. 1. CRC Press, Inc., Boca Raton, FL, pp. 161–200.

Einhellig, F. A. 1985. Effects of allelopathic chemicals on crop productivity. *In* P. Hedin, (ed.). *Bioregulators for Pest Control.* American Chemical Society, Washington, D.C., pp. 109–130.

Einhellig, F. A. 1986. Interactions among allelopathic chemicals and other stress factors of the plant environment. *In* G. R. Waller, (ed.). *Symposium on Allelochemicals: Role in Agriculture, Forestry, and Ecology*, American Chemical Society, Washington, D.C. (in press).

Einhellig, F. A. and P. C. Eckrich. 1984. *J. Chem. Ecol.* 10:161.

Einhellig, F. A. and L. Kuan. 1971. *Bull. Torrey Bot. Club.* 98:155.

Einhellig, F. A. and J. A. Rasmussen. 1978. *J. Chem. Ecol.* 4:425.

Einhellig, F. A. and J. A. Rasmussen. 1979. *J. Chem. Ecol.* 5:815.

Einhellig, F. A. and M. K. Schon. 1982. *Can. J. Bot.* 60:2923.

Einhellig, F. A., E. L. Rice, P. G. Risser, and S. H. Wender, 1970. *Bull. Torrey Bot. Club.* 97:22.

Einhellig, F. A., M. K. Schon, and J. A. Rasmussen. 1982. *J. Plant Growth Regul.* 1:251.

Einhellig, F. A., P. F. Nyberg, and G. R. Leather. 1985. *Intl. Soc. Chem. Ecol. Annu. Mtg. Abstr.* p. 13.

Einhellig, F. A., M. Stille Muth, and M. K. Schon. 1985. Effects of allelochemicals on plant–water relationships. *In* A. C. Thompson, (ed.). *The Chemistry of Allelopathy*, American Chemical Society, Washington, D.C., pp. 170–195.

Feucht, W. and P.P.S. Schmid. 1980. *Physiol. Plant.* 50:309.

Fisher, R. F., R. A. Woods, and M. R. Galvicic. 1978. *Can. J. For. Res.* 8:1.

Glass, A.D.M. 1973. *Plant Physiol.* 51:1037.

Glass, A.D.M. 1974. *J. Exp. Bot.* 25:1104.

Glass, A.D.M. and B. A. Bohm. 1971. *Planta.* 100:93.

Glass, A.D.M. and J. Dunlop. 1974. *Plant Physiol.* 54:855.

Hall, A. B., U. Blum, and R. C. Fites. 1983. *J. Chem. Ecol.* 9:1213.

Harper, J. R. and N. E. Balke. 1981. *Plant Physiol.* 68:1349.

Harvey, R. G. and J. J. Linscott. 1978. *Soil Sci. Soc. Am. J.* 42:721.

Horsley, S. B. 1977. Allelopathic interference among plants. II. Physiological modes of action. *In* H. E. Wilcox and A. F. Hamer, (eds.), *Proceedings of the 4th North American Forest Biology Workshop*, School of Continuing Education, College of Environmental Science and Forestry, Syracuse, NY., pp. 93–136.

Jacobson, A. and M. R. Corcoran. 1977. *Plant Physiol.* 59:129.

Jain, A. and H. S. Srivastava. 1981. *Physiol. Plant.* 51:339.

Jankay, P. and W. H. Muller. 1976. *Am. J. Bot.* 63:126.

Kadlec, K. D. 1973. Para coumaric and caffeic acid induced inhibition of growth, photosynthetic and transpiration rates in grain sorghum. M. A. Thesis, University of South Dakota, Vermillion.

Kanchan, S. D. and Jayachandra. 1980. *Plant Soil* 55:61.

Katznelson, J. 1972. *Plant Soil* 36:379.

Kobza, J. 1980. The effects of ferulic acid on the mineral nutrition of grain sorghum. M. A. Thesis, University of South Dakota, Vermillion.

Koeppe, D. E. 1972. *Physiol. Plant.* 27:89.

Lee, T. T. 1980. *Physiol. Plant.* 50:107.

Lee, T. T., A. N. Starratt, and J. J. Jevnikar. 1982. *Phytochemistry.* 21:517.

Lodhi, M.A.K. and G. L. Nickell. 1973. *Bull. Torrey Bot. Club.* 100:159.

Lorber, P. and W. H. Muller. 1980. *Compar. Physiol. Ecol.* 5:68.

McCahon, C. B., R. G. Kelsey, P. P. Sheridan, and F. Shafizadeh. 1973. *Bull. Torrey Bot. Club.* 100:23.

McCalla, T. M. and F. A. Norstadt. 1974. *Agric. Environ.* 1:153.

McClure, P. R., H. D. Gross, and W. A. Jackson. 1978. *Can. J. Bot.* 58:764.

Moreland, D. E. 1980. *Annu. Rev. Plant Physiol.* 31:597.

Moreland, D. E. and W. P. Novitzky. 1986. Effects of phenolic acids, coumarins, and flavonoids on isolated chloroplasts and mitochondria. *In* G. R. Waller, (ed.). *Symposium on Allelochemicals: Role in Agriculture, Forestry, and Ecology*, American Chemical Society, Washington, D.C. (in press).

Muller, C. H. 1966. *Bull. Torrey Bot. Club.* 93:332.

Muller, W. H. 1965. *Bot. Gaz.* 126:195.

Muller, W. H., P. Lorber, B. Haley, and K. Johnson. 1969. *Bull. Torrey Bot. Club.* 96:89.

Patterson, D. T. 1981. *Weed Sci.* 29:53.

Patrick, Z. A. and L. W. Koch. 1958. *Can. J. Bot.* 36:621.

Patrick, Z. A., T. A. Toussoun, and L. W. Koch. 1964. *Annu. Rev. Phytopathol.* 2:267.

Ray, S. D., K. N. Guruprasad, and M. M. Laloraya. 1980. *J. Exp. Bot.* 31:1651.

Rietveld, W. J., R. C. Schlesinger, and K. J. Kessler. 1983. *J. Chem. Ecol.* 9:1119.

Rice, E. L. 1984. *Allelopathy.* 2nd Ed., Academic, Orlando, FL.

Rose, S. L., D. A. Perry, D. Pilz, and M. M. Schoeneberger. 1983. *J. Chem. Ecol.* 9:1153.

Sato, T., F. Kiuchi, and U. Sankawa. 1982. *Phytochemistry.* 21:845.

Schon, M. K. and F. A. Einhellig. 1982. *Bot. Gaz.* 143:505.

Shindo, H. and S. Kuwatsuka. 1977. *Soil Sci. Plant Nutr.* 23:319.

Stenlid, G. 1970. *Phytochemistry.* 9:2251.

Tomaszewski, M. and K. V. Thimann. 1966. *Plant Physiol.* 41:1443.

Van Sumere, C. F., J. Cottenie, J. DeGreef, and J. Kint. 1971. *Rec. Adv. Phytochem.* 4:165.

Walters, D. T. and A. R. Gilmore. 1976. *J. Chem. Ecol.* 2:469.

Wang, T.S.C., S. Y. Cheng, and H. Tung. 1968. *Soil Sci.* 104:138.

Wang, T.S.C., K. L. Yeh, S. Y. Cheng, and T. K. Yang. 1971. Behavior of soil phenolic acids. *In* U.S. Natl. Comm. for IBP, (ed.). *Biochemical Interactions Among Plants.* National Academy of Science, Washington, D.C., pp. 113–120.

Weaver, T. W. and D. Klarich. 1977. *Am. Midl. Nat.* 97:508.

11

ALLELOCHEMICAL MECHANISMS IN THE INHIBITION OF HERBS BY CHAPARRAL SHRUBS

WALTER H. MULLER

Department of Biological Science
University of California, Santa Barbara
Santa Barbara, California

Biochemical inhibition, or allelopathy, has been observed in a variety of situations, but there has been little investigation of the specific effects that the allelochemicals have produced upon the susceptible plants. As Rice (1984) pointed out in his text on allelopathy, ". . . the surface has just been scratched in determining the mechanisms by which the different kinds of allelopathic compounds exert their actions." Although Molisch originally used "allelopathy" to refer to both inhibitory and stimulatory biochemical interactions, the term is almost always used in the former sense (i.e., harmful effects). The possible mechanisms discussed in this chapter concern only deleterious effects and are based on work done in my laboratory by various individuals over a number of years dealing with the influence of natural inhibitors produced by chaparral shrubs.

In the soft chaparral found in the Mediterranean climate of coastal Southern California, one frequently encounters greatly decreased growth of herbs that are in the immediate vicinity of *Salvia leucophylla*, a strongly aromatic shrub of the area. Particularly during periods of water stress, the shrub patches are surrounded by a zone almost completely barren of herbs and an area of decreased herb growth exterior to this bare zone. In aerial views the characteristic bare area of inhibition is quite clear at the zones of contact between shrub thickets and grassland, and extends 1–2 m beyond the crowns of the shrubs (Muller et al., 1964). Such views also show that the interiors of old shrub thickets are rather bare, indicating the occurrence of autointoxication as a result of long periods of exposure to a toxic material. The diminished herb growth that extends an additional 3–8 m farther from the shrub can be seen by closer viewing of the area. In this zone of inhibition there is not only a stunting of grasses and herbs, but also the complete exclusion of some species (Muller, 1970). The *Salvia* shrubs cause reduced growth of herbs and also bring about changes in community composition, because some herbs (e.g., *Avena fatua* and *Bromus rigidus*) are so sensitive to the phytotoxins as to be completely excluded from this zone of inhibition. Exclosure experiments showed that there was some effect by rodents in the bare zone but none in the zone of inhibition, and only a slight effect by the grazing of birds (mainly white-crowned and golden-crowned sparrows). Also, trenching across the zones indicated that there were no soil differences and no water zonation of any significance. Not only were there no differences in the levels of nitrogen, phosphorus, and potassium, but the electrical conductivity values in all zones were low, indicating no salinity stress upon any of the plant species. Further investigations demonstrated that volatile emanations from the leaves of *Salvia leucophylla* were responsible for the reduced growth of herbs.

The toxic volatile materials were monoterpenoids (derivatives of isoprene), of which camphor and cineole were the most effective (Muller and Muller, 1964). Heisey and Delwiche (1984) found that a volatile monoterpene alcohol (terpinen-4-ol) from vinegar weed (*Trichostema lanceolatum*, an annual herb of the California grassland) was more inhibitory than cineole, but less than camphor, to other plants in the vicinity. Asplund (1968) found that the most toxic in a group of structurally and metabolically related monoterpenoids were those having a ketone functional group, such as camphor. It appears likely that plants vary in their susceptibility to the allelochemicals, and that the presence of certain functional groups significantly influences toxicity.

Other shrubs of the chaparral, such as *Adenostoma fasciculatum* and *Arctostaphylos glandulosa* var. *zacaensis*, produce water-soluble phenolics that are inhibitory toward various herbs. The text by Rice (1984) is an excellent review of allelopathy and contains a number of references to the work done with phenolics especially. We have tested leachates from *A. fasciculatum*, obtained by simulating rain drip over leafy branches, and found that using this to moisten herb roots inhibited both growth and respiration. The leaf leachates were flash evaporated so that 1, 3, and 5X concentrations could be used. When placed in such leachates, the respective rates of oxygen uptake by excised *Avena fatua* roots were about 75, 15, and 12% of the control rates (Table 11.1). Figure 11.1 indicates the subsequent root growth when germinating seeds of *A. fatua* are moistened with the 1X leachate; primary and total (primary plus secondary) root growth are both less than the water-moistened controls. Figure 11.2 shows the reduction in root-growth rate brought about by the treatment. When the leachate is used at

TABLE 11.1.
Oxygen uptake (μM) (O_2/mg/hr) by excised roots of *Avena fatua* as influenced by
the leaf leachate of *Adenostoma fasciculatum*.

	O_2 Uptake	SD[a]
Control (water)	0.241	0.025
1X Leachate	0.180	0.021
3X Leachate	0.036	0.009
5X Leachate	0.028	0.015

[a]SD = standard deviation.

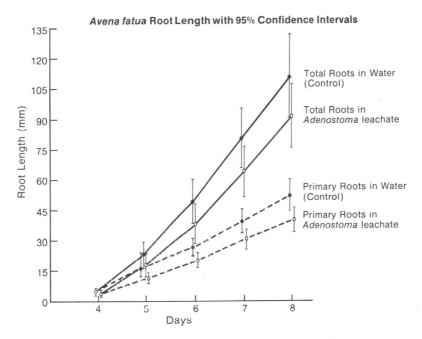

Figure 11.1 Root growth (mm) of germinating seeds of *Avena fatua* as influenced by leaf leachate from *Adenostoma fasciculatum*. Total root growth—top two curves; primary root growth—lower curves. Controls were moistened with water; others were moistened with leachate.

a higher concentration (3X), the inhibition of total root growth is much more pronounced (Figure 11.3). One of the phenols found in *A. fasciculatum* is hydroquinone, and its inhibitory effect on oxygen uptake by excised primary roots of *A. fatua* is indicated in Figure 11.4. It is clear that such leachates can produce detrimental effects upon herbs growing in the vicinity of *A. fasciculatum*.

Microbial activity also appears to be involved in the allelopathic effects found in the field. In a series of investigations Kaminsky (1981) demonstrated that

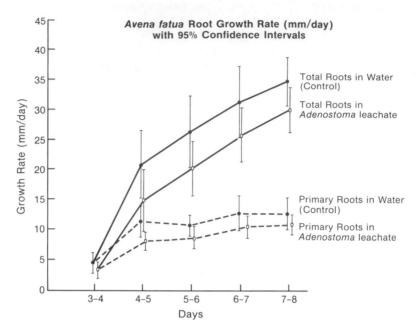

Figure 11.2 Root growth rate (mm/day) of same roots as in Figure 11.1.

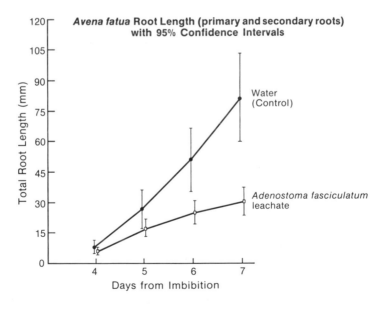

Figure 11.3 Roots similar to those in Figure 11.1, but moistened with leachate concentrated 3X; total roots (primary plus secondary).

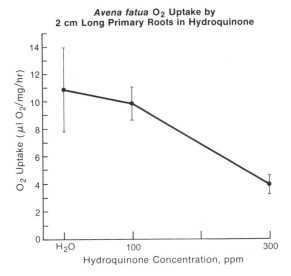

Avena fatua O_2 Uptake by
2 cm Long Primary Roots in Hydroquinone

Figure 11.4 Oxygen uptake by primary roots of *Avena fatua* as influenced by hydroquinone.

microorganisms in the soil under *A. fasciculatum* produce substances capable of inhibiting the germination and growth of seedlings. The microbially produced soil toxicity is linked to the presence of *A. fasciculatum* because only soils under that shrub developed toxicity on drying and rewetting—a treatment that results in a dramatic increase in microbial activity, as shown by various soil bacteriologists (e.g., Stevenson, 1956).

Most of our research on allelopathy, however, has been with *Salvia* and terpenoids and this chapter is based primarily on that work. We will update and expand the relatively little coverage available in review articles concerning the manner in which natural inhibitors such as terpenoids influence susceptible plants (i.e., the possible mechanisms involved).

A number of herbaceous plants were used in the studies with quite similar results: *Bromus rigidus* (ripgut grass), *Avena fatua* (wild oat), *Cucumis sativus* (cucumber), and *Allium cepa* (onion). The grass and wild oats are usually found adjacent to *Salvia* thickets and are very sensitive to terpenoids. Cucumber and onion provide more uniform germination and growth, exhibit an appearance similar to that of inhibited herb species common to the area where *Salvia* is found, and are less sensitive to the monoterpenoids. *Allium cepa* is especially useful because its karyotype (chromosomal complement) and root development are well-known; abnormalities are thus readily apparent.

Bioassays were generally based upon a demonstrable reduction in the radicle (or root) growth of susceptible species when germinating seeds were exposed to vapors from *Salvia* leaves or from specific terpenoids. To insure that the seeds were exposed only to vapors, the moistened seeds were placed within sealed storage dishes that also contained a small beaker with 2.0 g of crushed *Salvia* leaves (or the specific terpenoid). Seeds of grasses that are common to the *Salvia* area are so susceptible to the phytotoxins that they usually do not even germinate under the conditions of the bioassay. With *Cucumis* the treated plants produce short, bulky radicles (about 4–5 mm in length) as compared with the much longer (40–50 mm), more slender radicles of the control (normal) plants, which also produce numerous lateral roots (W. H. Muller, 1965). The volatiles from *S. apiana* were considerably less inhibitory than those from *S. leucophylla* and *S. mellifera*. By sealing the dishes with parafilm, it was possible to use a hypodermic syringe to remove samples of the enclosed atmosphere for gas chromatographic analysis. Chromatograms of atmospheres from the bioassay chambers, and from ether extracts of leaves, indicated that the most prominent terpenoids were α-pinene, camphene, β-pinene, cineole, and camphor (Muller and Muller, 1964). The areas beneath the peaks constituted a rough measure of the terpenes in each atmosphere and the amounts corresponded pretty well to the degree of inhibition of *Cucumis* by each species of *Salvia*—approximately 40, 10, and 8% of normal growth, respectively, in numerous assays (root length measurements). The investigations then focused on *S. leucophylla*, and subsequent discussion of *Salvia* vapors or leaves refers to this species.

When *Cucumis* seedlings (48 hours after germination) are exposed to *Salvia* vapors, elongation of the hypocotyls is severely curtailed (another 3 days); growth is only about 25% of the control even with exposure to volatiles from only 1.0 g of macerated *Salvia* leaves, as indicated in Table 11.2. Clearly, stem cells are susceptible as well as root cells to the effects of terpenoids. It was also noted that those cells that first develop when a seed germinates are more sensitive to the inhibitory effects of *Salvia* volatiles. When hypocotyls or radicles commence enlarging in the absence of terpenoids, subsequent exposure to these vapors is less detrimental to growth, although still inhibitory (W. H. Muller, 1965). Perennial plants such as *Stipa lepida* and *Poa scabrella* are not affected, but seedlings of these plants do not become established in the zone of inhibition (Muller et al., 1964). These allelochemicals are effective primarily during the initial phases of seedling growth and establishment.

Anatomical studies have shown that cell elongation, cell division, and tissue maturation are all affected. Root and hypocotyl cells of plants exposed to ter-

TABLE 11.2.
Growth of *Cucumis sativus* hypocotyls exposed, after an initial 48-hour germination period, to an atmosphere containing volatiles from macerated *Salvia* leaves. Total germination time of 120 hours.

Leaf material (g)	0.0	1.0	2.0
Hypocotyl length (mm)[a]	110.0	25.9	3.9
Percent of control	—	23.6	3.6

Source. Muller and Hauge (1967).
[a] Average length of 30 plants per group.

penes are much shorter and bulkier than normal; for example, about 49×39 μm vs. the normal 69×25 μm (averaged data from measurements of 30 cells from each of three treated roots and 30 cells from each of three normal roots; W. H. Muller, 1965). The stubby, compressed radicles of the bioassays result in part from this reduced elongation and increased lateral expansion of root cells.

Counts of mitotic figures in several thousand cells indicated that approximately 10% of the cells in normal root tips (*Allium cepa*) were undergoing nuclear division, whereas no such divisions were found in severely inhibited roots. Significant inhibition occurred within 24 hours of exposure, and mitosis essentially ceased by 72 hours, as indicated in Table 11.3 (from Lorber and Muller, 1980a). Because of the mitotic cycles known to occur in onion, these root tips were fixed between 3 and 4 P.M. each day and then stained.

More detailed examinations of chromosomes were possible by using lower concentrations of toxins, so that some mitotic figures were present even though the root was considerably inhibited in its growth (Lorber and Muller, 1980a). When roots were about 10 mm long, *Allium cepa* bulbs were placed on vermiculite that had been exposed to vapors from *S. leucophylla* leaves in a closed system where the atmosphere was circulated through macerated leaves and vermiculite and then returned to the leaf chamber. Within 24 hours the roots exhibited a swelling of tissue adjacent to the meristem, similar in appearance to colchicine-induced swellings known in various species. Contraction and fragmentation of chromosomes are common effects of colchicine treatment, and this was also found after exposure to *Salvia* vapors. Metaphase chromosomes of a normal *Allium* root cell (as seen in squash preparations stained with acetocarmine) are typically large and rather elongated. After 24 hours exposure to *Salvia* vapors, the chromosomes in metaphase cells of inhibited roots are shorter and much thicker. Continued exposure produces further contraction, and breakage regularly occurs at the kinetochores after 48 hours. Metaphase chromosomes in root tips treated for 72 hours appeared as poorly defined structures that were often fragmented. The longer the time of exposure, or the more concentrated the

TABLE 11.3.

Cell division in *Allium cepa* roots grown on vermiculite aerated with vapors from macerated leaves of *Salvia leucophyla.*[a]

	Control	Hours Exposed to *Salvia* vapors			
		8	24	48	72
Number of mitotic cells	103	98	33	11	3
Mitotic index	10.3	9.8	3.3	1.1	0.3

[a]All figures are averages from at least three experimental trials. Three root tips were counted at each sampling in each separate experiment. The values for each category are based upon an examination of 1000 cells.

terpenoids to which the cells are exposed, the greater are the abnormalities with regard to chromosomes. Even contraction without fragmentation is considered to be detrimental to the normal growth and development of a cell, producing nonspecific interference.

Analysis by gas chromatography of extracts of vermiculite samples aerated with volatiles from *S. leucophylla* leaves indicated the presence of the usual five terpenes, which were identified by comparison with peaks produced by reagent grade compounds. *Allium cepa* bulbs placed on vermiculite not exposed to *Salvia* vapors produced normal roots with normal mitotic figures.

In the early stages of these investigations it was noted that numerous lateral roots were present on the primary roots of normal (control) plants within 48 hours, but none appeared on the bulky radicles of treated plants. Besides a reduction in the number of lateral root initials produced, laterals that developed did not extend to or through the epidermis. Transverse sections of normal *Cucumis* roots frequently contain lateral roots rupturing out through the epidermis. Such laterals are not found in the inhibited roots, although occasional initials are found if the root is exposed to emanations from small amounts of *Salvia* leaves (Muller and Hauge, 1967). Even these few initials were usually so inhibited in cell division and growth that they seldom showed much more development than that sufficient to enable their recognition. Terpenoids from these leaves effectively inhibit lateral root production, which results in a poorly developed root system. The effect would be even more drastic with regard to the several species of grasses normally found adjacent to *Salvia* thickets, because such grasses are much more susceptible than are *Cucumis* seedlings. *Bromus rigidus*, *B. mollis*, *B. rubens*, *Avena fatua*, *Stipa pulchra*, and *Festuca megalura* all proved to be extremely sensitive.

Differential staining procedures and microscope examination have shown excessive deposition of cutin on the outer walls of root epidermal cells and the accumulation of lipid droplets in the cortical cells of inhibited roots. Increased cutin deposits occur with increased exposure to *Salvia* volatiles (Muller and Hauge, 1967), and large aggregates of lipid globules are found in most root cells inhibited by terpenoids (Lorber and Muller, 1976). Monoterpenoids are readily soluble in waxes and lipids. For example, when paraffin chips were introduced into a flask containing terpenoid vapors, about 70% of these terpenes were taken up by the paraffin within 5 minutes (C. H. Muller, 1965). Soil colloids also adsorb volatile terpenoids from the atmosphere and retain the resulting toxicity for several months (Muller and del Moral, 1966). This suggests that the roots of germinating herb seeds are in contact with toxic terpenoids, which causes a marked reduction in root growth. The increased amounts of cutins and lipids that also result may enhance the inhibitory effects by providing easier access of the allelochemicals into the cells by way of the cutin and the plasmodesmata (tiny cytoplasmic strands) which extend to the cuticular layers.

Ultrastructure studies utilizing electron micrographs have demonstrated that cells exposed to volatile monoterpenoids from *S. leucophylla* are considerably altered from the normal situation (Lorber and Muller, 1976). In all of the un-treated (normal) material, cell and organelle membranes are intact and no ab-normalities are apparent. However, after exposure for 3 days, the absence of intact organelles is striking and those that are present show severe damage. Mi-tochondria have disrupted membranes and fragmented cristae and dictyosomes are absent. The cytoplasm appears much less dense than in control tissues, with far fewer ribosomes in evidence. In many locations the double membrane of the nucleus has separated, leaving small vacuole-like areas between the two mem-branes; these spaces range in size up to more than 1 μm in diameter. The endo-plasmic reticulum exhibits a spherical or clustered appearance not found in the untreated material, and ribosomes are prominent on the surface of these spheroid forms. Also prominent in the cells of inhibited tissue are large aggre-gates of lipid globules found throughout the cytoplasm.

It is possible that the lipid globules may result from abnormal metabolism due to blocking of metabolic pathways by toxins or poor utilization of food and its conversion to storage products, or as by-products from organelle and membrane decomposition. The absence of a variety of intact organelles and the presence of membrane fragments indicate that structural breakdown and decomposition are occurring within inhibited root cells. Lipids are a known structural component of membranes and it is possible that membrane degradation could result in the freeing of lipids within the cytoplasm of affected cells. In studies of chloroplast senescence Strunk and Wartenberg (1960) noted that degradation of the granal structure was accompanied by an increase in the development of lipid globules and suggested that granal membranes were acting as a lipid source. Various other workers have also shown the accumulation of lipid globules in diseased tissues. Although we believe that the accumulated lipids are degradation prod-ucts, it is also true that cell metabolism is inhibited by terpenoids, as might be suspected by the damaged state of the mitochondria. Additionally, the conver-sion of storage products (e.g., lipids) to usable carbohydrates might cease as the inactivation of mitochondria progresses, allowing for the buildup of free lipids in the cells. Thus it is not possible to state with certainty the sources of these lipid globules.

In addition to the various anatomical and cytological effects that have been noted, *Salvia* volatiles (and especially cineole, one of the monoterpenoids from *Salvia*) have a marked effect on seedling metabolism. Placing roots of 2-day old seedlings in Warburg flasks for measurement, Muller et al. (1968) found that oxygen uptake per gram dry weight by excised inhibited *Cucumis* roots was only about 25% of that of excised normal roots. It was also shown that oxygen uptake by excised normal *Avena* roots was decreased by the presence of cineole or *Sal-via* leaves in the side arm of the Warburg flask. There was also a decrease in oxygen utilization by excised normal *Cucumis* hypocotyls under similar treat-ment. In addition, the respiration of entire *Bromus rigidus* seedlings was only about 20% of the controls when 0.1 g of *S. leucophylla* leaves was present in the side arm. The metabolism of cells in general was inhibited by the allelochemicals that were in the atmosphere of the flask because there was no contact between the material in the side arm and the excised roots in the main body of the flask except by volatilization of the phytotoxins.

Oxygen uptake by mitochondrial suspensions of *Avena fatua* was inhibited by cineole when succinate or fumarate were used as substrates, but α-pinene and β-pinene had no effect (nor on oxygen uptake by excised roots). Camphene and dipentene were somewhat inhibitory, but only in relatively high concentrations. The main effective allelochemicals in the mixture of terpenoids produced by *Salvia* leaves are cineole and camphor. Measurements made with a biological oxygen monitor system (Yellow Springs Instrument Co.) using a Clark electrode showed that inhibition increased with the concentration of cineole used (Muller et al., 1969).

In another investigation (Lorber and Muller, 1980b), mitochondrial samples were prepared for thin-layer chromatographic analysis after respiratory activity had been determined using ^{14}C-succinic acid added with Na-succinate and ADP. Autoradiographic results indicate that succinate utilization was retarded in cineole-treated mitochondria. Fumarate accumulation occurred, and only traces of other ^{14}C-labeled Krebs cycle acids were recovered from these treated preparations. Control suspensions rapidly metabolized both succinate and fumarate, and provided substantially larger quantities of other labeled acids. Polarographic tracings indicated active mitochondria in the suspensions with reasonably high respiratory control (RC) ratios and ADP:O values, both of which are good indices of mitochondrial integrity, and that 5×10^{-3} M cineole suppressed oxygen uptake by more than 60%. Although the RC ratio was lower in cineole-treated mitochondria, the ADP:O ratio remained stable, indicating that the efficiency of oxidative phosphorylation did not deteriorate appreciably although the rate of phosphorylation decreased with exposure to cineole. Electron micrographs of the isolated material revealed no apparent change in the ultrastructure of cineole-treated mitochondria. The short exposure of about 2 hours to cineole did not have the marked effect on isolated mitochondria that prolonged exposure had on the mitochondria in intact plants, as mentioned previously.

The high solubility of terpenoids in lipids suggests that cineole may be absorbed easily into lipid membrane structures, thereby affecting membrane permeability. The destruction of membrane integrity when cells are exposed to terpenoids has been demonstrated. Approximately one-third of mitochondrial dry weight consists of lipids, with much of this material present in the outer membrane. If terpenoids decrease membrane permeability, substrate concentration is lowered and the RC ratio is decreased as exposure time is increased. This may explain the increased inhibition with time reported by Muller et al. (1969).

CONCLUSIONS

The monoterpenoids emanating from *Salvia* leaves interfere with cell metabolism, elongation, and division, and the differentiation of susceptible herbs. It is also likely that the disruption of membranes implicates earlier effects on cellular permeabilities as well as metabolism. Soil colloids adsorb such terpenoids and retain the resulting toxicity for months. Microbial action on such toxins is also a possibility. The germinating herb seeds are probably in contact with toxic mono-terpenoids for a considerable period of time and this results in a marked reduction of root growth and respiration. It is quite likely that the adsorbed terpenes dissolve in the cutin of the root epidermal cells and then through lipids into the protoplasts. The increased deposition of cutin and lipids, by cells under the influence of monoterpenoids, could enhance this pathway and result in relatively high concentrations of these allelochemicals within the cells.

If the susceptible plants manage to survive, they would be poorly established and quite likely to succumb to environmental stress, such as that which occurs as the result of the low rainfall in the areas showing such allelopathic responses. In fact the zone of inhibition around *Salvia* shrubs is much more pronounced during years of poor rainfall (Muller, 1970). Chemical inhibition (or allelopathy) may be mitigated or intensified by environmental factors. In this instance the death of the herb seedling is most probably a result of desiccation made more likely by the poor root system that develops under allelopathic stress.

REFERENCES

Asplund, R. O. 1968. *Phytochem.* 7:1995.

Heisey, R. M. and C. C. Delwiche. 1984. *Am. J. Bot.* 71:821.

Kaminsky, R. 1981. *Ecol. Monog.* 51:365.

Lorber, P. and W. H. Muller. 1976. *Am. J. Bot.* 63:196.

Lorber, P. and W. H. Muller. 1980a. *Comp. Physiol. Ecol.* 5:60.

Lorber, P. and W. H. Muller. 1980b. *Comp. Physiol. Ecol.* 5:68.

Muller, C. H. 1965. *Bull. Torrey Bot. Club.* 92:38.

Muller, C. H. 1970. *Recent Adv. Phytochem.* 3:105.

Muller, C. H. and R. del Moral. 1966. *Bull. Torrey Bot. Club.* 93:130.

Muller, C. H., W. H. Muller, and B. L. Haines. 1964. *Science.* 143:471.

Muller, W. H. 1965. *Bot. Gaz.* 126:195.

Muller, W. H. and R. Hauge. 1967. *Bull. Torrey Bot. Club.* 94:182.

Muller, W. H. and C. H. Muller. 1964. *Bull. Torrey Bot. Club.* 91:327.

Muller, W. H., P. Lorber, and B. Haley. 1968. *Bull. Torrey Bot. Club.* 95:415.

Muller, W. H., P. Lorber, B. Haley, and K. Johnson. 1969. *Bull. Torrey Bot. Club.* 96:89.

Rice, E. L. 1984. Allelopathy. 2nd ed. Academic, Orlando, FL.

Stevenson, K. L. 1956. *Plant Soil* 8:170.

Strunk, C. and H. Wartenberg. 1960. *Phytopathol. Z.* 38:109.

CHEMISTRY AND POTENTIAL USES OF ALLELOPATHY

12

THE FUNCTION OF MONO AND SESQUITERPENES AS PLANT GERMINATION AND GROWTH REGULATORS

NIKOLAUS H. FISCHER

*Department of Chemistry, Louisiana State University,
Baton Rouge, Louisiana*

During the last 2 decades considerable evidence has been accumulated which suggests that plant natural products play a major role in plant–plant interactions (Rice, 1984). Most observations in allelopathic studies are based on laboratory bioassays and unambiguous experiments under field conditions are generally not available. In spite of the lack of field data it cannot be denied, however, that secondary plant metabolites play distinct ecological roles (Harborne, 1982).

Although it has been known for more than 60 years that terpenoids affect the germination and growth of plants (Sigmund, 1924), earlier allelopathic studies were mainly restricted to phenolic acids and other phenolic plant constituents (Rice, 1984). This may have occurred because the isolation and identification of phenolic compounds is generally easier than procedures in the terpenoid series. Furthermore, the common belief seems to be that terpenoid plant lipids usually show low solubility in water. Therefore, transport of these compounds in rain washes from the plant surface into the soil was considered unlikely. It has been suggested recently, however, that one possible transport mechanism of plant lipids in water washes might involve micelle formation with natural tensides, such as fatty acids and sterols, which are frequently present in copious amounts on a plant's surface (Fischer and Quijano, 1985).

This chapter presents an overview of research on the germination and growth regulation of plants by mono- and sesquiterpenes. As more biological and ecological studies on terpenoid compounds become available, their important role of interference in natural and agricultural plant communities will undoubtedly be demonstrated.

MONOTERPENES

Presently, several hundred naturally occurring monoterpenes are known (Devon and Scott, 1972). They represent 10-carbon compounds with acyclic, monocyclic, and bicyclic carbon skeletons, which are biogenetically derived from two isoprenoid units (Croteau, 1981). Monoterpenes are common constituents in higher plants and they represent the major components in many essential oils. They are often found in copious amounts compartmentalized in glandular hairs (trichomes) of the plant surface (Kelsey et al., 1984). Although the function of monoterpenes in plants is not well understood, there exists considerable evidence that they play an ecological role in the interaction of plants with other organisms, including other plants (Harborne, 1982). The function of monoterpenes as germination and growth regulators was reviewed by Evenari (1949).

Over 60 years ago, Sigmund (1924) screened essential oils and a number of pure monoterpenes and observed significant inhibitions of seed germination and plant growth in wheat, rape, and vetch. In Muller's pioneering study of the role of allelopathy in the fire cycle of the California chapparal, the bicyclic monoterpenes (Figure 12.1) α-pinene (1), β-pinene (2), camphene (3), camphor (4), and cineole (5) from *Salvia leucophylla*, *S. apiana*, and *S. mellifera* were shown to be growth inhibitors on test seedlings. (Muller and Muller, 1964). The bicyclic terpenes camphor (4) and 1,8-cineole (5) were the major and most active constituents in *S. leucophylla* and *S. mellifera* (Muller, 1965) as well as *Artemisia californica* (Halligan, 1975). The seed germination inhibitory monoterpene mixture of eight *Artemisia* species (sagebrush) from Western Montana also contained camphor as the major constituent (Kelsey et al., 1978). Germination inhibitions of radish seeds by a series of monoterpenes (1, 2, 4–10) were investigated by Asplund (1968, 1969). Most compounds exhibited low activity, with the exception of (+)- and (−)- camphor and (+)-pulegone, which were at least one order of magnitude more toxic than the other tested monoterpenes and about twice as toxic as HCN.

Volatile monoterpenes, released from the leaves of *Salvia leucophylla*, cause dramatic anatomical and physiological changes in herb seedlings when exposed to the vapors. Tests on *Cucumis sativus* root-tip cells showed systemic disturbances including accumulation of lipid globules in the cytoplasm, reduction in number and variety of organelles that included mitochondria, and disruption of membranes surrounding nuclei, mitochondria, and dictyosomes (Lorber and

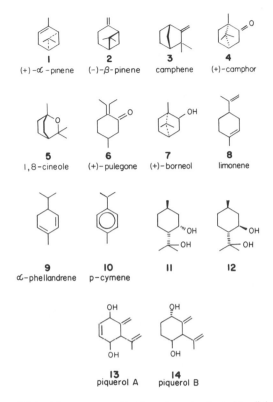

1
(+)-α-pinene

2
(-)-β-pinene

3
camphene

4
(+)-camphor

5
1,8-cineole

6
(+)-pulegone

7
(+)-borneol

8
limonene

9
α-phellandrene

10
p-cymene

11

12

13
piquerol A

14
piquerol B

Figure 12.1   Monoterpenes that have been implicated in allelopathy.

Muller, 1976). The volatile terpene mixture also produced swelling in the root
tips of *Allium cepa* and reduced mitotic activity. Furthermore, chromosome hy-
percontraction and breakage were observed (Lorber and Muller, 1980a). At a
5×10^{-3} molar concentration, cineole (**5**) suppressed the activity of mitochondria isolated from etiolated coleoptiles of *Avena fatua* and decreased the rate of
oxidative phosphorylation (Lorber and Muller, 1980b).

Gum trees of the genus *Eucalyptus* are frequently surrounded by grass-free
zones, which has led to a search for possible allelopathic agents in *Eucalyptus*
species. *Eucalyptus camaldulensis* produces 1,8-cineole, α- and β-pinene, and
α-phellandrene as volatile inhibitors (del Moral and Muller, 1970). From the
leaves of *E. citriodora* (lemon-scented gum), two germination and growth inhibitors against lettuce (*Lactuca sativa* L.) were isolated by Nishimura et al. (1982)
and shown to be racemic *p*-menthane-3,8-*cis*-diol (**11**) and its trans isomer (**12**).
It was determined that the cis isomer **11** exhibited higher germination inhibitory
effects than the trans isomer **12**, and the synthetic (+)-*cis*-enantiomer was more
active than its optical antipode. In a more recent paper, Nishimura et al. (1984)
demonstrated that the synthetic (+)-*cis*-enantiomer **11** was inhibitory to seed
germination and hypocotyl growth in lettuce, garden cress, (*Lepidium sativum*

L.), green foxtail (*Setaria viridis* L.), and barnyardgrass (*Panicum Crus-galli* L.), but no effects were observed on *E. citriodora* itself and rice (*Oryza sativa* L.). The allelopathic potential of two monoterpene diols, piquerols A(**13**) and B(**14**), from *Piqueria trinervia*, Cav. of the family Asteraceae (the weed Tabardillo in Mexico and Central America), was studied by Gonzalez de la Parra et al. (1981). The authors found that the germination and growth of a number of test plants from the habitat of *P. trinervia* was significantly inhibited by both compounds, but piquerol A was more active. However, an interesting difference regarding the action of the two diols was that piquerol B more strongly inhibited stem growth and diol A was most inhibitory to roots.

In a study of old-field succession in Tennessee, possible allelopathic influences of *Sassafras albidum* upon annual herbs, which are effectively excluded from sassafras understory flora, were investigated by Gant and Clebsch (1975). Besides eugenol and safrole, the monoterpenes citral, pinene, α-phellandrene, and (+)-camphor were found in the leaves, litter, soil, and roots of sassafras stands. A distinct correlation between the α-phellandrene concentration and the reduction in radicle growth of *Acer negundo* and *Ulmus americana* was observed.

SESQUITERPENES

The structural variety among sesquiterpenes, which contain three isoprenoid units with 15 carbons, is considerably larger than in the monoterpene series (Ruecker, 1973; Loomis and Croteau, 1980). This increase in skeletal types is expressed in a dramatic increase in the number of known compounds. Among the sesquiterpene lactones alone, over 2000 compounds are known today (Fischer et al. 1979; Seaman, 1982; Hoffmann and Rabe, 1985) and publications on new methylene lactones appear at an accelerating rate.

Reports related to the allelopathic potential of sesquiterpenes, in particular sesquiterpene lactones, appear with increasing frequency. Three reviews on the plant growth regulation of sesquiterpenes were published during the last decade (Gross, 1975; Julio and Pinto, 1976; Stevens, 1984). In this chapter the literature on plant germination and growth regulation effects by sesquiterpenoids is reviewed through early 1985.

Germination and Growth Regulation by Sesquiterpene Hydrocarbons and Derivatives

Two allelopathic sesquiterpene hydrocarbons, β-bisabolene (**15**) and β-caryophyllene (**16**), from *Artemisia absinthium*, were described by Grummer (1961) (Figure 12.2). In a bioassay-guided search for potential allelopathic compounds from Louisiana common ragweed (*Ambrosia artemisiifolia* L.) we recently isolated a mixture of highly active sesquiterpene hydrocarbons consisting of β-bisabolene (**15**), bergamotene (**17**), α-guayene (**18**), α-bulnesene (**19**), and β-pat-

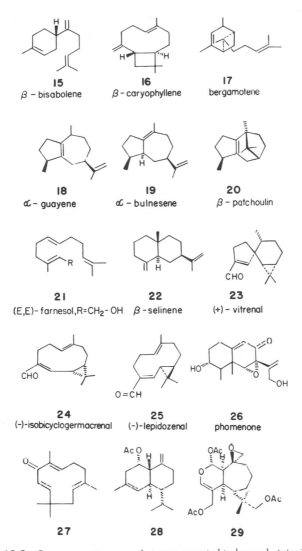

Figure 12.2 Some sesquiterpenes that were reported to have phytotoxic activity.

choulin (**20**) (Fischer and Quijano, 1985). Seeds pretreated with a dichloromethane solution of the mixture given above caused strong germination inhibitions in onion, oat, rye grass, and the weed *Amaranthus palmeri* S. Wats. (Bradow, 1985).

The essential oil fraction of a methanol extract of water nutgrass tubers (*Cyperus serotinus* Rottb.) inhibited the germination of lettuce at 300 ppm and the growth of lettuce and rice. The volatile oil contained farnesol (**21**, R =

$-CH_2OH$), its acetate (**21**, R $=$ CH_2OAc), methyl farnesate (**21**, R $=$ $-COOCH_3$), and β-selinene (**22**). All four compounds possess antigibberellin and antiauxin activities. Root elongation and the formation of adventitious roots were distinctly different between the open-chain farnesol and the bicyclic β-selinene (**22**) (Komai et al., 1981). Although a number of sesquiterpenes were present in the tubers of purple nutsedge (*Cyperus rotundus* L.), growth inhibitory effects on white clover (*Digitaria sanguinalis*) and Rumex were caused by a fraction containing a mixture of *p*-coumaric, ferulic, vanillic, *p*-hydroxybenzoic, and protocatechuic acid (Komai and Ueki, 1981).

Matsuo and co-workers have published a series of papers related to the chemistry of the growth-inhibitory sesquiterpene aldehydes (+)-vitrenal (**23**), (−)-isobicyclogermacrenal (**24**), and (−)-lepidozenal (**25**), which were isolated from the liverwort *Lepidozia vitrea* Steph. (Matsuo et al., 1984a,b), but no details on the biological activity were given.

The mycotoxin phomenone (**26**), which is produced by *Phoma destructiva* Plowr. and causes wilting and necroses of leaflets of tomato, is also growth inhibitory to shoots and rootlets of tomato (Capasso et al., 1984). Structure–activity relationship studies on the eremophilane **26** and several synthetic derivatives suggested that the presence of an epoxide function appears to be necessary for growth inhibitory activity.

The cross-conjugated humulene derivative zerumbone (**27**) and its epoxide derivative cause adventitious rooting in the hypocotyl cuttings of mung bean (*Phaseolus mungo*) with greater activity than indole acetic acid (IAA) at 10 ppm and below (Kalsi et al., 1978). Another α,β-unsaturated ketone, isopatchoulenone, was also distinctly more active than IAA (Kalsi et al., 1979). Similar strong effects were observed with the cadinane derivatives *epi*-khusinol acetate (**28**) and khusinoloxide when tested on *Phaseolus aureus* (Kalsi and Talwar, 1981). The *ent*-2,3-*seco*-alloaromadendrane plagiochilin A (**29**) and congeners from the liverwort *Plagiochila ovalifolia* strongly inhibited the growth of rice seedlings at the 50-ppm level (Matsuo et al., 1981). Structure–activity studies suggested that the biological activity is due to the acetyl hemiacetal moiety in **29**.

Germination and Growth Regulation by Sesquiterpene Lactones

Sesquiterpene lactones commonly occur in the family Asteraceae (sunflower family) but are also found in the Umbelliferae and Magnoliaceae (Fischer et al., 1979; Seaman, 1982) as well as in certain liverworts (Asakawa, 1982). The biogenetic relationships of sesquiterpene lactones were discussed previously (Fischer et al., 1979). The major skeletal types of sesquiterpene lactones mentioned in this chapter are given in Figure 12.3. The typical feature of most sesquiterpene lactones is the presence of an α-methylene-γ-lactone moiety (**A**), and some lactones also contain a cyclopentenone group (**B**). Both are potent receptors for biological nucleophiles, in particular, thiol groups that form the respective addition products **C** and **D**. This can result in the inhibition of key enzymes that contain essential thiol groups. As a consequence these highly bitter substances

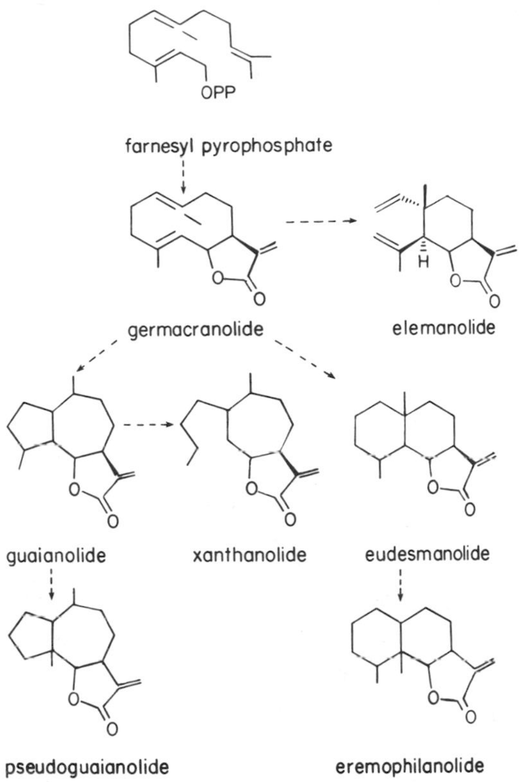

farnesyl pyrophosphate

germacranolide   elemanolide

guaianolide   xanthanolide   eudesmanolide

pseudoguaianolide   eremophilanolide

Figure 12.3   Biogenetic relationships of major skeletal types of sesquiterpene lactones.

exhibit a wide spectrum of biological activities (Rodriguez et al., 1976; Stevens, 1984) which includes cytotoxicity and antitumor properties (Cassady and Suffness, 1980), antimicrobial (Lee et al., 1977), insecticidal (Smith et al., 1983), molluscicidal (Marston and Hostettmann, 1985), and antimalarial activity (Klayman, 1985). Furthermore, they are known causes for livestock poisoning (Ivie et al., 1975a,b) and contact dermatitis in humans (Arlette and Mitchell, 1981). The function of sesquiterpene lactones as growth regulators was reviewed by Gross (1975) and more recently by Stevens (1984).

We tested 10 sesquiterpene lactones for their germination regulation activity (Fischer and Quijano, 1985). Their effects on the seed germination of 11 crop plants and the weed *Amaranthus palmeri* were tested at concentrations near 0.1 mM or below (< 50 ppm). The lactones depressed the germination of some species and stimulated others. For instance, the germacranolide costunolide (**30**) promoted germination of sorghum, carrot, and cucumber but inhibited ryegrass, wheat, and *Amaranthus palmeri* (Figure 12.4). In contrast, the biogenetic derivative 11,13-dihydroparthenolide (**31**) from the ragweed *Ambrosia artemisiifolia*, which has a saturated γ-lactone, significantly promoted germination of wheat, clover, carrot, and cucumber but inhibited sorghum and *A.*

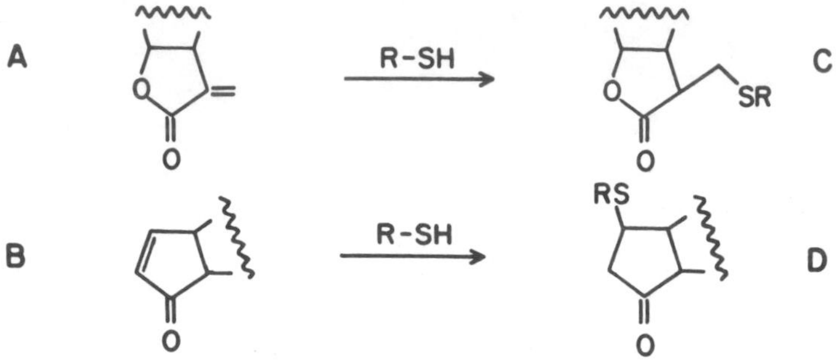

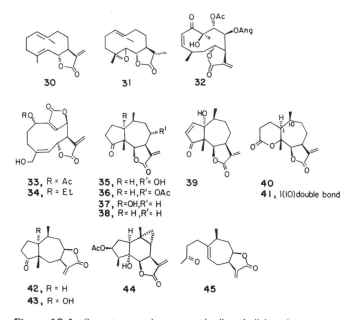

Scheme 12.1   Possible mechanism of action of methylene lactones and cyclopentenones.

palmeri. The germacranolide calein A (**32**), which also contains an α,β-unsaturated ketone, exhibited at 5×10^{-5} molar concentration a significant inhibitory effect on *A. palmeri* but no pronounced activity on crop seeds was observed. The dilactone melampodin B (**33**) promoted germination of ryegrass but the analog cinerenin (**34**) inhibited ryegrass and wheat and promoted carrot. The pseudoguaianolides confertiflorin (**36**) and desacetylconferfiflorin (**35**), which are constituents of the Texas ragweed *Ambrosia confertiflora* DC., exhibited distinct differences in activity. Compound **36** inhibited sorghum and ryegrass but promoted wheat and had little effect on the other seeds. Parthenin (**39**) inhibited

Figure 12.4   Sesquiterpene lactones with alleged allelopathic activity.

clover and carrot but promoted onion, wheat, cucumber, and *A. palmeri*. It had previously been demonstrated that parthenin, which is a major constituent of the aggressive weed *Parthenium hysterophorus* L., inhibited the development of radicals and hypocotyls of *Phaseolus vulgaris* (Garciduenas et al., 1972) and *Eleusine coracana* (Kanchan, 1975). Parthenin (39) and the minor constituent damsin (38) were found to be present in allelopathic root exudates, leaf washings, pollen, and trichome leachates of *P. hysterophorus* (Kanchan and Jayachandra, 1980).

Two constituents of *Parthenium hysterophorus*, parthenin (39) and coronopilin (37), were autotoxic to seedlings and older plants (Picman and Picman, 1984). The authors suggest that the autotoxic effects of water-soluble plant metabolites (lactones and phenolic acids) in *P. hysterophorus* not only play a role as allelopathic agents and in the defense against herbivores and diseases, but are also autotoxic to secure population regulation. The root and seedling growth inhibitory activity of parthenin (39) on *Crotalaria mucronata*, *Cassia tora*, *Ocimum basilicum*, *O. americanum*, and barley was reported (Khosla and Sobti, 1981a,b).

Anaya and del Amo (1978) found that aqueous extracts of leaves and roots of the semiarid and subtropical ragweed *Ambrosia cumanensis* H.B.K., a pioneer species in secondary succession in Vera Cruz, Mexico, inhibited the germination and growth of several associated pioneer species and was also autoinhibitory. Seven sesquiterpene lactones, including the common *A. cumanensis* constituents psilostachyin B (41), C (40), confertin (42), and peruvin (43), inhibited root and shoot growth of most test species at the 250-ppm levels and germination varied with pronounced inhibitions and promotions (del Amo and Anaya, 1978). Peruvin (43) and damsic acid were isolated from hog weed (*Ambrosia artemisiifolia*) and both inhibited rice seedling growth and lettuce germination at a concentration of 250 ppm (Watanabe et al., 1981). The germination and growth of velvet leaf (*Abutilon theophrasti* Medic.), a weed causing severe problems in corn and soybean fields, was inhibited by axivalin (44) and tomentosin (45), two lactones isolated from seeds of povertyweed (*Iva axillaris* Pursh) (Spencer et al., 1984).

The function of sesquiterpene lactones as growth regulators has been studied and in several cases structure–activity relationships have clearly emerged to allow conclusions about the mechanism of action of lactonic growth regulators. Earlier studies by Shibaoka et al. (1967a,b) showed that heliangine (46) from Jerusalem artichoke (*Helianthus tuberosus* L.) and related sesquiterpene lactones promote the adventitious root formation on hypocotyls of cuttings taken from *Phaseolus mungo* seedlings and inhibit the elongation of *Avena* coleoptile sections (Figure 12.5). The authors also demonstrated that lactones in which the α-methylene-γ-lactone moiety had been reduced to the saturated lactone showed no effect on root formation. Since all active compounds reacted with cysteine, it was concluded that the root growth activity was related to the reaction of the α-methylene-γ-lactone with biological nucleophilic centers. Indeed, cysteine deac-

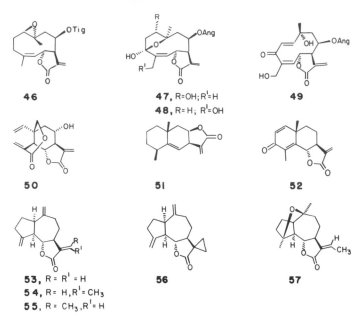

Figure 12.5 Sesquiterpene lactones with demonstrated growth regulator activity.

tivated heliangine in its root formation activity (Shibaoka et al., 1967a,b). More recently, Spring et al. (1981, 1982) investigated the growth inhibition in the sunflower (*Helianthus annuus* var. *giganteus*), which led to the isolation of two known furanoheliangolides, niveusin C (**47**) and B (**48**) (Ohno and Mabry, 1980), and a new germacranolide (**49**). Compounds **47–49** caused a linear reduction of growth in *Avena* coleoptiles. Growth assays on *Avena sativa* L. and *Helianthus annuus* with an inductive displacement transducer showed inhibition of the indole acetic acid (IAA)-induced straight growth of stem segments of *H. annuus* and *Avena sativa* coleoptile segments. In the presence of dithiothreitol (DTT) the inhibitory effect of the lactones in the *Avena*-segment test was neutralized, which was attributed to the binding of DTT to the methylene lactones. Spring and Hager (1982) proposed the following mechanism of action for the sesquiterpene lactone inhibitions of auxin-receptor sites—the binding of an auxin to an auxin receptor liberates a thiol that is irreversibly blocked by the alkylating methylene lactone. Consequently, the IAA–receptor–inhibitor complex is not able to initiate elongation growth.

The antiauxin activity of two germacranolides, argophyllin A and B from *Helianthus agrophyllus*, are comparable to heliangine (**46**) (Watanabe et al., 1982). Kupchan and co-workers demonstrated that the elemanolide vernolepin (**50**) from *Vernonia hymenolepis* A. Rich. inhibited the extension growth of wheat coleoptile sections at a 1.8×10^{-5} molar concentration. When increasing amounts of the auxin were simultaneously administered the inhibitory effect was reduced (Sequeira et al., 1968; Kupchan, 1975).

The phytotoxicity of the eudesmanolide alantolactone (51) was demonstrated by its inhibitory effect on seed germination, seedling growth, rate of respiration, and degradation of starch and protein in mung beans (*Phaseolus mungo*) (Dalvi et al., 1971). In contrast, the cross-conjugated ketolactone santonin (52) promoted, at 10 ppm, adventitious root formation in *Phaseolus mungo*, which was comparable to the activity of IAA (Kalsi et al., 1978). The effect of alantolactone (51) and the related isoalantolactone on the respiration of *Chlorella pyrenoidosa* was studied by Kwon et al. (1973) and Kwon and Woo (1975). The lactones greatly enhanced the rate of respiration but inhibited cell growth.

Kalsi and co-workers investigated a number of naturally occurring sesquiterpene lactones and their chemically modified derivatives to learn about the structural influences on plant growth activity. In a large-scale screening of various essential oils for plant growth regulatory activity it was found (Kalsi et al., 1977) that costus root oil (*Saussurea lappa*) was active, which was due to the presence of costunolide (30) and the guaianolide dehydrocostus lactone (53). Compound 53, its synthetic methyl derivatives 54 and 55, as well as the cyclopropane derivative 56 exhibited strong promoting activity in hypocotyl rooting bioassays of mung bean (*Phaseolus aureus* Roxb.). In this test, compounds 54 and 55 were slightly more active than dehydrocostus lactone (53) (Kalsi et al., 1977). Comparison of the Z-isomer 54 with the E-isomer 55 revealed that the two geometric isomers exhibit different root initiation activities, which also change with the lactone concentration (Kalsi et al., 1981a). Other synthetic derivatives of dehydrocostus lactone included 4(15)- and 10(14)-epoxides, alcohols, and their modification products, with lactone groups as in 54–56. Among the 14 modified guaianolides, 10 compounds showed, at concentration levels of 10 ppm, increased activity in rooting bioassays of mung bean with increasing activity at higher concentrations (Kalsi et al., 1979, 1981b). The chemical modifications resulted in a product (57) with rooting activity higher than IAA at the 10-ppm concentration level (Kalsi et al., 1981b). The guaianolide ketones 58 and 59 as well as the 3-keto-derivative of 56 showed distinct enhancement in the root formation of *Phaseolus areus*, when compared with their 3-deoxy derivatives 53, 54, and 56 (Figures 12.5 and 12.6). The keto derivatives exhibit maximum activity at 10 ppm or below and show toxicity at higher concentrations. The presence of epoxide functions had no major effect on the biological activity (Kalsi et al., 1984). The structurally related guaianolides solstitolide (60), repin, acroptilin, and centaurepensin from Russian knapweed (*Centaurea repens*) and yellow starthistle (*C. solstitialis*) increased root elongation of lettuce seedlings at 10 ppm and inhibited growth at higher concentrations (Stevens and Merrill, 1985). It had also been demonstrated that several guaianolides from sagebrush (*Artemisia tridentata* var. *vaseyana*) inhibited the growth and stimulated the respiration of cucumber (*Cucumis sativus*) at 10^{-4} molar level and above (McCahon et al., 1973).

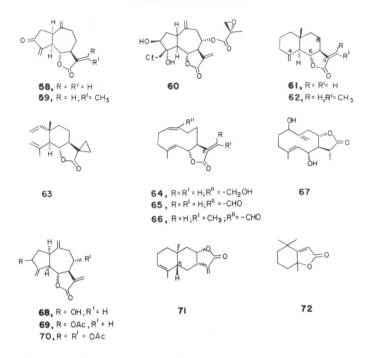

Figure 12.6 Additional sesquiterpene lactones with plant growth regulating activity.

In a series of growth tests of β-cyclocostunolide (**61**) and derivatives, the root-forming activity again increased with the presence of a methyl group, as in **62**, or a cyclopropane ring at C-11 of the eudesmanolide. The germacranolide costunolide (**30**) and its C-11-cyclopropane homolog as well as the Cope-rearrangement product **63** also showed distinct rooting enhancements on hypocotyl cuttings of *P. aureus* seedlings (Kalsi et al., 1983). Comparison of the rooting promotion activities of β-cyclocostunolide (**62**) with its C-6 isomer β-frullanolide showed a distinct increase for the second lactone (Kalsi et al., 1985). The synthetic melampolides **64–66** exhibited rooting enhancement activities of increasing magnitude within this series. As in previous examples, the C-11-cyclopropane homolog of **65** was most active (Kalsi et al., 1985).

The root-growth inhibitory constituents of the pyrethrum flower (*Chrysanthemum cinerariaefolium* Vis.) were tartridin-type sesquiterpene lactones, several of which have a saturated methylene lactone moiety as exemplified by (11R)-11,13-dihydrotartridin B (**67**). In tests on Chinese cabbage (*Brassica rapa* L. var. *pervidis* Bailey) germination was not affected, but root growth was significantly inhibited by **67** and the related lactones at the 10-ppm level (Sashida et al., 1983).

The liverworts (Hepaticeae) are rich in terpenoids and are known to produce sesquiterpene lactones of the configurational series commonly found in higher plants, but a number of genera elaborate lactones of the antipodal series (Asakawa, 1982). In studies related to the search for natural plant growth inhibitors from liverworts, a series of active terpenoids, in particular sesquiterpene lactones, were isolated (Asakawa 1981, 1984; Matsuo et al., 1981). The guaianolides zaluzanin C (**68**) and D (**69**) and the 8 α-acetoxyzaluzanin D (**70**) were isolated from *Conocephalum conicum* and *Wiesnerella denudata*. Complete inhibitory activities toward the germination and root growth of rice in the husk was 100 and 50 ppm, respectively, for compounds **68** and **69**, and 200 and 50 ppm for the diacetate **70**. The eudesmanolide 8 α-acetoxy-β-cyclocostunolide (**61**, 8 α-acetoxy) and its 3,4-double bond isomer exhibited similar activities (Asakawa and Takemoto, 1979; Awakawa et al., 1980). Other guaianolides and 3 β-hydroxycostunolide (**30**, 3 β-OH) from *Porella japonica* were also active at the 100–200 ppm level (Asakawa et al., 1981). Several pungent *ent*-eudesmanolides from *Diplophyllum albicans* and *Chiloscyphus polyanthus*, represented by *ent*-diplophyllolide (**71**), inhibit the germination and root elongation of rice husk at 100–200 ppm (Asakawa et al., 1979).

Finally, the potent growth inhibitory effect of the terpene lactone dihydroactinidiolide (**72**) from spikerush (*Eleocharis coloradoensis*) should be mentioned (Stevens and Merrill, 1981).

CONCLUSIONS

A summary of the studies of mono- and sesquiterpenes and their effects on the germination and growth of plants provided several interesting facts, but many questions remain to be answered. Activity could be related to the presence of a α-methylene-γ-lactone moiety for several sesquiterpene lactones. However, for most active compounds, predictions on stimulatory and/or inhibitory activities are presently not possible. This might be due, in part, to the wide range of concentrations (10–1000 ppm) applied in the bioassays. Also, the use of a great variety of test plants does not permit easy comparison of data. However, one fact has been demonstrated in several studies: there exists a distinct activity–concentration relationship. In many instances, stimulatory effects were observed at low concentrations and phytotoxicity increased with increasing regulator concentration. Therefore, in future studies it might be useful to include the following experiments in any growth regulatory study:

1. Activity–concentration studies to learn about the toxicity of a compound.
2. Systematic structural modifications of an active compound to determine structural moieties that are essential for bioactivity.
3. Study of anatomical and physiological effects on test plants caused by an active compound to provide information related to sites of action and physiological processes.

These studies will require extensive cooperation between chemists and biologists. The efforts will undoubtedly be rewarded by the discovery of new and novel bioactive compounds to be used as models for synthetic herbicides and growth hormones.

ACKNOWLEDGMENT

Financial support by the Department of Entomology and the Louisiana Agricultural Experiment Station, Louisiana State University Agricultural Center, is acknowledged.

REFERENCES

Anaya, A. L. and S. del Amo. 1978. *J. Chem. Ecol.* 4:289.

Arlette, J. and J. C. Mitchell. 1981. *Contact Dermatitis.* 7:129.

Asakawa, Y. 1981. *J. Hattori Bot. Lab.* 50:123.

Asakawa, Y. 1982. *In* W. Herz, H. Grisebach, and G. W. Kirby (eds.), *Progress in the Chemistry of Organic Natural Products*, Vol. 42. Springer, New York, pp. 1–269.

Asakawa, Y. 1984. *Rev. Latinoam. Quim.* 14:109.

Asakawa, Y. and T. Takemoto. 1979. *Phytochemistry.* 18:285.

Asakawa, Y., M. Toyota, T. Takemoto, and C. Suire. 1979. *Phytochemistry.* 18:1007.

Asakawa, Y., R. Matsuda, and T. Takemoto. 1980. *Phytochemistry.* 19:567.

Asakawa, Y., M. Toyota, and T. Takemoto. 1981. *Phytochemistry.* 20:257.

Asplund, R. O. 1968. *Phytochemistry.* 7:1995.

Asplund, R. O. 1969. *Weed Sci.* 17:454.

Bradow, J. M. 1985. *In* A. C. Thompson (ed.), *The Chemistry of Allelopathy.* American Chemical Society Washington, D.C., pp. 285–299.

Capasso, R., N. S. Iacobellis, A. Bottalico, and G. Randazzo. 1984. *Phytochemistry.* 23:2781.

Cassady, J. M. and M. Suffness. 1980. *In* J. M. Cassady and J. D. Douros (eds.), *Medicinal Chemistry, Vol. 16: Anticancer Agents Based on Natural Product Models.* Chapter 7. Academic, London.

Croteau, R. 1981. *In* J. W. Porter and S. L. Spurgeon (eds.), *Biosynthesis of Isoprenoid Compounds.* Vol. 1. Wiley-Interscience, New York, p. 225.

Dalvi, R. R., B. Singh, and D. K. Salunkhe. 1971. *Chem. Biol. Interact.* 3:13.

Del Amo, S. and A. L. Anaya. 1978. *J. Chem. Ecol.* 4:305.

Del Moral, R. and C. H. Muller. 1970. *Am. Midl. Nat.* 85:254.

Devon, T. K. and A. I. Scott. 1972. *Handbook of Naturally Occurring Compounds (Terpenes).* Vol. 2. Academic, New York.

Evenari, M. 1949. *Bot. Rev.* 15:153.

Fischer, N. H. and L. Quijano. 1985. *In* A. C. Thompson (ed.), *The Chemistry of Allelopathy.* American Chemical Society, Washington, D.C., pp. 133–147.

Fischer, N. H., E. J. Olivier, and H. D. Fischer. 1979. *In Progress in the Chemistry of Organic Natural Products*, Vol. 38. Springer, New York, pp. 47–390.

Gant, R. E. and E.E.C. Clebsch. 1975. *Ecology.* 56:604.

Garciduenas, M. R., X. A. Dominguez, J. Fernandez, and G. Alanis. 1972. *Rev. Latinoam. Quim.* 3:52.

Gonzalez de la Parra, M., A. L. Anaya, F. Espinosa, M. Jiménez, and R. Castillo. 1981. *J. Chem. Ecol.* 7:509.

Gross, D. 1975. *Phytochemistry.* 14:2105.

Grummer, G. 1961. *In* F. L. Milthorpe (ed.), *Mechanism in Biological Competition.* Academic, New York, pp. 219–228.

Halligan, J. P. 1975. *Ecology.* 56:999.

Harborne, J. 1982. *Introduction to Ecological Biochemistry.* 2nd ed. Academic, New York.

Hoffmann, H.M.R. and J. Rabe. 1985. *Angew. Chem. Int. Ed. Engl.* 24:94.

Ivie, G. W., D. A. Witzel, W. Herz, R. Kannan, J. O. Norman, D. D. Rushing, J. H. Johnson, L. D. Rowe, and J. A. Veech. 1975a. *J. Agric. Food Chem.* 23:841.

Ivie, G. W., D. A. Witzel, and D. D. Rushing. 1975b. *J. Agric. Food Chem.* 23:845.

Julio, A. M. and C. Pinto. 1976. *Bol. Fac. Farm. Coimbra.* 1:9.

Kalsi, P. S. and K. K. Talwar. 1981. *Phytochemistry.* 20:511.

Kalsi, P. S., V. K. Vij, O. S. Singh, and M. S. Wadia. 1977. *Phytochemistry.* 16:784.

Kalsi, P. S., O. S. Singh, and B. R. Chhabra. 1978. *Phytochemistry.* 17:576.

Kalsi, P. C., B. R. Chhabra, and O. S. Singh. 1979. *Experientia.* 35:481.

Kalsi, P. S., P. Kaur, and B. R. Chhabra. 1979. *Phytochemistry.* 18:1877.

Kalsi, P. S., M. L. Sharma, R. Handa, K. K. Talwar, and M. S. Wadia. 1981a. *Phytochemistry.* 20:835.

Kalsi, P. S., D. Gupta, R. S. Dhillon, G. S. Arora, K. K. Talwar, and M. S. Wadia. 1981b. *Phytochemistry.* 20:1539.

Kalsi, P. S., V. B. Sood, A. B. Masih, D. Gupta, and K. K. Talwar. 1983. *Phytochemistry.* 22:1387.

Kalsi, P. S., G. Kaur, S. Sharma, and K. K. Talwar. 1984. *Phytochemistry.* 23:2855.

Kalsi, P. S., S. Khurana, and K. K. Talwar. 1985. *Phytochemistry.* 24:103.

Kanchan, S. D. 1975. *Curr. Sci.* 44:358.

Kanchan, S. D. and A. Jayachandra. 1980. *Plant Soil.* 55:67.

Kelsey, R. G., T. T. Stevenson, J. P. Scholl, T. J. Watson, and F. Shafizadeh. 1978. *Biochem. Syst. Ecol.* 6:193.

Kelsey, R. G., G. W. Reynolds, and E. Rodriguez. 1984. *In* E. Rodriguez, P. L. Healey, and I. Mehta (eds.), *Biology and Chemistry of Plant Trichomes.* Plenum, New York, pp. 187–241.

Khosla, S. N. and S. N. Sobti. 1981a. *Indian J. Forestry.* 4:56.

Khosla, S. N. and S. N. Sobti. 1981b. *Pesticides.* 15:8.

Klayman, D. L. 1985. *Science.* 228:1049.

Komai, K. and K. Ueki. 1981. Shokubutsu no Kagaku Chosetsu. 16:32. (*Chem. Abstr.* 97:123942u.)

Komai, K., Y. Sugiqaka, and S. Sato. 1981. Kinki Daigaku Nogakubu Keyo. 14:57. (*Chem. Abstr.* 95:162961c.)

Kupchan, S. M. 1975. *In* V. C. Runeckles (ed.), *Recent Advances in Phytochemistry.* Vol. 9. Plenum, New York, pp. 167–188.

Kwon, Y. M. and W. S. Woo. 1975. Yakhak Hoe Chi. 19:118. (*Chem. Abstr.* 84:100251k.)

Kwon, Y. M., W. S. Woo, L. K. Woo, and M. J. Lee. 1973. Han'guk Saenghwahakhoe Chi. 6:85. (*Chem. Abstr.* 163964d.)

Lee, K. H., T. Ibuka, R. Y. Wu, and T. A. Geissman. 1977. *Phytochemistry.* 16:1177.

Loomis, W. D. and R. Croteau. 1980. *In* P. K. Stumpf (ed.), *Biochemistry of Plants.* Vol. 4. Academic, New York, pp. 363–418.

Lorber, P. and W. H. Muller. 1976. *Am. J. Bot.* 63:196.

Lorber, P. and W. H. Muller. 1980a. *Comp. Physiol. Ecol.* 5:60.

Lorber, P. and W. H. Muller. 1980b. *Comp. Physiol. Ecol.* 5:68.

Marston, A. and K. Hostettmann. 1985. *Phytochemistry.* 24:639.

Matsuo, A., K. Nadaya, M. Nakayama, and S. Hayashi. 1981. Nippon Kagaku Kaishi. 1981:665. (*Chem. Abstr.* 95:39086u.)

Matsuo, A., H. Nozaki, N. Kubota, S. Uto, and M. Nakayama. 1984a. *J. Chem. Soc. Perkin Trans.* I:203.

Matsuo, A., S. Uto, H. Nozaki, and M. Nakayama. 1984b. *J. Chem. Soc. Perkin Trans.* I:215.

McCahon, C. B., R. G. Kelsey, R. P. Sheridan, and F. Shafizadeh. 1973. *Bull. Torrey Bot. Club.* 100:23.

Muller, C. H. 1965. *Bull. Torrey Bot. Club.* 92:38.

Muller, W. H. and C. H. Muller. 1964. *Bull. Torrey Bot. Club.* 91:327.

Nishimura, H., K. Kaku, T. Nakamura, Y. Fukazawa, and J. Mizutani. 1982. *Agric. Biol. Chem.* 46:319.

Nishimura, H., T. Nakamura, and J. Mizutani. 1984. *Phytochemistry.* 23:2777.

Ohno, N. and T. J. Mabry. 1980. *Phytochemistry.* 19:609.

Picman, J. and A. K. Picman. 1984. *Biochem. Syst. Ecol.* 12:287.

Rice, E. L. 1984. *Allelopathy*, 2nd ed. Academic, Orlando, FL.

Rodriguez, E., G.H.N. Towers, and J. C. Mitchell. 1976. *Phytochemistry.* 15:1573.

Ruecker, G. 1973. *Angew. Chem. Int. Ed. Engl.* 12:793.

Sashida, Y., H. Nakata, H. Shimomura, and M. Kagaya. 1983. *Phytochemistry.* 22:1219.

Seaman, F. C. 1982. *Bot. Rev.* 48:121.

Sequeira, L., R. J. Hemingway, and S. M. Kupchan. 1968. *Science.* 161:789.

Shibaoka, H., M. Shimokoriyama, S. Iriuchijima, and S. Tamura. 1967a. *Plant Cell Physiol.* 8:297.

Shibaoka, H., M. Mitsuhashi, and M. Shimokoriyama. 1967b. *Plant Cell Physiol.* 8:161.

Sigmund, W. 1924. *Biochem. Z.* 146:389.

Smith, G. M., K. M. Kester, and N. H. Fischer. 1983. *Biochem. Syst. Ecol.* 11:377.

Spencer, G. F., R. B. Wolf, and D. Weisleder. 1984. *J. Nat. Prod.* 47:730.

Spring, O. and A. Hager. 1982. *Planta.* 156:433.

Spring, O., K. Albert, and W. Gradmann. 1981. *Phytochemistry.* 20:1883.

Spring, O., K. Albert, and A. Hager. 1982. *Phytochemistry.* 21:2551.

Stevens, K. L. 1984. *In* W. D. Nes, G. Fuller, and L. S. Tsai (eds.), *Isopentenoids in Plants.* Dekker, New York, pp. 65-80.

Stevens, K. L. and G. B. Merrill. 1981. *Experimentia.* 37:1133.

Stevens, K. L. and G. B. Merrill. 1985. *In* A. C. Thompson (ed.), *The Chemistry of Allelopathy.* American Chemical Society, Washington, D.C., pp. 83-98.

Watanabe, S., A. Kobayashi, and K. Yamashita. 1981. *Agric. Biol. Chem.* 45:2919.

Watanabe, K., N. Ohno, H. Yoshioka, J. Gershenzon, and T. J. Mabry. 1982. *Phytochemistry.* 21:709.

13

POLYACETYLENES AS ALLELOCHEMICALS

KENNETH L. STEVENS

Western Regional Research Center, United States Department of Agriculture, Agricultural Research Service, Berkeley, California

The invasion of weeds into range, crop, and pasture land has been a troublesome problem plaguing mankind for millennia. In many instances, it appears as if weeds have an added advantage in the competitive struggle over domestic crops by incorporating allelochemics into their strategem of survival and propagation. Many cases have now been documented that attest to the role of allelopathy in the propagation of weeds. The number and variety of phytotoxic compounds that have been isolated, characterized, and shown to be involved in the ecological distribution of plants is rather large and continues to grow at a phenomenal rate. Included in the list of allelochemics are a class of compounds called polyacetylenes.

As early as 1961 N. A. Sørenson reported on the role of polyacetylenes and noted that these natural products were, in general, physiologically active against other organisms. Since his paper was concerned mainly with the chemotaxonomic importance of polyacetylenes, he did not elaborate on their biological activity. One observation by Sørenson, which has been substantiated by a number of other workers, is the chemical reactivity of polyacetylenes. Many of the compounds are unstable above 30°C, polymerize in light, and readily react with themselves if they are concentrated. It was observed that many of the polyacetylenes have a half-life of 1–2 days in the plant, as determined by radiolabeling.

Kawazu et al. in 1969 were the first to report on the phytotoxicity of *cis*-dehydromatricaria ester (*cis*-DME) (**1**) which had been isolated from *Solidago altissima*. Following this early work, Kobayashi (1974) examined *Erigeron annus* as well as *S. altissima* and outlined not only the role of *cis*-DME (**1**) but also lachnophyllum ester (LE) (**2**) and matricaria ester (ME) (**3**) in

$$CH_3-(C\equiv C)_3-CH=CHCOOCH_3$$

1

$$CH_3CH_2CH_2-(C\equiv C)_2-CH=CHCOOCH_3$$

2

$$CH_3-CH=CH-(C\equiv C)_2-CH=CHCOOCH_3$$

3

plant growth suppression. He also looked at a number of geometric isomers of these naturally occurring esters. These compounds, along with 2-*cis*-8-*cis*-ME, 2-*trans*-8-*cis*-ME, and 2-*cis*-LE, were examined and found to be toxic to young rice plants by inhibiting the growth of both the hypocotyl and root (Numata et al., 1973). They were relatively inactive in suppression of seed germination. Figure 13.1 shows the activity of *cis*-DME, *cis*-LE, and *cis*-ME toward root inhibition of young rice plants. Approximately a 50% inhibition was obtained at 10–20 ppm for DME and LE.

Kobayashi (1974) found that the active compounds appear to be concentrated in the roots but are also found in the aerial parts of the plants. This is in accord with the earlier work of Sorensen (1961). Since these phytotoxic compounds are predominately found in the roots, it was stated that the observed allelopathic reactions of *S. altissima* and *E. annus* took place underground. To lend cre-

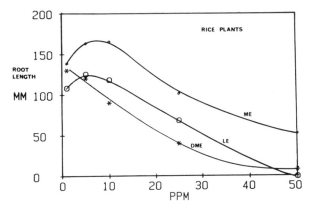

Figure 13.1   Root inhibition of young rice plants by *cis*-DME, *cis*-LE and *cis*-ME[3].

dence to this hypothesis, they examined the soil around the roots and isolated not only *cis*-DME but also *trans*-DME at concentrations (5–6 ppm) that would affect the surrounding plant community.

A number of Composites found in Japan were examined for the presence of DME, ME, and LE. Table 13.1 summarizes their results.

The investigation only revealed their presence or absence with no information on the amounts actually present. As can be seen from Table 13.1, DME is found only in *A. vulgaris* and *S. altissima*, which in turn do not contain LE or ME. However, both LE and ME are distributed almost equally among the other Composites investigated, that is, both occur together in all except *S. virgaurea*.

A later publication (Kobayashi et al, 1980) confirmed much of the earlier work and concluded that these C_{10} polyacetylenes found in *S. altissima* and *Erigeron* species are probably allelopathic substances of ecological importance. Specifically *cis*-DME in *S. altissima* and *cis*-ME, *trans*-LE, and *cis*-LE in *Erigeron* spp. show strong inhibitory effects on other plants.

A new polyacetylene **4** isolated from the roots of *S. altissima* was also phytotoxic (Ichihara et al., 1976).

$$CH_3CH=\overset{\overset{\displaystyle CH_3}{|}}{C}-COO-CH_2CH=CH-(C\equiv C)_2-CH=CHCOOCH_3$$

4

This compound is an oxidation product of ME (**3**) with oxidation occurring at the ω carbon with subsequent esterification by a C_5 acid. This particular acyl moiety is quite prevalent in the Compositae but usually occurs as the C_5 side chain on sesquiterpene lactones (Fischer et al., 1979). As with ME, this C_{16} polyacetylene is also phytotoxic to the growth of barnyard millet (*Panicum crusgalli* L. var. *frumentaceum* trin.) in which 20% inhibition was observed at 1 ppm. It was suggested that this matricaria ester derivative may act as an allelopathic substance.

TABLE 13.1.

Presence or absence of DME, LE, and ME in selected compositae found in Japan

	DME	LE	ME
Artemissia vulgaris	+	–	–
Solidago altissima	+	–	–
S. virgaurea	–	–	+
Erigeron annus	–	+	+
E. philadelphicus	–	+	+
E. canadensis	–	+	+
E. linifolius	–	+	+

In a later publication, Ichihara et al. (1978) reported the isolation from the same plant, dehydromatricaria lactone (5). This compound, like ME, also inhibited the growth of barnyard millet; a 66% reduction at 10 ppm was observed. The presence of these ME derivatives in *S. altissima* undoubtedly contribute to the allelopathic nature of the plant.

Towers and Wat (1978) published a review on the biological activity of polyacetylenes and concluded that ". . . all the physiological activities of these compounds which have been reported so far are detrimental to living organisms, including plants." They also stated that many of the biological activities associated with the Composites and often ascribed to the sesquiterpene lactones are probably due to the presence of polyacetylenes. Indeed, the wide distribution of polyacetylenes in the Composite family (Bohlmann et al., 1973) and the activity of those compounds tested in biological systems would seem to lend support to this hypothesis.

A recent review by Numata (1982) summarizes the allelopathic effects of ME derivatives and states that the inhibitory activity against *Oryza sativa* (rice) for both the second sheath and the root was LE > *cis*-DME > *cis*-ME > *trans*-ME > control. *Trans*-ME has almost no inhibitory effect. The application of 5 ppm of gibberelin (GA$_3$) and 5 ppm of auxin (indoleacetic acid, IAA) to rice seedlings with 20 ppm of a C$_{10}$-polyacetylene methyl ester failed to nullify the phytotoxicity of the polyacetylene. It was concluded that the inhibitory effect is not based on a hormonal mechanism.

With the possible exception of the two polyacetylenes phenylheptatriyne (PHT) (6) and α-terthienyl (7)*, which have been found in the roots of the tropical weed *Bidens pilosa* and the common marigold *Tagetes erecta*, respectively, there appears to be no data to suggest that other polyacetylenes are phytotoxic. It must be recognized, however, that the vast majority of polyacetylenes described in the literature have never been tested for biological activity and in particular for phytotoxicity.

$$CH_3-(C\equiv C)_2-CH= \underset{O}{\boxed{}} =O$$

5

$$\boxed{}-(C\equiv C)_3-CH_3$$

6

7

*α-Terthienyl may be viewed as a dodecane containing six triple bonds, with subsequent addition of three equivalents of hydrogen sulfide.

Our investigation of allelopathic weeds, *viz*, *Centaurea repens* L. (Compositae, Russian knapweed) has led to the isolation and identification of several polyacetylenes in both the aerial parts and the roots.

ISOLATION AND PURIFICATION

Roots from plants of Russian knapweed (*C. repens* L.) were collected throughout the growing season from early spring to late summer near Discovery Bay (Highway 4 between Brentwood and Stockton, California). The severed roots were immediately placed on dry ice, then ground in a rotating plate mill with dry ice to approximately $1/8$-in. particle size. The dry ice/root mixture was then extracted with ether/Skelly-F (1:2) in a large beaker which was allowed to warm to room temperature. Light was excluded to prevent any photolysis of the polyacetylenes. After the ether/Skelly-F warmed to room temperature, the mixture was filtered and the volume was reduced on a rotary evaporator at approximately 25°C. The polyacetylenes were separated from the mixture by a combination of column chromatography on silica gel with benzene and preparative TLC (silica gel), using a variety of solvents, that is, ether/Skelly-F (1:9), benzene, and cyclohexane/acetone (4:1). Polyacetylenes were collected from the preparative plates by extraction with ether. In all instances, care was taken to exclude light and keep temperatures below 30°C.

RESULTS

Structural identification of each component was made by UV and NMR spectroscopy. Table 13.2 shows the five pure polyacetylenes we isolated from *C. repens* along with their respective UV and NMR spectra. In addition to those listed in the table, compounds **13** and **14** were isolated as a mixture. Lack of material coupled with their instability precluded separating the two isomers; however, they were readily identified by comparing the UV and NMR spectra of the mixture with that reported by Bohlmann et al. (1965).

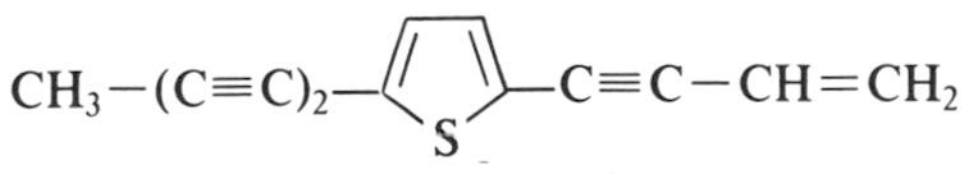

$$CH_3-(C\equiv C)_2-\underset{S}{\langle\hspace{1em}\rangle}-C\equiv C-CH=CH_2$$

13

$$CH_3-C\equiv C-\underset{S}{\langle\hspace{1em}\rangle}-(C\equiv C)_2-CH=CH_2$$

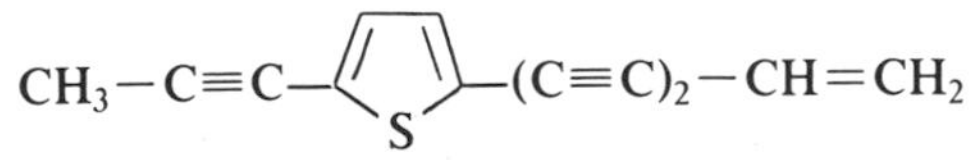

14

TABLE 13.2.
UV and NMR data of polyacetylenes 8-12

$CH_3-(C{\equiv}C)_2-$ [thienyl] $-C{\equiv}C-R$

Compound	UV (Ether) $(\lambda_{max})^a$	NMR($CDCl_3$), 90 MHz (δ Values Relative to TMS)
$R = -CH-CH_2OH$ | Cl **8**	344, 324, 270, 250, 206	2.04(s,3H,Me),3.92(d,1H,J=6,H_A), 3.97(d,1H,J=5,H_B), 4.88(dd,1H,J=5,6,H_C), 7.12(s,2H,A_r-H)
$R = -CH-CH_2Cl$ | OH **9**	342, 322, 250, 236, 210	2.00(s,3H,Me),3.70(d,1H,J=6,H_A), 3.71(d,1H,J=4,H_B), 4.79(dd,1H,J=4,6,H_C), 7.02(d,1H,J=3.5,A_r-H)
$R = -CH-CH_2OAc$ | Cl **10**	342, 324, 270, 250, 236, 210	2.00(s,3H,Me),2.08(s,3H,COMe), 4.37(d,1H,J=6,H_A), 4.38(d,1H,J=5,H_B), 4.91(dd,1H,J=5,6,H_C), 7.08(s,2H,A_r-H)
$R = -CH-CH_2Cl$ | OAc **11**	342, 322, 250, 238	2.00(s,3H,Me),2.10(s,3H,COMe), 3.73(d,2H,J=6,H_A,H_B), 5.78(t,1H,J=6,H_C), 7.06(s,2H,A_r-H)
$R = -CH-CH_2OA_c$ | OAc **12**	342, 322, 248, 210	2.00(s,3H,Me),2.07(s,3H,COMe), 2.09(s,3H,COMe), 4.22(dd,1H,J=5,12,H_A), 4.34(dd,1H,J=7,12,H_B), 5.80(dd,1H,J=5,7,H_c), 7.02(s,2H,A_r-H)

^{a}Intensities were similar for all compounds, that is, 342 (2×10^4), 322 (2.2×10^4), 250 (8×10^3), 210 (1.7×10^4).

All of the isolated polyacetylenes are C_{13} derivatives, each containing one thienyl group which can formally be envisioned as the addition of hydrogen sulfide across two triple bonds. With the exception of **13** and **14**, each polyacetylene varies only with different substituents on C_1 and C_2. All of the compounds obtained from Russian knapweed have been found in other Composites; however, these particular substances have not been found in *C. repens*. It is significant that even though Russian knapweed has been reported to be allelopathic (Fletcher and Renny, 1963), no matricaria esters or ME derivatives have been found in the plant. Compound **8** has been found in a number of *Echinops* spp. (Bohlmann et al., 1973) and *Pluchea dioscorides*. Compound **9** has been found in the roots of *Eclipta prostata* L. (Bohlmann and Zdero, 1970), and **10** has been found in both *Echinops* spp. and *Centaurea cristata* (Bohlmann et al., 1966). Although a number of *Centaurea* species have been investigated for polyacetylenes, compounds **8-10** represent new compounds to this genus. In 1965, Bohlmann et al. reported the presence of the diacetate **12** in *Echinops sphaerocephalus* L.; however, it was only inferred to be present and was not isolated nor were any data given for the compound. Its presence in Russian knapweed thus confirms it as a natural product. Both the unsaturated compounds **13** and **14** as well as the oxidation products (chlorohydrins and chlorohydrin acetates) have been isolated in this study; however, the presumed intermediate epoxide was not found in spite of its presence in other *Centaurea* spp.

In all cases, with the exception of **13** and **14**, the compounds are optically active, which rules out the possibility of them being artifacts (i.e., hydrolysis products of **13** and **14**). As a typical example, **9** showed $[\alpha]589 = + 17.5°$; $[\alpha]578 = + 18.9°$; $[\alpha]546 = + 20.6°$; $[\alpha]436 = + 37.9°$. The asymmetry of the molecules is also apparent from the NMR spectra. For instance, the methylene group shows a different chemical shift for each of the protons. The diacetate **9** is the most obvious with δ 4.22 and δ 4.34 for each of the methylene protons with coupling constants of 12 Hz (geminal), 5 Hz (vicinal), and 7 Hz (vicinal).

BIOLOGICAL ACTIVITY

Pregerminated seeds were placed on 20 ml of 0.5% agar gels that had previously been slurried with the prescribed amount of polyacetylene in ether. The polyacetylenes were added to the still fluid gel at approximately 30°C to avoid thermal

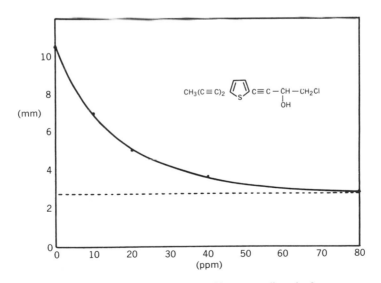

Figure 13.2 Root inhibition of lettuce seedlings by **9**.

decomposition. Controls had only solvent incorporated into the gels. The pregerminated seedlings on the agar gels were then placed in the dark at 20°C and root growth was measured to the nearest millimeter.

The polyacetylene **9** is the only compound isolated from Russian knapweed that has shown phytotoxic activity. Figure 13.2 shows the activity of **9** against lettuce seedlings after 1 day in the dark. The average root length of the control was approximately 10.5 mm (starting length was 3 mm), while at 10 ppm the root length was approximately 7 mm. Fifty percent reduction in root length occurs at 12 ppm. It is interesting to note that the chlorohydrin isomer **8** does not

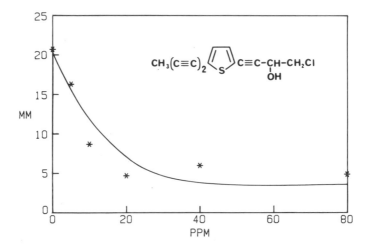

Figure 13.3 Root inhibition of barnyardgrass seedlings by **9**.

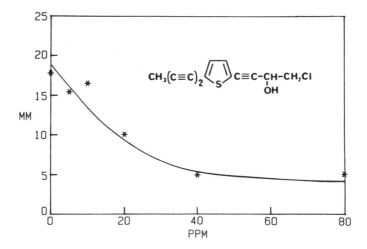

Figure 13.4 Root inhibition of alfalfa seedlings by **9**.

show any activity. Compound **9** was also tested against barnyard grass (Figure 13.3), alfalfa (Figure 13.4), and red millet (Figure 13.5). In all cases the results were approximately the same.

Matricaria ester (**3**) was also tested in our bioassay system against lettuce seedlings and the root length was measured. Owing to the relative stability of ME, the test could be conducted over 2 or 3 days rather than the 1 day necessary for the thiophene analogs of the polyacetylenes. As Figure 13.6 illustrates, matricaria ester has approximately the same activity as **9** with 50% reduction in root length at 10 ppm.

Examination of the soil around the roots of Russian knapweed throughout its growing season showed a concentration between 4 and 5 ppm of **9**. The concentration did not vary significantly throughout the growing season. Since the assay method involved the isolation of **9**, the 4–5 ppm represents a minimum amount, owing to the instability of the compound. Extrapolating the data from the bioas-

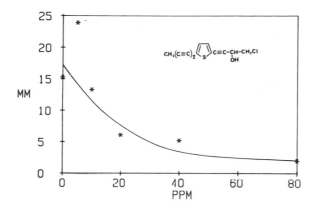

Figure 13.5   Root inhibition of red millet seedlings by **9**.

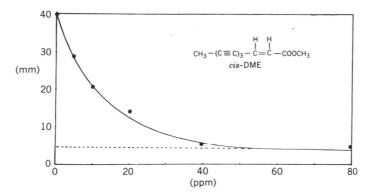

Figure 13.6   Root inhibition of lettuce seedlings by ME.

say to field conditions, one would expect about a 30% reduction in root length elongation at 4 ppm. This is certainly sufficient to have an appreciable effect on the surrounding plant community.

REFERENCES

Bohlmann, F. and C. Zdero. 1970. *Chem. Ber.* 103:834.

Bohlmann, F., K.-M. Kleine, and C. Arndt. 1964. Chem. Ber. 97:2125.

Bohlmann, F., C. Arndt, K.-M. Kleine, and H. Bornowski. 1965. *Chem. Ber.* 98:155.

Bohlmann, F., K.-M. Kleine, and H. Boronowski. 1965. *Chem. Ber.* 98:155.

Bohlmann, F., K.-M. Rode, and C. Zdero. 1966. *Chem. Ber.* 99:3544.

Bohlmann, F., T. Burkhardt, and C. Zdero. 1973. *Naturally Occurring Acetylenes.* Academic, London. 547 pp.

Fischer, N. H., E. J. Olivier, and H. D. Fischer. 1979. *In Progress in the Chemistry of Organic Natural Products.* Vol. 38. Springer-Verlag, Wien, New York, pp. 47–430.

Fletcher, R. A. and A. J. Renny. 1963. *Can. J. Plant Sci.* 43:475.

Ichihara, K., T. Kawai, M. Kaji, and M. Noda. 1976. *Agric. Biol. Chem.* 40:353.

Ichihara, K., T. Kawai, and M. Noda. 1978. *Agric. Biol. Chem.* 42:427.

Kawazu, K., A. Nakamura, S. Nishino, H. Koshimizu, and T. Mitsui. 1969. Ann. Mtg. Agric. Chem. Soc. Japan (Abstr). 130 pp.

Kobayashi, A. 1974. *Chem. Regul. Plants.* 9:95.

Kobayashi, A., M. Morimoto, Y. Shibata, K. Yamashita, and M. Numata. 1980. *J. Chem. Ecol.* 6:119.

Numata, M. 1982. *In* W. Holzner and M. Numata (eds.), *Biology and Ecology of Weed.* Dr. W. Junk, The Hague, pp. 169–173.

Numata, M., A. Kobayashi, and N. Ohga. 1973. *In* M. Numata (ed.), *Fundamental Studies in the Characteristics of Urban Ecosystem.* pp. 59–64.

Sørensen, N. A. 1961. *In Polyacetylenes and Conservatism of Chemical Characters in the Compositae.* Proceedings of the Chemical Society, London. 91 pp.

Towers, G.H.N. and C.-K. Wat. 1978. *Rev. Latinoam. Quim.* 9:162.

14

QUALITATIVE AND QUANTITATIVE DETERMINATION OF THE ALLELOCHEMICAL SPHERE OF GERMINATING MUNG BEAN

CHUNG-SHIH TANG

Department of Agricultural Biochemistry,
University of Hawaii, Honolulu, Hawaii

BAOCHEN ZHANG

Northwest Plateau Institute of Biology, Academia Sinica,
Xining, Qinghai, China

One of the early events in the germination of a seed is the release of organic and inorganic exudates into its moistened surroundings (Simon and Kaja Harun, 1972; Abdel Samad and Pearce, 1978). The outward movement of certain secondary metabolites has been regarded as the removal of endogenous inhibitors which might account for the imposition and maintenance of dormancy (Bewley and Black, 1982). The phenomenon has also been considered as an evolutionary advantage from an ecological or plant protection point of view; certain seed exudates were known to adversely affect microbial activity (Rice, 1984) and the germination of competing plant species (Gressel and Holm, 1964; Junttia, 1975; Elmore, 1980).

In recent years, allelopathy of higher plants has received wide attention because of its relevance to both agricultural and nonagricultural ecology (Rice,

1984). It is becoming increasingly clear that plants are capable of maintaining a chemical sphere of influence by releasing allelochemicals into their environment (Whittaker and Feeny, 1971). To define this sphere, it is necessary to identify the responsible compounds exuded by the donor plants under natural conditions, to determine the biological activities of the exudates against acceptor organisms, and finally, to assess the concentration of allelochemicals in the microenvironment where the interaction takes place. It remains an overly complicated problem to measure the allelochemical sphere of a growing plant under natural conditions, such as the root–soil interface; however, it is possible to detail its presence under simple and controlled conditions. In this chapter we describe the exudation of inhibitory phenolic compounds from mung bean (*Vigna radiata* L.) into its aqueous environment during germination and propose the formation of an allelochemical sphere under defined conditions.

When a mung bean seed was planted in the center of a petri dish (50-mm

Figure 14.1 A germinating mung bean seed releases allelochemicals inhibitory to the cogerminating lettuce seeds within the area restricted by a section of Teflon tubing.

diameter) containing 1.5% (w/v) agar, a light brown ring of exudates inhibitory to lettuce (*Lactuca sativa* L.) seed germination and seedling growth was observed. The inhibitory effects were enhanced (Figure 14.1) if diffusion of the exudates in the agar was confined within a cylindrical ring cut from Teflon tubing (17-mm diameter). To investigate the chemical nature of the inhibition, we collected the active fraction using a continuous hydrophobic exudate trapping system similar to the root exudate trapping system developed by Tang and Young (1982).

EXPERIMENTAL

Mung bean seeds (*Vigna radiata* L., cv. Berken) were purchased from W. Atlee Burpee Co. (Warminster, PA, 18974). Seeds of uniform size and appearance were selected and treated in a 0.5% sodium hypochlorite solution for 15 minutes, followed by thorough rinsing in distilled water. Lettuce seeds (*Lactuca sativa* L., cv. Anuenue) were obtained from the Seed Laboratory, Department of Horticulture, University of Hawaii.

Collection of Inhibitory Exudates from the Germinating Mung Bean

The continuous hydrophobic exudate trapping apparatus for collecting inhibitors from the germinating mung bean is shown in Figure 14.2 (cf. Chapter 7). In this apparatus, seeds were irrigated periodically by water released from a Soxhlet type time-release reservoir. The water flooded the seedbed briefly and then drained into a lower chamber to which an XAD-4 resin column was attached. Hydrophobic exudates leached from the germinating seeds were continuously trapped by the column while the water was transported back to the Soxhlet reservoir using a peristaltic pump. In a typical experiment, 20 g of mung beans were incubated in the dark at 24°C, the total volume of circulating water was maintained at 150 ml, and the Soxhlet reservoir was adjusted to 80 ml. At a flow rate of 4 ml/min, the seedbed was irrigated every 20 minutes, and an equivalent of 11.5 l of water passed through the germinating seeds and the resin column in 48 hours.

The XAD-4 column was disconnected and eluted with MeOH. After removal of MeOH under reduced pressure at 40°C, the "XAD-4 fraction" was diluted to concentrations equivalent to gram mung bean per milliliter (i.e., exudates trapped from 1 g of mung bean and then diluted to 1 ml aqueous solution). The circulating water was also collected and freeze-dried to provide the "aqueous residue fraction," which contained inorganic ions and hydrophilic organic exudates. For bioassay, 50 lettuce seeds were germinated in the dark under 24°C in a 50-mm diameter petri dish containing 1 ml of test solution and a Whatman No. 1 filter paper disk (50-mm diameter). The rates of germination were recorded at 12 hours and radicle lengths at 48 hours. Distilled water was used as the control.

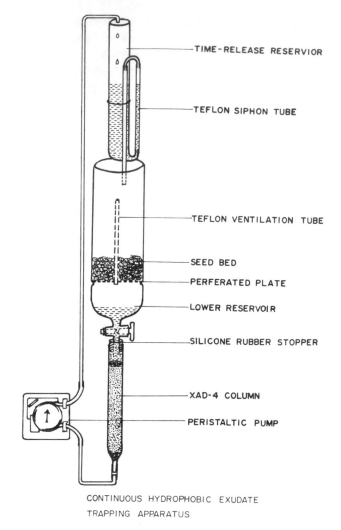

TIME-RELEASE RESERVOIR

TEFLON SIPHON TUBE

TEFLON VENTILATION TUBE

SEED BED

PERFERATED PLATE

LOWER RESERVOIR

SILICONE RUBBER STOPPER

XAD-4 COLUMN

PERISTALTIC PUMP

CONTINUOUS HYDROPHOBIC EXUDATE
TRAPPING APPARATUS

Figure 14.2   The continuous hydrophobic exudate trapping apparatus.

For HPLC, a 4.6 × 250-mm Lichrosorb C-18 column was used. The gradient elution used H_2O/MeOH as the mobile phase at a flow rate of 2.5 ml/min, linearly programed from 7 to 15% MeOH in 2 minutes, 15 to 75% in 6 minutes, and 75 to 80% in 17 minutes. Three major peaks representing more than 80% of the total absorbance at 280 nm were detected: peak I, retention time (RT) = 16.6 minutes, yield 2.3 mg/g mung bean, dry weight basis; peak II, RT = 20.0 minutes, yield 3.2 mg/g; peak III, RT = 21.2 minutes, yield 3.3 mg/g. For collection of these peaks, a semipreparative column (Ultrasil-ODS, 10 × 250 mm) was used with a flow rate of 3.5 ml/min under a similar solvent gradient

program. The ^{1}H NMR spectra were obtained in methanol-d_4 at 300 MHz. Ultraviolet spectra were measured in MeOH on a recording spectrophotometer. To
determine the rate of release of the inhibitory exudates, the XAD-4 columns
were collected and replaced every 12 hours. Amounts of the individual inhibitors
exuded in each interval were measured by HPLC.

RESULTS AND DISCUSSION

Distribution and Exudation of Inhibitors

The lettuce seed bioassay (Table 14.1) showed that the inhibitors were quantitatively recovered in the XAD-4 fraction, the aqueous residue fraction was nontoxic. The seed coats were separated from the briefly imbibed seeds. A comparison of exudates collected from the seed coats and the seeds without seed coats,
using HPLC, indicated that more than 80% of the inhibitors were localized in
the seed coat, which contains approximately 7% of these compounds by dry
weight. This high local concentration contributed to the rapid release of these
reserved metabolites during the germination. The rate of exudation reached its
maximum at 24 hours, and by 48 hours from the onset of imbibition, over 90%
of the inhibitors were released into the immediate environment of the germinating seeds.

Identification of Inhibitors

The inhibitory fraction consisted of three major compounds. Vitexin (II) and
isovitexin (III) were identified based on the comparison of retention times and
UV and NMR spectral characteristics of the authentic compounds. Both compounds have been reported in the seeds of several *Vigna* species (Ishikura et al.,
1981). Compound I was a novel *C*-glucosylflavonoid. The pale yellow powder
had an absorption maximum at 289 nm (logE_{289} = 4.2) and slight shoulders at
380 and 410. In dilute acids (e.g., HCl) compound I converted readily and irreversibly to a near 1:1 mixture of vitexin and isovitexin, as indicated by both
HPLC (Figure 14.3) and continuous time recording UV spectra (Figure 14.4).
The ^{1}H NMR of I in methanol-d_4 at 22°C showed typical absorption for sugar
(broad peak, 3.25–4 ppm), and poorly resolved peaks at 4.2, 4.8, 6.1, 6.8, and
7.6 ppm (Figure 14.5a). Under −51°C, these peaks were partially resolved in
multiple peaks (Figure 14.5b), indicating the existence of 3-*C*-glucosyl-
2,4,6,4′-tetrahydroxydibenzoylmethane (Ia), its tautaomers (Ib–d), and the
diastereomers of Ic and Id (Figure 14.6). The ^{13}C NMR also showed complicated
spectra supporting this interpretation. Despite the difficulty of assigning the
NMR peaks owing to the multiplicity and variability of structures in a given
equilibrium, the UV spectrum of I and its acid catalyzed conversion to II and III
firmly established the identity of this *C*-glucoside (Figure 14.6).

TABLE 14.1.
Inhibition of lettuce seed germination and radicle elongation by the exudates collected from the germinating mung bean

Exudate	Concentration g Mung Bean/ml	Percent of Germination ($n = 8$)		Radicle Length (mm, $n = 80$)	
		Means ± SE	Percent of Control[a]	Means ± SE	Percent of Control[a]
Control	0	83.0 ± 5.7	100[a]	8.80 ± 0.45	100[a]
XAD-4 fraction	0.25	69.8 ± 3.2	83[b]	6.50 ± 0.44	66[b]
	0.50	49.3 ± 6.6	59[c]	3.55 ± 0.34	41[c]
	1.00	2.5 ± 2.9	3[d]	1.75 ± 0.24	20[d]
Aqueous residue	0.25	83.0 ± 8.2	100[a]	8.45 ± 0.44	96[a]
Fraction	0.50	86.0 ± 4.1	103[a]	9.10 ± 0.35	101[a]
	1.00	80.0 ± 8.2	96[a]	8.70 ± 0.36	99[a]

[a]Means in a column followed by the same letter are not significantly different at 1% level using Duncan's multiple range test.

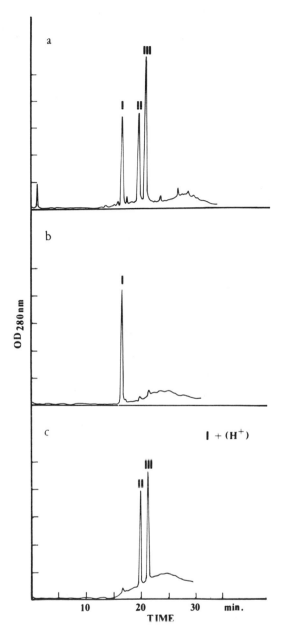

Figure 14.3 (a) HPLC chromatogram of the XAD-4 fraction of exudates collected from the germinating mung bean. (b) Isolated peak I, a novel C-glucosylflavonoid. (c) Peak I was converted to a near 1 : 1 ratio of peaks II and III after treating with dilute HCl.

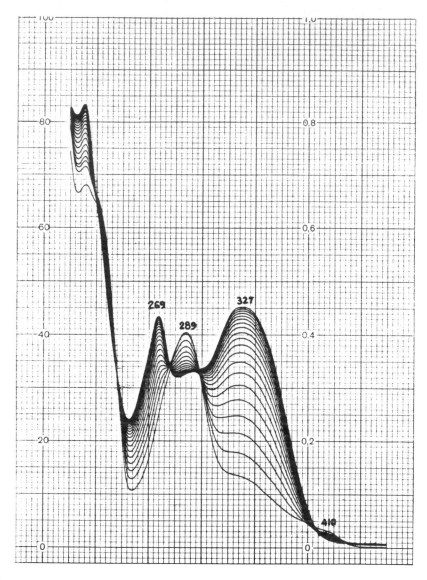

Figure 14.4 Conversion of peak I (λ_{max} = 289 nm) to a mixture of II and III; both II and III have similar UV spectra with λ_{max} at 269 and 372 nm. One drop of 0.02 N HCl was added to a 3-ml MeOH solution of I; spectra were recorded every 2 minutes.

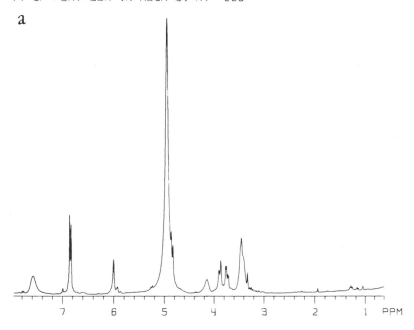

a

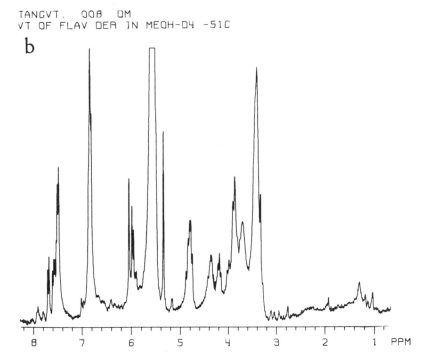

b

Figure 14.5 Variable temperature ^{1}H NMR spectra of peak I in MeOH-d$_4$ at 300 MHz. (a) 22°C. (b) −51°C.

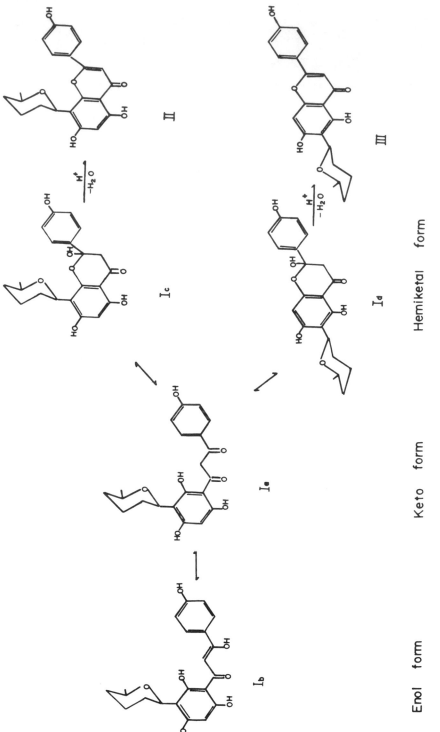

Enol form   Keto form   Hemiketal form

Figure 14.6 Multiple structures (a–d) of peak I, 3-C-glucosyl-2,4,6,4′-tetrahydroxybenzoylmethane (Ia); II, vitexin; III, isovitexin.

238

C-Glucosylflavonoids as Germination and Growth Inhibitors

The XAD-4 fraction effectively inhibited both the germination and seedling growth of lettuce. At a concentration of 1.0 g mung bean equivalent per 1 ml, the radicle length of lettuce seedling was reduced to 20% of the control (Table 14.1). The XAD-4 fraction was also inhibitory to mung bean itself to a lesser extent, radicle growth was reduced to 54% at the 1.0-g equivalent concentration. This autotoxicity is demonstrated in Figure 14.7. Mung bean seedlings in the left container are larger and healthier than those in the right because of the specific removal of flavonoids by the XAD-4 column through continuous trapping. Without the column, however, seedlings in the right container are stunted and the roots appear brownish.

The individual C-glucosides were tested for their inhibitory activities against

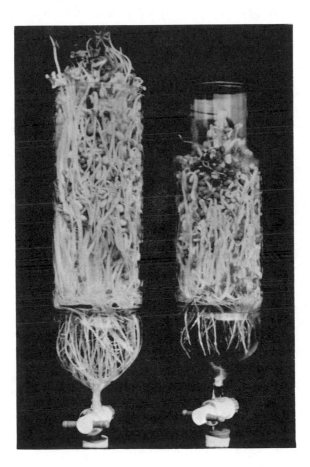

Figure 14.7 Mung bean seedlings germinated in the continuous exudate trapping apparatus (cf. Figure 14.2). The left side had an XAD-4 column attached, but the right side had the column removed during the circulation, indicating autotoxic exudates (i.e., the glucosides) were adsorbed by XAD-4 resins.

lettuce (Table 14.2). All compounds showed weak activities at levels equivalent to 0.25 and 0.50 g. Isovitexin was the most toxic among the three, at 0.5 g mung bean equivalent per 1 ml, radicle growth of lettuce seedlings was reduced to 52% of that of the control at 48 hours.

Allelochemical Spheres of the Germinating Mung Bean (Figure 14.8)

The average dry weight of a mung bean seed used in the present experiment was 50 mg and its volume at 48 hours after imbibition was 110 mm^3 (diameter, approximately 3 mm, calculated as a spherical body). Using these figures and the data in Table 14.1, it is possible to quantitatively correlate the thickness of an allelochemical sphere and its inhibitory activity. For example, since 0.25 g mung bean would produce 1 ml of exudates, which reduced lettuce germination approximately 17% and the growth rate of radicles 34% when compared with those treated with distilled water (see Table 14.1), a single mung bean of 50 mg would afford 200 mm^3 of aqueous exudates around the seed with the same inhibitory power. This layer of aqueous exudate is the allelochemical sphere maintained by this mung bean. The volume (V) of the mung bean plus the allelochemical sphere is 110 mm^3 + 200 mm^3 = 310 mm^3, and the diameter (r) is 4.2 mm as calculated from $V = 4\pi r^3/3$ or $r = (3V/4\pi)^{1/3}$. The thickness of the sphere is therefore 4.2 − 3.0 = 1.2 mm. Similarly, in a 0.7-mm sphere, germination is 41% and radicle elongation is 59% of the control. A very small sphere of 0.4-mm thickness would only permit 3% of the lettuce to germinate, and of those that germinate, the radicle length would be one-fifth of the control.

In reality, the spheres described above may be greatly expanded if inert, space-filling particles such as sand are packed around the seeds. However, water availability directly affects their size and potency. Competing acceptor seeds in the sphere are more severely inhibited when the water supply is reduced or the

TABLE 14.2.

Inhibition of lettuce seed germination and radicle elongation by the isolated flavonoids

Compound	Concentration g Mung Bean/ml	Germination (Percent of Control)	Radicle Length (Percent of Control)
Dibenzoylmethane	0.25	88	77
	0.50	90	75
Vitexin	0.25	92	87
	0.50	84	75
Isovitexin	0.25	90	65
	0.50	92	52

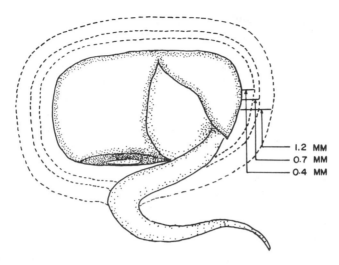

1.2 MM
0.7 MM
0.4 MM

Figure 14.8 Allelochemical spheres of germinating mung bean against cogerminating lettuce seeds. See text for the activities of each sphere.

population density of the donor seeds is increased. Thus both crowding and water reduction enhance allelopathic interactions. An allelochemical sphere is further affected by other environmental factors, including adsorption by soil particles, reaction with chemicals in soil, and degradation by microorganisms. In the case of mung bean, its allelochemical composition is pH dependent; the C-glucosyl-dibenzoylmethane converts readily to'vitexin and isovitexin under mild acidic conditions. The biological significance of this change is not clear.

CONCLUSIONS

As stated in the second edition of *Allelopathy* by E. L. Rice (1984): "Probably one of the most critical points in the life cycle of many plants is seed germination . . . . It seems surprising therefore that little research has been done (on allelopathy and prevention of decay before germination) in the past decade in this important area." Seeds and fruits often contain preexisting secondary metabolites which inhibit microbial activity and seed germination. These compounds have been regarded as agents preventing microbial infection (see Chapter 8, Rice, 1984) and possibly maintaining dormancy (Bewley and Black, 1982). While the importance of these inhibitors as preservatives requires further research, their role as allelochemicals against cogerminating seeds has also been largely neglected, and their possible significance in evolution has not been explored.

Based on the data obtained, we introduced and established the concept of allelochemical spheres. Application of this concept should not be limited to the germinating seeds. The plant root system, for example, is a continuous source of bioactive metabolites to the root–soil interface; therefore, the rhizosphere is an allelochemical sphere to the susceptible acceptors in that environment. Unfortunately, it is difficult, if not impossible at present, to describe this sphere quantitatively as we have demonstrated here with the germinating mung bean.

As mentioned previously, the allelochemical sphere changes in size and nature according to the chemical–physical environment. In addition, the sphere is also acceptor dependent; to an organism (i.e., microorganisms, plants, nematodes, etc.) susceptible to these metabolites, the size of the sphere is larger, and the physiological impact greater than to a less susceptible organism.

The allelochemicals isolated from the germinating mung bean are C-glucosylflavonoids. They are compounds effectively adsorbed by XAD-4 and are partially soluble in water. Their inhibitory activity is far from striking; each compound caused approximately 25% reduction of the radicle elongation of lettuce at 10^{-3} M levels. Nevertheless, the massive (approximately 8.8 mg/g seeds) and rapid (within 48 hours) release of these partially soluble inhibitors helps maintain an effective allelochemical sphere in the agar medium (Figure 14.1) and under certain natural conditions.

ACKNOWLEDGMENTS

Supported in part by USDA Agreement No. 83-CRSR-2-2293. We thank R. S. H. Liu for helpful discussions; A. C. Waiss, Jr. and R. M. Horowitz for samples of vitexin and isovitexin; the NMR Facility at the Chemistry Department, University of Hawaii; and the Regional NMR Facility at the California Institute of Technology for use of their NMR spectrometer.

REFERENCES

Abdel Samad, I. M. and R. S. Pearce. 1978. *J. Exp. Bot.* 29:1471.

Bewley, J. D. and M. Black, 1982. *Physiology and Biochemistry of Seeds. 2. Viability, Dormancy and Environment Control.* Springer-Verlag, New York.

Elmore, C. D. 1980. *Weed Sci.* 28:658.

Gressel, J. B. and L. G. Holm. 1964. *Weed Res.* 4:44.

Ishikura, N., M. Iwata, and S. Miyazaki. 1981. *Bot. Mag. Tokyo.* 94:197.

Junttia, O. 1975. *Physiol. Plant.* 33:22.

Rice, E. L. 1984. *Allelopathy.* 2nd ed. Academic, Orlando, FL.

Simon, E. G. and R. M. Kaja Harun. 1972. *J. Exp. Bot.* 23:1076.

Tang, C. S. and C. C. Young. 1982. *Plant Physiol.* 69:155.

Whittaker, R. H. and P. P. Feeny. 1971. *Science.* 171:757.

15

CAFFEINE AUTOTOXICITY IN *COFFEA ARABICA* L.

GEORGE R. WALLER

*Department of Biochemistry, Oklahoma State University,
Stillwater, Oklahoma*

DURGA KUMARI

*Department of Forest Products, University of Minnesota,
St. Paul, Minnesota.*

JACOB FRIEDMAN

*Department of Botany, The George S. Wise Faculty of Life Sciences,
Tel Aviv University, Tel Aviv, Israel*

NURIT FRIEDMAN

*Department of Botany, The George S. Wise Faculty of Life Sciences,
Tel Aviv University, Tel Aviv, Israel*

CHANG-HUNG CHOU

*Institute of Botany, Academia Sinica,
Taipei, Taiwan, Republic of China*

Coffee plantations (Figure 15.1) are managed in different ways, ranging from intensive cultivation to maintenance as natural forest-like ecosystems, but little or no attention has been given to adjusting their management for the allelopathic effects that are produced by secondary metabolites such as caffeine (Figure 15.2). Caffeine is present to the greatest extent in fruits ($1.5 \pm 0.3\%$) in *Coffea arabica*, the plant from which most commercial coffee is made. Other compounds present (Chou and Waller, 1980a,b), such as theophylline, theo-

Figure 15.1 (*a*) Research workers in the field at a coffee plantation in Coatepec, Veracruz. (*b*) Closeup of a coffee plant in Coatepec, Veracruz. (Courtesy A. L. Anaya.)

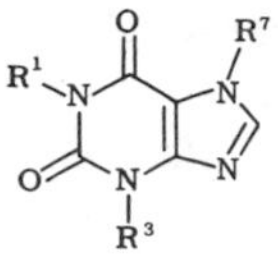

	Trivial name	R^1	R^3	R^7
Xanthine		H	H	H
1,3-Dimethylxanthine	Theophylline	CH_3	CH_3	H
3,7-Dimethylxanthine	Theobromine	H	CH_3	CH_3
1,7-Dimethylxanthine	Paraxanthine	CH_3	H	CH_3
1,3,7-Trimethylxanthine	Caffeine	CH_3	CH_3	CH_3

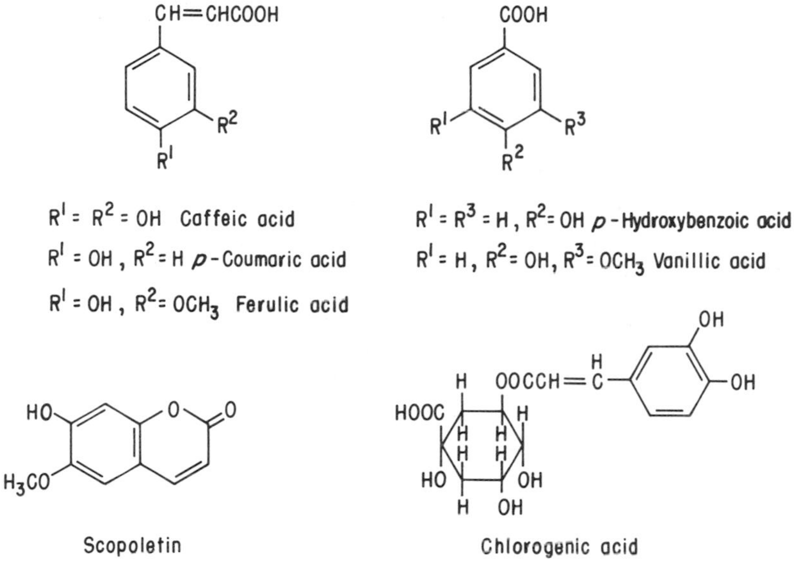

$R^1 = R^2 = OH$ Caffeic acid

$R^1 = OH$, $R^2 = H$ p-Coumaric acid

$R^1 = OH$, $R^2 = OCH_3$ Ferulic acid

$R^1 = R^3 = H$, $R^2 = OH$ p-Hydroxybenzoic acid

$R^1 = H$, $R^2 = OH$, $R^3 = OCH_3$ Vanillic acid

Scopoletin

Chlorogenic acid

Figure 15.2 The chemical structure of allelochemical compounds present in coffee plant parts.

bromine, paraxanthine, scopoletin, and caffeic, coumaric, ferulic, p-hydroxy-benzoic, vanillic, and chlorogenic acids, are also known as allelochemicals (Figure 15.2).

The allelopathic effects of *Coffea arabica* trees have been studied by Anaya-Lang and Del Amo (1978), Anaya et al. (1978, 1982, 1985), Chou and Waller (1980a,b), Ramos et al. (1983), and Friedman and Waller (1983a,b); however, much research remains to be done. Anaya's group performed research on coffee plantations in Coatepec, Veracruz, Mexico which are characterized by the presence of shade trees that resemble the deciduous temperate forests, with three well-defined strata: the herbaceous layer, the shrub layer represented by coffee plants, and the tree layer. The main objective of this group was to assess the

allelopathic interactions among the species present in this community, in particular the coffee plants. Anaya-Lang et al. (1978) reported that aqueous extracts
of fresh and dried roots of coffee plants had an allelopathic effect on weeds associated with coffee plantations. Four varieties of coffee, all *C. arabica*, inhibited
the radicle growth of most test weeds, but there were varietal differences (Anaya
et al., 1982). Inhibition was also produced by suspension of soils beneath coffee
trees (Figure 15.3).

One of the weed species tested was a member of the Commelinaceae family.
The coffee plantations in Coatepec have a sparse to dense ground cover of herbaceous vegetation dominated by Gramineae and Compositae for the sunny plantings and by Commelinaceae in the shade (Figure 15.4). Anaya's group (Ramos
et al., 1983) reported on leachates from fresh, air-dried, and oven-dried material
and litter from some Commelinaceae, collected at different times of the year,
which reduced the radicle length of *Brassica campestris*, *Bidens pilosa*, and *Rumex* sp.

The scarcity of weeds around many coffee trees is due at least in part to a slow
leaching of caffeine from the tree canopy as well as from litter and lost coffee
beans. Such beans lose their viability after about 6 months and then release the
caffeine seven times as fast as viable seeds (Friedman and Waller, 1983b). The
coffee and shade tree litter in combination with that from weeds should produce
decomposed material to supply the soil with enough toxins to inhibit the germi-

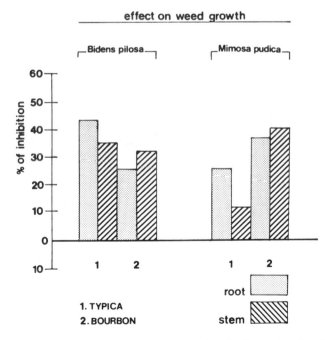

Figure 15.3 Effect on weed growth of aqueous suspension of soil beneath coffee plants. (Anaya et
al, 1982).

Figure 15.4   Commelinas dominating and covering the soil in a shade coffee plantation in Coate-pec, Veracruz. (Courtesy of A. L. Anaya.)

nation and growth of various other plant species as well as to be autotoxic to coffee trees.

The emission of inhibitors can occur by leaching, root exudation, volatiliza-tion, and decomposition of plants, and inhibition can occur only after uptake of allelochemicals, usually by the roots. Many allelopathic plants are perennials whose roots may often grow far below the upper soil layer where allelochemicals usually accumulate. However, when such chemicals are released from roots they poison the same or adjacent roots. Root tips are often the susceptible zone; thus the question arises of what distance between two zones can prevent autotoxic hazards. In coffee, where roots are highly susceptible to caffeine and are also close to the soil surface, the degeneration known to occur in old plantations may be (at least in part) due to autointoxication. The raison d'etre for caffeine pro-duction in coffee plants is protection against predators, regulations of plant spacing, and suppression of weeds. These probably override the cumulative neg-ative effects of caffeine, though it eventually reduces the life-span of the tree (Friedman and Waller, 1985).

Rizvi et al. (1980a) reported that seed extracts of *Coffea arabica* were most inhibitory toward the germination of *Amaranthus spinosus* (pigweed) out of 50 species of weeds and crops tested. The active principle of the seed extract was identified as caffeine. It completely inhibited the germination of *A. spinosus* at a concentration of 1200 ppm in water, but exerted no adverse effect on the germi-nation and growth of *Phaseolus mungo* (mung bean) (Rizvi et al., 1981). Seeds of *A. spinosus* treated with caffeine (sublethal dose) showed a 30% decrease in amylase activity. Kinetic studies of substrate saturation and K_m values indicated no effect on its catalytic property. Rizvi and Rizvi (1983) suggested that inhibi-tion of seed germination of *A. spinosus* was caused by reduced synthesis of amyl-ase and this reduction was not GA^3 mediated.

Results from the same group showed that caffeine has fungicidal activity against *Helminthosporium maydis*, a parasitic pathogen of corn, at a concentration of 1500 ppm (Rizvi et al., 1980a–c) and against *Saprolegnia ferax*, a fish pathogen, at 100 ppm (Prabhuji et al., 1983). They also demonstrated an insect-sterilizing property of caffeine for *Callosobruchus chinensis*, a stored-grain pest of beans (Rizvi et al., 1980c.)

Another plant–insect interaction involving caffeine (and other purine alkaloids and some synthetic methylxanthines) was shown by Nathanson (1984), who used it to inhibit the feeding by larvae of *Manduca sexta* (tobacco hornworm), *Tenebrio* spp. (mealworm), *Vanessa cardui* (butterfly), nymphs of *Ancopeltus fasciatus* (milkweed bug), and adult *Tribolium confusum* and *T. castaneum* (flour beetles). Caffeine was toxic at the concentration that occurs in *C. arabica* (0.4%) in the whole plant; but all insects were affected by methylxanthines, although the ED_{50} ranged from 0.007 to 3%. There was an inhibition of phosphodiesterase (PDE) activity and an increase in intracellular cyclic adenosine monophosphate (cyclic AMP) in the insects; Duffus and Duffus (1969) had previously shown that theophylline inhibits PDE activity in barley endosperm. At lower concentrations Nathanson demonstrated that methylxanthines are synergistic with certain other insecticides (e.g., chlorodimeform) and they are known to activate adenylate in insects. The natural insecticidal effects of caffeine occur through an increase in concentration of cyclic AMP in tissue and an inhibition of PDE activity.

Our interest in the metabolism of secondary metabolites has extended over the last quarter of a century. We believe that some secondary metabolites play a role as allelochemicals, and that these biochemical compounds may be essential to the survival of the plants. Studies on the metabolism of caffeine began in our laboratory in 1977 (Waller, 1983; Waller et al., 1983; Suzuki and Waller, 1984a–c) and have been extended to include the allelochemical effects of coffee plants and coffee soils (Chou and Waller 1980a,b; Waller et al., 1982; Friedman and Waller, 1983a,b; 1985).

EXPERIMENTS

Plant Material Used for the Initial Determination of Phytotoxic Activity

The fallen leaves of 2-year-old coffee plants were obtained from a greenhouse of the Department of Horticulture, Agricultural Experiment Station, Oklahoma State University, Stillwater, and young seedlings of about 10 cm in length were collected from the Kew Gardens, United Kingdom in 1979. This plant material was dried at room temperature and stored in a desiccator before use.

To estimate the phytotoxicity associated with the coffee tissues, a 5% aqueous extract of leaves or roots was bioassayed with lettuce (*Lactuca sativa* var. Great

Lakes), rye grass (*Lolium multiflorum*), and fescue (*Festuca* sp.) as test materials, using the techniques described by Chou and Lin (1976).

The identity of the allelopathic compounds present in coffee tissues was established by extraction, TLC, paper chromatography, spectrophotometric assay, and GC/MS.

Germination and Growth of Coffee Seedlings

Seeds of *C. arabica* cv. Bourbon, 3–6 months after collection, containing 32 ± 3% water (on a dry-weight basis), were obtained from Conafruit, Jalapa, Mexico. As described by Valio (1976), endocarp-free seeds were allowed to germinate at 27 ± 0.1°C in darkness. To furnish optimal conditions for germination and early growth, seeds were initially soaked and shaken in 0–20 mM aqueous solutions of caffeine, pH 5.2–6.0, for 48 hours at the same temperature and were later washed and set on the margins in the middle of four layers of strips (10 × 50 cm) of Whatman chromatographic paper, 0.3 mm thick. These were rolled and placed in glass tanks (30 × 30 × 10 cm), each with a different solution of caffeine. To avoid gross changes in caffeine concentration by either uptake or evaporation of water, 300 ml of solution per tank were used and the tanks were sealed. In each tank 150 seeds, divided into 3 groups (replicas) of 50 each, were set. The effect of exogenous theophylline (1,3-dimethylxanthine) was also similarly tested. Lengths of rootlet and hypocotyl were periodically measured. Extraction and quantitative evaluation of caffeine were conducted by the method of Chou and Waller (1980b), slightly modified. Caffeine content was determined in each of the five seedlings weekly, and was measured separately for the rootlet, hypocotyl, cotyledons, endosperm, and the germination medium; measurements were conducted during the course of germination, 6 weeks after seed wetting.

Coffee seeds are known to lose viability some months after collection, as already noted. The rate of release of caffeine was determined both in new, viable seeds of coffee (*C. arabica* cv. Bourbon) 3 months after their collection (in Xalapa, Mexico), having 38 ± 4% water (on a dry-weight basis), and showing 87 ± 6% germination during 10 days, as well as in one-year-old dead seeds of the same variety and origin, with 18 ± 8% water on a dry-weight basis. Seeds of both types contained 0.8% caffeine. Twenty-five seeds of each type were immersed in 50 ml of glass-distilled water, then the mixture was shaken for 24 hours at 27 ± 0.5°C. The leachate was replaced by distilled water every 24 hours for a period of 12 days, and the caffeine content of extracts was determined according to a slightly modified method of Chou and Waller (1980b).

The effect of caffeine on cell division was followed in cells of root tips of *C. arabica* L. cv. Bourbon seedlings. These were allowed to develop on filter paper in distilled water for 4 weeks at 27°C in darkness, and then were placed on filter paper immersed in 10 mM caffeine in a tightly covered glass tank (30 × 30 × 9 cm). Root tips were removed, fixed, and prepared for microscopic examination by Warmake's method (Warmake, 1935) with some modification.

Caffeine Addition to Coffee Seedlings

Seedlings 6 weeks old, initially grown in distilled water in the darkness, were transferred into caffeine solutions (0, 10, 20, 30, 40, and 50 mM) and set on top of rolled filter paper 15 cm long. These rolls were placed in sealed glass tanks (20 × 20 × 8 cm) with their lower part (1 cm) dipping into one of the various caffeine solutions. Rootlets were aerated by this method. Tanks with the seedlings were kept in darkness at 27°C for another 20 days and caffeine was determined separately for the roots and the shoots (hypocotyl and cotyledons).

Recovery of Caffeine from Coffee Soils from Mexican Plantations

Soil samples were randomly collected in coffee plantations near Xalapa at various depths (0–10, 11–20, 21–30, and 31–40 cm), some near the base of the tree under the canopy and others about 1 m apart around the circumference of this canopy. Samples were collected from several locations, including coffee plantations that had trees aged 6, 25, 40, and 75 years. Samples were air-dried and passed through a sieve (2 × 2 mm) to remove pebbles and fragments of coffee tree roots. The methods used to determine caffeine are the standard ones employed in this laboratory (Waller et al., 1983); they involve reflux with 0.012 N H_2SO_4, extraction with chloroform, TLC, spectrometric determination of quantity, and finally confirmation of identity by mass spectrometry. A modification of the method proposed by Lailach et al. (1968) and Wang (personal communication, 1982), which employed 10% hydrochloric-hydrofluoric acid (5% HCl and 5% HF) instead of the H_2SO_4, was also used. Both of these techniques were adopted for analysis of the soil from the base of a 25-year-old tree and immediately outside the canopy (1 m away). The HCl/HF method was only partially successful in breaking the binding of caffeine in the soil.

Soil Analysis from Stillwater-Grown *C. arabica* Trees

Soil from around the roots of each of the 20 coffee trees sacrificed (see below) was stored at −18°C. One kilogram of soil was extracted with 85% methanol for 24 hours; this treatment was then repeated. The extracts were combined, filtered through Whatman 42 paper, and evaporated until an aqueous solution remained. The aqueous residue was extracted successively with hexane, chloroform, and ethyl acetate. These extracts were evaporated to dryness and stored at 10°C until analyzed.

Another method of soil extraction used was as follows: 1 kg of soil was soaked with 1.5 L of HF/HCl solution (1.3% HF and 0.9% HCl) for 24 hours; the mixture was then refluxed for 1 hour and filtered through a coarse glass filter. The filtrate was extracted with chloroform three times and the extract was reduced to dryness with a rotary evaporator and stored at −18°C until analysis.

Leachates of Stillwater-Grown *C. arabica* Trees

Twenty trees that had borne fruit until the week before were killed by cutting off the aerial part of the plant, and the roots were carefully separated from the soil by washing. A leaching apparatus consisted of wire netting of 1/4-in. squares spread out over cinder blocks spaced about 6 ft apart (Figure 15.5). Pans were placed underneath the apparatus to collect rainfall. The top parts of the coffee plants and the roots were put on the netting and set out to dry for 28 days in normal atmosphere in Stillwater, Oklahoma State University Horticultural Farm during September and October of 1982. On the 29th day rain started; after each rainfall the leachates of the top part of the coffee plant and the roots were collected separately within 2–4 hours. A part was evaporated to dryness on a steam hot plate (maximum temperature, 60–80°C). The dry leachate residues were stored at 0°C before use. The remaining fresh leachate was extracted successively with hexane, chloroform, and ethyl acetate. The extracts were reduced to dryness on a rotary evaporator before analysis.

Gas Chromatograph/Mass Spectrometer Analysis

A Hewlett-Packard Model No. 5992-B gas chromatograph/mass spectrometer/ data system (GC/MS/DS) equipped with a fused silica column internally coated with SE-54 was operated at a flow rate of 2 ml He/min; the column temperature was held at 40°C for 5 min, then raised to 320°C at 4°/min; the spectrometer was operated at a source temperature of 250°C and 70 eV. An 0.1–0.2-μl sample was injected. Background was subtracted for each sample.

Figure 15.5 *Coffea arabica* plants arranged for natural leaching in Stillwater, OK. Right: shoots. Left: roots. *Note.* Pans were used to collect the rainfall and leachate.

Column Chromatography

A silica gel column, 2.5 × 60 cm, was used. The compounds were eluted by consecutive use of hexane (600 ml), chloroform (600 ml), and ethyl acetate (600 ml). The compounds from the leaf extract were eluted from the column in the following order: caffeine, scopoletin, theobromine, theophylline, and paraxanthine; only caffeine was eluted from the root extract.

RESULTS

Allelochemical Effects of Caffeine and Other Purine Alkaloids Isolated From Coffee Plants

The aqueous extracts of the fallen leaves and roots exhibited remarkable phytotoxic effects in seed germination and radicle growth of all species tested (Figure 15.6). Results expressed as percent of inhibition over the distilled water control are given in Figure 15.6, which shows that the inhibition of both seed germination and radicle growth is lower at lower concentrations of extracts, as might be expected. However, significant inhibition of radicle growth was still caused by a 1% extract of young seedlings. Furthermore, the ether-soluble fraction of an aqueous extract of fallen leaves was chromatographed and materials in the chromatograms were bioassayed. The material in all chromatographic spots, except in the segment from R_f 0.00 to 0.12, caused significant inhibition of radicle growth.

To clarify the phytotoxic effect of the purine alkaloids, aqueous solutions of each compound in a series of concentrations (100, 200, 300, and 400 ppm) were

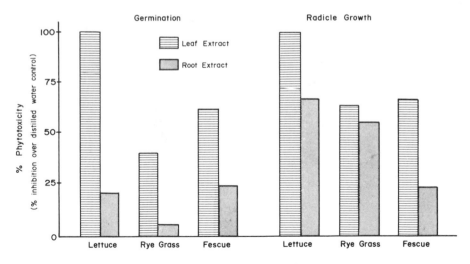

Figure 15.6 Inhibiting effects of 1% aqueous extracts for coffee leaves and roots on seed germination and radicle growth of lettuce, rye grass, and fescue.

bioassayed with lettuce seeds. The tested compounds exhibited significant phytotoxicity, as manifested by radicle growth, at concentrations below 100 ppm (Table 15.1). The phytotoxicity of scopoletin and chlorogenic, caffeic, ferulic, vanillic, p-coumaric, and p-hydroxybenzoic acids for lettuce has been previously reported (Fay and Duke, 1977; McPherson et al., 1971; Rice, 1984).

Coffee Seeds as an Allelopathic Agent

Since the soil surrounding coffee trees contains a bank of seeds of various ages, Friedman and Waller (1983b) studied the effect of seed senescence on the rate of emanation of germination inhibitors. The available data had shown that leakage of ionic solutes from germinating seeds increases with seed age, thereby increasing the electrical conductivity of the medium (Harman and Granett, 1972; Parish and Leopold, 1978). Young, viable seeds often reabsorb sugars, amino acids, and other metabolites by active uptake (Osborne, 1981), but as senescence progresses, the loss of cytoplasmic solutes is correspondingly greater and the extent of subsequent uptake is lower (Roberts, 1972). Because of these findings, as well as the limited data dealing with the escape of secondary metabolites from seeds, we tested the hypothesis that the escape of a secondary metabolite that also inhibits germination will increase with seed age. Experiments with seed types were replicated five times, as described above. The results on rates of release of caffeine are shown in Figure 15.7. The average leakage of caffeine is expressed as a percentage of the amount initially present in the seed. During a period of 24 hours, the viable seeds showed an average rate of caffeine efflux comprising

TABLE 15.1.
Inhibitory effects of four alkaloids in aqueous solution of radicle growth of lettuce after 48 hours at 25°C

Compound	Phytotoxicity (Percent Inhibition Compared to Distilled Water) at Four Concentrations (ppm)				Statistical Value (F)
	100	200	300	400	
Distilled water (control)	0	0	0	0	
Caffeine	54	78	90	84	101.3[a]
Theobromine	41	51	32	64	10.2[a]
Theophylline	58	69	81	76	76.1[a]
Paraxanthine	62	66	66	85	40.9[a]

Source. Chou and Waller (1980a).
[a]Statistical significance below 1% ($F + 6.0$) level. All compounds significant inhibition below 1% level at 100 ppm or below.

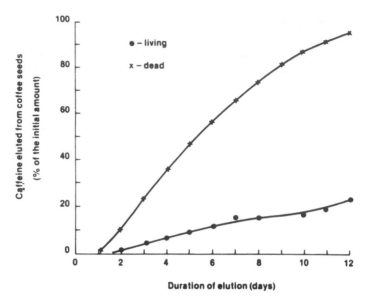

Figure 15.7   Average rate of caffeine release from young viable seeds (●) and from old dead seeds (x) of coffee (*Coffea arabica* cv. Bourbon) shaken in distilled water. (Friedman and Waller, 1983b.)

1.8% of the total caffeine; the average rate of loss from the old, nonviable seeds was more than five times that value.

Our incidental observations in coffee plantations in Xalapa, Mexico have established that only a small number of lost, uncollected coffee seeds germinate. Thus we conclude that in nature most of the caffeine in seeds of coffee is discharged into the soil around the tree within 1–2 yr after seed dispersal. Rate of caffeine efflux will surely increase as the last seed crop ages. Hence, if the allelopathic potential of seeds is to be properly evaluated, both old nonviable and young seeds in the soil should be considered.

Avoidance of Caffeine Autotoxicity by Coffee Seedlings

Phytotoxins with autotoxic activity are often present in the outer parts of seeds or diaspores. If these toxins are not sufficiently leached out by rainfall or metabolized by soil microflora, they hold germination in check and ensure that this will occur only after the amounts of rainfall are sufficient for the establishment of seedlings (Friedman and Waller, 1983b). Other toxins, however, are stored within the seeds and are difficult to leach; these are natural protectants (e.g., caffeine in *C. arabica* and strychnine in *Strychnos nux vomica*). When either of these compounds is applied exogenously to the seed that stores it, germination is inhibited, even when the concentration of the exogenous inhibitor is much lower than that found endogenously in the imbibed seed (Evenari, 1949). This suggests that such seeds have some means to avoid the effect of their own phytotoxins.

Fowden and Lea (1979) describe mechanisms by which plants avoid self-poisoning by their secondary metabolites, especially by nonprotein amino acids.

When seeds of coffee were allowed to germinate in aqueous solutions of caffeine of various concentrations, elongation of hypocotyls was reduced in all cases and growth of rootlets was almost completely inhibited by 10 mM caffeine (Figure 15.8, Table 15.2). Root tips darkened and 4–5 days later deteriorated. A similar although milder inhibition was caused by theophylline. Suppression of growth occurred, although the concentration of the endogenous caffeine in the imbibed coffee seeds was 40–60 mM. This indicated that embryos of germinating coffee seeds must be able to avoid autotoxic hazards from their endogenous caffeine.

Inhibitors may be sequestered in specific tissues, dead or alive. For example, in the fruit of *Ammi majus* (Bishop's weed, Umbelliferae), 8-methoxypsoralen is stored mainly in the outer dead layers of the fruit (Friedman et al., 1982). Autoinhibition occurs only after the inner fruit shell is pierced, that is, an impenetrable barrier keeps the embryo away from autotoxic agents. Germinating coffee seeds contain an average of 40 mM caffeine. However, mitosis in root tips is arrested by 10 mM caffeine, in a manner similar to the response of roots of onion [(Kihlman, 1949) (Figure 15.9)]. The caffeine of the seeds (1–2% dry weight) is stored within the endosperm and the dormant embryo is nearly devoid of the

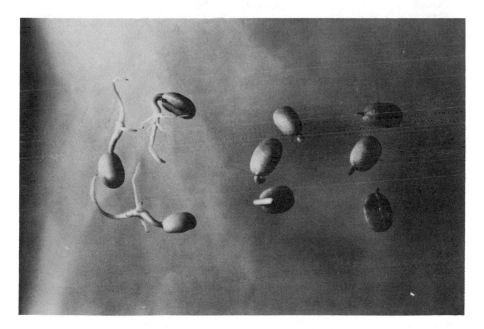

Figure 15.8 Seedlings of *Coffea arabica* L. cv. Bourbon germinated in distilled water (left) and in 4.25 mM (center) and 25 mM (right) caffeine, 3 weeks after wetting, in the dark. In all caffeine-treated seeds, emergence of the hypocotyl did occur but rootlet growth was arrested (× 2.5 natural size). (Friedman and Waller, 1983a.)

TABLE 15.2.

Growth of rootlet and of hypocotyl of 4-week-old seedlings of coffee (*Coffea arabica* L. cv. Bourbon) germinated in various concentrations (5, 10, 20 mM) of caffeine or theophylline[a]

		Concentration (mM)		
		5	10	20
Caffeine	Rootlet	72 ± 8	3 ± 2	3 ± 3
	Hypocotyl	152 ± 7	61 ± 19	39 ± 12
Theophylline	Rootlet	59 ± 7	27 ± 4	14 ± 7
	Hypocotyl	158 ± 4	48 ± 11	54 ± 6

Source. Friedman and Waller (1983a).
[a]Data presented as percentage of growth (length in mm) of the control (distilled water); average of 50 seedlings.

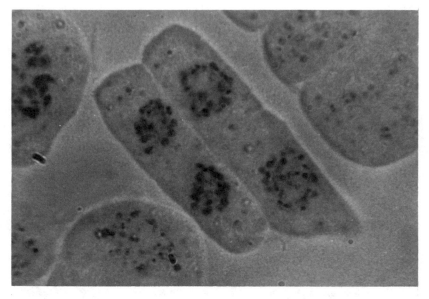

Figure 15.9 Inhibition of cell-plate formation in root tips of coffee (*Coffea arabica* L. cv. Bourbon) seedlings after exposure to 10 mM caffeine for 248 hours (× 1200). (Friedman and Waller, 1983a.)

alkaloid. During the first stages of germination no cell division occurs in the root tip; it starts only after the root tip is pushed outside (1–3 mm away from the caffeine-rich endosperm) as a result of cell expansion and elongation of the hypocotyl. During this stage the cotyledons of the embryo are about 2 mm in diameter and mitosis proceeds until the third week of germination. At the end of this stage caffeine disappears from the endosperm and is accumulated within the cotyledons (Figure 15.10) when mitosis is arrested. At the final stage of germination most of the caffeine of the seedling is in cotyledons (85%) and the hypocotyl (13%) with only a small amount in the roots. Studies by Baumann and Gabriel (1984) using sterile conditions did not support the loss of 13% in the hypocotyl, presumably because our system was nonsterile. Caffeine is thought to be sequestered from the sites where mitosis occurs, but how this is effected is not yet known.

Despite starvation, caffeine is preserved, which indicates that the alkaloid is not easily utilized either for energy or as a nitrogen source. Uptake of the alkaloid into the cotyledons, despite its potential hazard, suggests that caffeine is an important protective agent for the plant.

Uptake and Translocation of Exogenous Caffeine by Coffee Seedlings

Caffeine exogenously applied to seedlings of coffee remarkably inhibited cell division and growth of rootlet, so it was of interest to follow the fate of applied caffeine in both the rootlet and the shoot. Data presented in Figure 15.11 show a remarkable increase in caffeine concentration in the shoots, but no such increase in the rootlets. A slight increase in concentration of endogenous caffeine in the roots occurred for the 40- or 50-mM solutions and is probably the result of a considerable amount of dead parts of root tips that appeared at these concentrations.

To ascertain that exogenous caffeine was taken up by the plants and that the increase of endogenous caffeine was not due to increased biosynthesis induced by the exogenous alkaloid, determinations of caffeine in the exogenous solutions

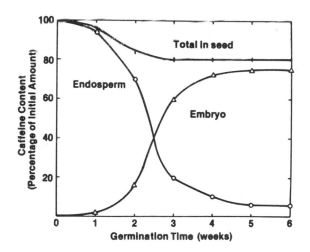

Figure 15.10 Caffeine content in the endosperm and in the embryo during the germination of seeds of coffee (*Coffea arabica* L. cv. Bourbon). (Friedman and Waller, 1983a.)

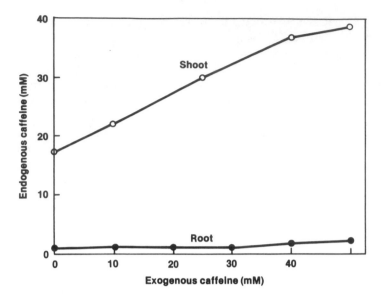

Figure 15.11   Comparison of exogenous with endogenous caffeine in *Coffea arabica* L. cv. Bourbon. Caffeine was determined weekly in each of five seedlings and in the rootlet. (Waller et al., 1983a.)

were conducted. These analyses showed decreased amounts relative to those observed in the seedlings. Also, autoradiograms made of young seedlings after rootlets were immersed in [2-¹⁴C]caffeine for 24 hours showed uptake and translocation to all parts of the seedling. It is evident that coffee rootlets can absorb caffeine and that the absorbed alkaloid is translocated into the shoots. This occurs even though such shoots grown in distilled water contain 17–18 mM of endogenous caffeine. Accumulation of caffeine in the shoot, against the gradient concentration, could be the plant's means to remove the alkaloid from the susceptible root tip.

Isolation of Caffeine in Soils from Coffee Plantations

The nature of the interaction between biochemical molecules and soil matrices depends upon the chemistry of the compounds and the properties of the matrix near the soil surface. The interaction between plants and soil microorganisms is less complicated but little investigated; the allelochemical effect is governed by the soil environment and climatic conditions, which vary by geographical regions. We had not previously worked with soil before collecting samples in Mexico, around Xalapa, the center of coffee production in that country. To gain an understanding of how the caffeine is bound in the soil, experiments on extracting the alkaloid from it were performed.

Attempts to extract endogenous caffeine by the method used for fruit or seed were not successful when applied to soil (Figure 15.12a). However, soil treated

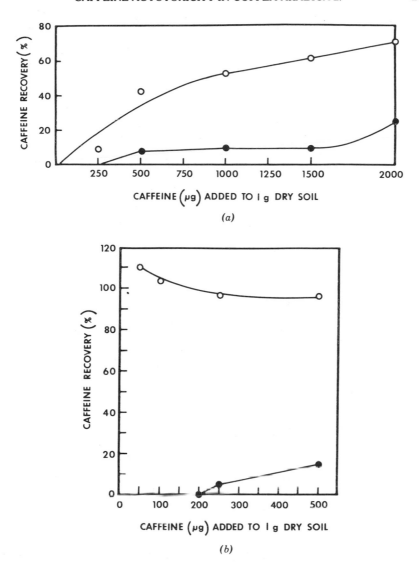

Figure 15.12 Caffeine recovery (%) from soil samples previously treated with different amounts of caffeine (a) By H$_2$SO$_4$ method: soil under coffee canopy, -O-; soil 2 m away from coffee tree, -●-; (b) By HF-HCl method: soil under coffee canopy, -O-; soil 2 m away from coffee tree -●-. (Waller et al., 1983.)

with caffeine in varying amounts and then incubated for 24–48 hours at 28°C gave higher recoveries of caffeine from canopy soil than the samples outside the canopy. Treatment of canopy soil with refluxing HF and HCl for 20 minutes gave a recovery of nearly 100% (Figure 15.12b). Again, no endogenous caffeine was found by this method, but caffeine should be used to ascertain what the survival of the alkaloid in the Mexican soil may be. It is remarkable that soil from under coffee trees allows complete recovery of added caffeine. An understanding of the differences between the soils from the canopy and outside the canopy is needed. There is a possibility that soil microbes in canopy soil have adapted to metabolizing caffeine, rather than its being tightly bound, but this does not seem likely.

These preliminary analytical data on the soils suggest that the binding of caffeine is very complicated. They imply that the canopy soil already contains much caffeine while the outside soil is far from being saturated. The term "saturation" refers to the amount of caffeine bound to the soil matrices (by covalent or ionic bonds or sorption) and to humic and fulvic acids. For the sake of simplicity, two different groups of sites that immobilize caffeine can be recognized, those that have a very strong binding capacity (covalent or strong ionic bonding) and those that have a weaker binding capacity (weak ionic and sorption bonding). In canopy soil under old trees the first sites are regarded as saturated with caffeine (and its derivatives), but these alkaloids are not removable by our methods; any added amount of alkaloids is more weakly held on the second group of sites and therefore is recoverable by our extraction. Of course the soil outside the canopy of coffee trees has the same types of sites of bonding. When the added caffeine is so strongly bound (by the first mechanisms proposed) that it is very difficult to remove by the HCl/HF procedure, we can calculate how much caffeine a soil might take up before the second type of binding is reached. This value is approximately 2–4 mg/g. We suggest that this level of caffeine in soil might produce the "tired soil." It is not clear what level of caffeine should be regarded as producing "tired soil." In fact, most trees are developed from plants that are first grown from seed in nursery beds and transplanted to the field after a year, which has a significant amount of caffeine present in the soil.

Determination of Natural Products from Soil and Leachates of Coffee Trees Grown in Stillwater

A study of naturally occurring compounds from coffee tree leachates (Figure 15.6) and from soil yielded a variety of compounds, some of which are known to have allelopathic effects. Analysis by GC/MS/DS showed the presence of caffeine, N-phenyl-1-naphthylamine, 1-methylnaphthalene, naphthalene, n-alkanes from C_{12} to C_{28}, furfural, 1,3,5-trimethylbenzene, an antioxidant (2,6-di-t-butyl-1,4-benzoquinone), and 3-methyl-2,4-pentanedione in the leachates from the tops of the plant, as shown in Table 15.3. Table 15.4 illustrates the results of GC/MS/DS analysis of leachates of the roots. Again the most predominant compounds were caffeine, N-phenyl-1-naphthylamine, and some higher molecular weight hydrocarbons. The chloroform extracts from soil that had been used for growing coffee trees in Stillwater contained methyl esters of myristic, pentadecanoic, palmitic, heptadecanoic, stearic, nonadecanoic, eicosanoic, docosanoic, and dehydroabietic acids, as shown in Figure 15.13. The composition of a hexane extract of 85% methanol extracts from the same soil is shown in Figure 15.14; it contained vanillin, 2-hydroxybenzothiazole, caffeine, and several high molecular weight hydrocarbons. From these preliminary data it is possible to conclude that biochemicals can reach the soil from either the roots or the

TABLE 15.3.
Some compounds occurring in *C. arabica* leaf leachate

Compound	Molecular Weight	Retention Time from GC/MS (R_t, min)
Furfural	96	3.4
1,3,5-Trimethylbenzene	120	4.2
Dodecane	170	5.9
Naphthalene	128	6.2
Tridecane	184	7.2
3-Methyl-2,4-pentanedione	114	7.5
1-Methylnaphthalene	142	7.8
Tetradecane	198	8.6
Pentadecane	212	9.9
2,6-Di-*t*-butyl-1,4-benzoquinone	220	10.2
Hexadecane	226	11.2
Heptadecane	240	12.4
Octadecane	254	13.6
Caffeine	194	15.7–16.9
(the major compound)		
Nonadecane	268	17.0
Eicosane	282	18.3
Methyleicosane	296	19.6
Docosane	310	20.9
N-Phenyl-1-naphthylamine	219	21.4
Tricosane	324	22.3
Tetracosane	338	23.6
Unknown hydrocarbon	352	24.1
(a major compound)		
Pentacosane	352	25.0
Hexacosane	366	26.3
Heptacosane	380	27.7
Octacosane	394	28.0
1,4-Dihydro-2,3-benzocarbazole	218	30.0
(tentative identification)		

aerial parts of the plants. It may be that the soil microorganisms are capable of converting hydrocarbons to the corresponding fatty acids, a part of which may be esterified. Methyl esters of fatty acids may have an allelochemical effect. Results obtained from fatty acids in soils from coffee plantations of Tanzania, Brazil, and Java by Haesler and Wanner (1977) showed that esterification had not occurred in the soil, but these authors did not look for, nor expect, this reaction in soil or microorganisms. Indeed methyl esters may be a small, common component in soils since Wang et al. (1971) observed methyl esters and free fatty

TABLE 15.4.
Some compounds occurring in *C. arabica* root leachate

Compound	Molecular Weight	Retention Time from GC/MS (R_t, min)
Caffeine	194	15.7–16.9
N-Phenyl-1-naphthylamine	219	25.8
Tricosane	324	26.3
Tetracosane	338	27.8
Pentacosane	352	29.2
Hexacosane	366	30.6

acids in Taiwanese soils. Recently, Al Saadawi et al. (1983) observed that free fatty acids with 14–22 carbons are important allelochemicals from soil under strands of *Polygonum aviculare* (knotweed) and *Cynodon dactylon* (bermuda grass).

Extraction of organic compounds of low molecular weight from soil can be a difficult process and completeness of extraction is not easily assured. Accumulation of toxic residues from the plant or metabolites occurs in the soil (Waller, et al., 1984).

Metabolism of Caffeine, Theobromine, and Theophylline in *C. arabica* Fruits

The biosynthesis and biodegradation of the purine alkaloids in developing fruits were shown by Suzuki and Waller (1984a,b) to follow the pathways summarized in Figure 5.15 (see pages 266 and 267). Caffeine is always the predominant alkaloid in *C. arabica* and the degradation to dimethylxanthine(s) is slow but steady, whereas further degradation to CO_2 proceeds somewhat faster. Among purine nucleosides and/or nucleotides, xanthosine (Negishi et al., 1985a–c; Suzuki and Waller, 1985a) is the most effective biosynthetic precursor of caffeine. The control mechanisms for caffeine in caffeine-producing plants is not known. It is significant that caffeine is accumulated throughout the entire plant in concentrations of 0.05 (roots) to 1.6% (mature seed) of the dry weight and that it can be degraded to CO_2; but (1) much is lost when the berries are harvested for commercial coffee, (2) part remains in association with the coffee tree to begin the new crop, and (3) part becomes litter from the coffee tree that falls in the soil, where it produces an allelochemical effect. It will be interesting to learn if caffeine can serve as an allelopathic agent in tea (Suzuki and Waller, 1985b,c) and other caffeine-producing plants.

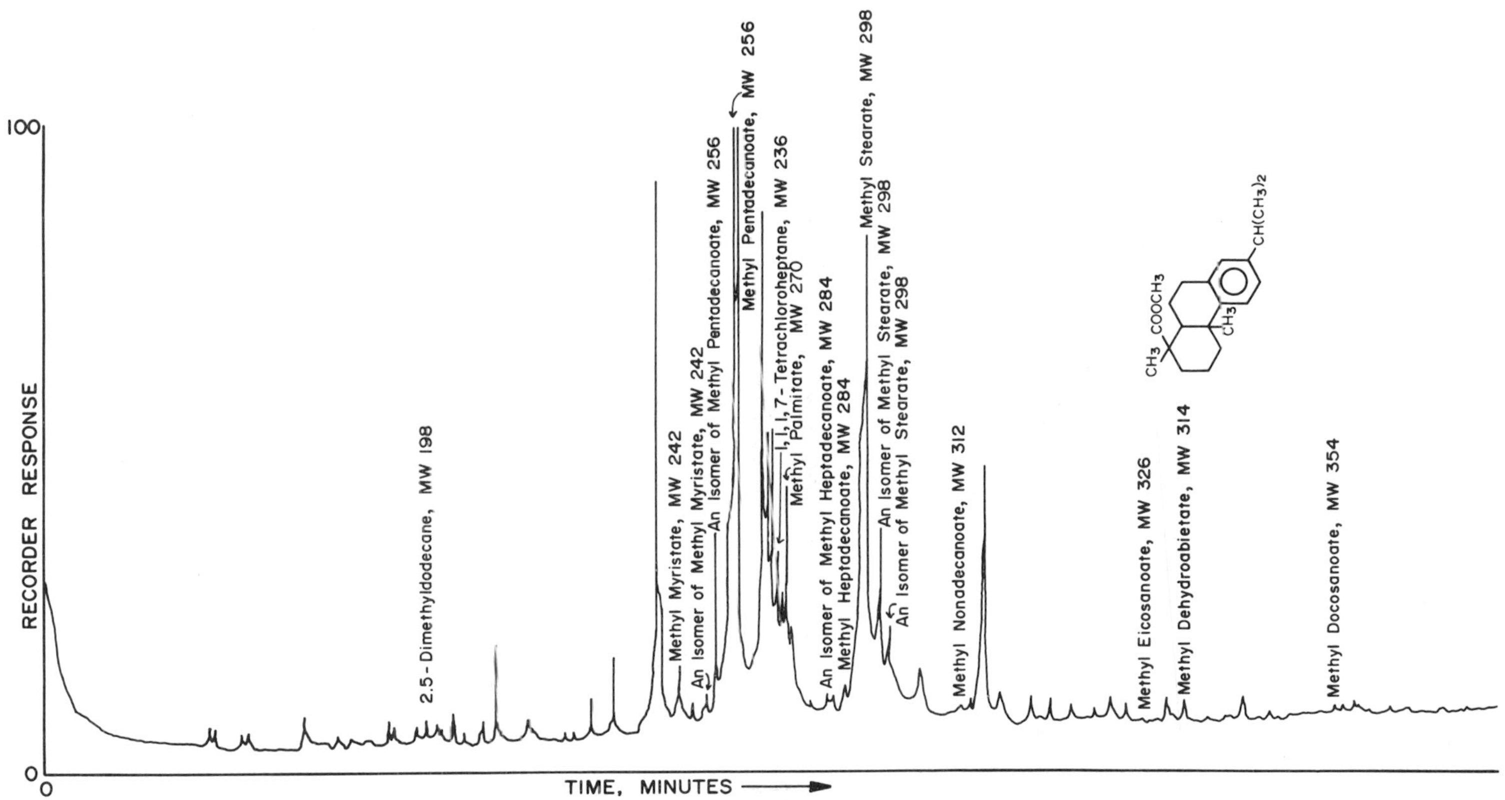

Figure 15.13 Natural products from coffee soil—chloroform extract of a 10% HCl-HF extract of the soil. Coffee trees were grown in the greenhouse in Stillwater, OK for 4 years.

263

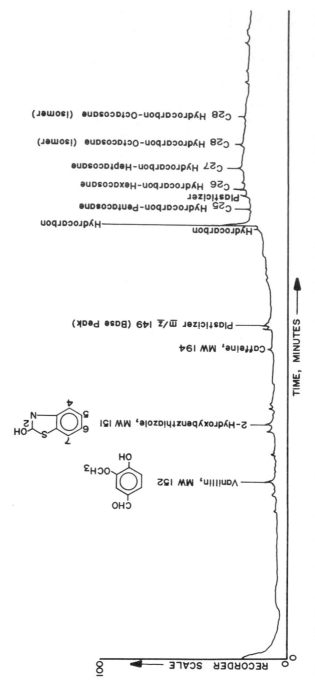

Figure 15.14 Natural products from coffee soil—hexane extract of an 85% methyl alcohol extract of the soil. Coffee trees were grown in the greenhouse in Stillwater, OK for 4 years.

264

CONCLUSIONS

The results show that autotoxicity caused primarily by caffeine may occur in nature. The growth of coffee seedlings exposed to 10 mM caffeine was inhibited in the rootlets, where mitosis and cell plate formation were blocked. Since concentrations of endogenous caffeine in the imbibed seed were 40–60 mM, 4–6 times as high as in the seedlings, we conclude that coffee embryos have specific means of avoiding caffeine autotoxicity. Leachates of coffee trees and roots and extracts from soil were analyzed with a GC/MS/DS system and shown to contain (among other compounds): (1) caffeine and methyl esters of myristic through docosanoic acids (which have allelopathic activity) from soil; and (2) furfural, N-phenyl-1-naphthylamine, n-alkanes, and caffeine from leaves, stems, and roots. Caffeine from Mexican canopy soil indicated that around old coffee trees considerable amounts of caffeine may be released from the tree litter and accumulate in the vicinity of roots over the years. The annual amount of litter (mostly leaves) produced, plus about 10% of lost fruits, in old coffee trees is 150–200 g dry matter/m^2/year (Epifanio, 1981). We reason that this litter may release 1–2 g caffeine/m^2/year, with some additional amounts of other purine alkaloids and compounds. The antimicrobial activity of caffeine may reduce catabolism of the alkaloid in the soil and thus prolong its retention and increase caffeine accumulation. Since most roots of coffee trees develop in the upper soil layer immediately under the tree's own litter, autotoxicity can occur. Soil that is "saturated" with caffeine can also continue building up other naturally occurring compounds from the action of microorganisms. Some of these products are toxic, particularly if formed in relatively high concentrations. This toxicity of allelochemicals may provide an explanation for the worldwide phenomenon of early degeneration of coffee plantations at 10–25 years of age (Wellman, 1961). Caffeine, the compound most associated with the production of coffee, is the factor that allows for survival of coffee plants in a hostile environment, but it may also be responsible for shortening the lives of mature trees.

ACKNOWLEDGMENTS

We are indebted to Otis C. Dermer, who critically read the manuscript. This work was supported in part by National Science Foundation Grant PCM-78-23160 and is published as journal article No. J-4830 of the Oklahoma Agricultural Experiment Station, Oklahoma State University, Stillwater, OK.

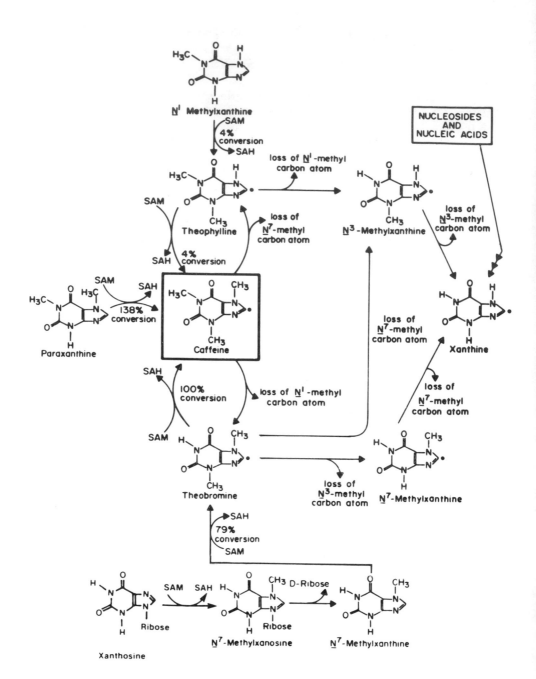

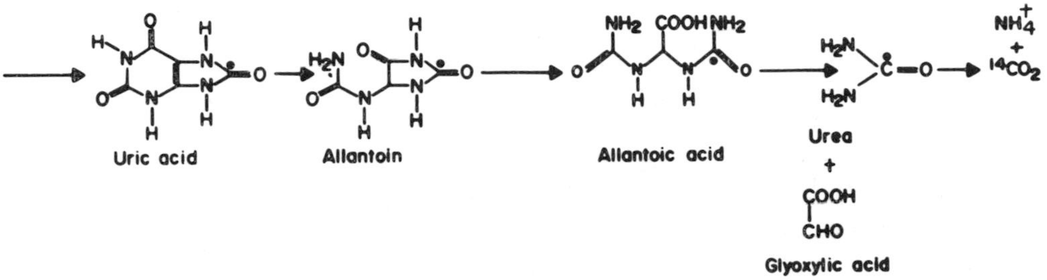

Figure 15.15 Pathways of metabolism of caffeine by *Coffea arabica* L. fruit; (●) represents the ^{14}C label from [8-^{14}C] theophylline in the biodegradation part. The rates of conversion are taken from Roberts and Waller (1979). *Note*. The left half of the diagram relates to the biosynthesis and the right half represents the biodegradation of caffeine (Suzuki and Waller, 1984a,b; 1985a,b; Negishi et al., 1985a–c).

REFERENCES

Al Saadawi, I. S., E. L. Rice, and T.K.B. Karns, 1983. *J. Chem. Ecol.* 9:761.

Anaya-Lang, A. L. and S. Del Amo. 1978. *J. Chem. Ecol.* 4:289.

Anaya-Lang, A. L., G. Roy-Ocotla, and L. Ortiz-Ortega. 1978. *2nd Int. Congr. Ecol. Jerusalem, Israel*, September 10–16. p. 8 (abstract).

Anaya, A. L., G. Roy-Ocotla, L. M. Ortiz, and L. Ramos. 1982. *In* E. Jimenez Avila and A. Gomez-Pompa (eds.), *Estudios Ecologicos en el Agroecosistema Cafetalero.* Instituto Nacional de Investigaciones sobre Recursos Bioticos, Xalapa, Veracruz, Cid Editorial Continental, S. A., Mexico D. F., pp. 83–92.

Anaya, A. L., L. Ramos, J. G. Hernandez, and R. Cruz. 1985. *Abstracts, 190th Am. Chem. Soc. Mtg.* September 9–14, Chicago, IL. AGFD-200.

Baumann, T. W. and H. Gabriel. 1984. *Plant Cell Physiol.* 25:1431.

Chou, C. H. and H. L. Lin. 1976. *J. Chem. Ecol.* 2:353.

Chou, C. H. and G. R. Waller. 1980a. *J. Chem. Ecol.* 6:643.

Chou, C. H. and G. R. Waller. 1980b. *Bot. Bull. Acad. Sinica* 21:25.

Duffus, C. M. and J. H. Duffus. 1969. *Experientia* 25:581.

Epifanio, J. A. 1981. *Ecologia del Agroecosistema Cafetalero.* Ph.D. Thesis. Universidad Nacional Autonoma de Mexico, Mexico D. F., Mexico. 166 pp.

Evenari, M. 1949. *Bot. Rev.* 15:153.

Fay, P. K. and W. B. Duke. 1977. *Weed Sci.* 25:224.

Fowden, L. and P. J. Lea. 1979. *In* G. A. Rosenthal and D. H. Janzen (eds.), *Herbivores, Their Interaction with Secondary Plant Metabolites.* Academic, New York, pp. 135–160.

Friedman, J., E. Rushkin, and G. R. Waller. 1982. *J. Chem. Ecol.* 8:55–65.

Friedman, J. and G. R. Waller. 1983a. *J. Chem. Ecol.* 9:1099.

Friedman, J. and G. R. Waller. 1983b. *J. Chem. Ecol.* 9:1107.

Friedman, J. and G. R. Waller. 1985. *Trends Biochem. Sci.* 10:47.

Haesler, von R. and H. Wanner. 1977. *Plant Soil* 48:397.

Harman, G. E. and A. L. Granett. 1972. *Physiol. Plant Pathol.* 2:271.

Kihlman, B. A. 1949. *Hereditas* 35:109.

Lailach, G. E., T. D. Thompson, and G. W. Brindley. 1968. *Clays Clay Miner.* 16:285.

McPherson, J. K., C. H. Chou, and C. H. Muller. 1971. *Phytochemistry* 10:2925.

Nathanson, J. A. 1984. *Science* 226:184.

Negishi, O., T. Ozawa, and H. Imagawa. 1985a. *Agric. Biol. Chem.* 49:251.

Negishi, O., H. Ozawa, and H. Imagawa. 1985b. *Agric. Biol. Chem.* 49:887.

Negishi, O., T. Ozawa, and H. Imagawa. 1985c. *Agric. Biol. Chem.* 49:2221.

Osborne, D. 1981. *In* K. V. Thiman (ed.), *Senescence in Plants.* CRC Press, Boca Raton, FL, pp. 13–38.

Parish, D. J. and A. C. Leopold. 1978. *Plant Physiol.* 61:365.

Prabhuji, S. K., G. C. Srivastava, S.J.H. Rizvi, and S. N. Mathur. 1983. *Experientia.* 39:177.

Ramos, L., A. L. Anaya, and J. N. Pascual. 1983. *J. Chem. Ecol.* 9:1079.

Rice, E. L. 1984. *Allelopathy*, 2nd ed. Academic, Orlando, FL. 422 pp.

Rizvi, S.J.H., D. Mukerji, and S. N. Mathur. 1980a. *Indian J. Exp. Biol.* 18:777.

Rizvi, S.J.H., V. Jaiswal, D. Mukerji, and S. N. Mathur. 1980b. *Naturwissenschaften* 67:459.

Rizvi, S.J.H., S. K. Padney, D. Mukerji, and S. N. Mathur. 1980c. *Z. Angew. Entomol.* 90:378.

Rizvi, S.J.H., D. Mukerji, and S. N. Mathur. 1981. *Agric. Biol. Chem.* 45:1255.

Rizvi, S.J.H. and V. Rizvi. 1983. *Int. Congr. Plant Protect.* (Brighton). 1:234. (Extended abstract.)

Roberts, E. H. 1972. *Viability of Seeds.* Chapman and Hall, London.

Roberts, M. F. and G. R. Waller. 1979. *Phytochemistry* 17:1083.

Suzuki, T. and G. R. Waller. 1984a. *J. Agric. Food Chem.* 32:845.

Suzuki, T. and G. R. Waller. 1984b. *J. Sci. Food Agric.* 35:66.

Suzuki, T. and G. R. Waller. 1985a. *Abstracts, 190th Am. Chem. Soc. Mtg.* September 9–14, Chicago, AGFD 109.

Suzuki, T. and G. R. Waller. 1985b. *Abstr. 13th Int. Congr. Biochem.* Amsterdam, The Netherlands, p. 537.

Suzuki, T. and G. R. Waller. 1985c. *Ann. Bot.* 56:537.

Valio, I.F.M. 1976. *J. Exp. Bot.* 27:983.

Waller, G. R. 1983. *In Proceedings of the Seminar on Allelochemicals and Pheromones,* Monograph 5. Institute of Botany, Academia Sinica, Taipei, ROC, pp. 1–25.

Waller, G. R., J. Friedman, C.-H. Chou, T. Suzuki, and N. Friedman. 1982. *In Proceedings of the Seminar on Allelochemicals and Pheromones,* Monograph 5. Institute of Botany, Academia Sinica, Taipei, ROC, pp. 239–260.

Waller, G. R., D. C. MacVean, and T. Suzuki. 1983. *Plant Cell Reports* 2:109.

Waller, G. R., C. R. Ritchey, E. E. Krenzer, G. Smith, and M. Hamming. 1984. *Abstracts, 32nd Annu. Conf. Mass Spectrom. Allied Topics.* San Antonio, TX. MPB14.

Wang, T.S.C., P. T. Hwang, and C.-Y. Chen. 1971. *Soil Sci. Soc. Am. Proc.* 35:584.

Warmake, H. E. 1935. *Stain Technol.* 10:101.

Wellman, F. L. 1961. *Coffee: Botany, Cultivation and Utilization.* Interscience Publishers, New York, pp. 345–347.

16

ALLELOPATHIC ACTIVITY OF RYE (*SECALE CEREALE L.*)

JANE P. BARNES AND ALAN R. PUTNAM

*Department of Horticulture and Pesticide Research Center,
Michigan State University, East Lansing, Michigan*

BASIL A. BURKE

ARCO Plant Cell Research Institute, Dublin, California

Winter rye (*Secale cereale* L.) is an annual grass species that has proved useful in a variety of cropping systems by contributing organic matter, reducing soil erosion, "smothering" weeds, and enhancing water penetration and retention (Benoit et al., 1962; Blevins et al., 1971; Overland, 1966; Shear, 1968). It germinates well in untilled soils and tolerates a wide range of soil moisture levels, pH's, and fertility conditions (Nuttonson, 1958). It also germinates well at low temperatures and is winter-hardy. Rye develops an extensive root system which improves soil tilth and enables it to use considerable soil moisture. The branching, slender, and fibrous adventitious roots are functional throughout the life of the plant. The root system of a single rye plant at the milky ripe stage extends downward as much as 1.5 m, with a lateral spread of over 0.9 m (Nuttonson, 1958).

Over-wintered rye produces considerable biomass early in the growing season. For this reason it has often been used as a green manure crop in sandy or low-fertility soils. In a comparison of residue production from cereals, rye out-produced wheat (*Triticum vulgare* L.), oats (*Avena sativa* L.), and barley (*Hor-*

deum vulgare L.) from 21 to 70% (Phillips and Young, 1973). The massive production of biomass by rye has the potential to influence the growth of succeeding plant species through the release of allelopathic chemicals from its residues. As early as 1925, Cubbon found a rye crop to inhibit the growth of grape (*Vitis vinifera* L.) plants and suggested that chemicals might be implicated.

PRIOR EVIDENCE FOR ALLELOPATHY IN RYE

Numerous studies indicate that rye interferes with the growth of other plants. Hill (1926) found that the addition of green rye to heavy soils depressed corn (*Zea mays* L.) growth, whereas growth was increased in light soils. Roots were more toxic than tops. Faulkner (1943) first suggested seeding the land with rye to help eliminate weeds. Osvald (1953) found that rye root exudates reduced germination of wild oats (*Avena fatua* L.). Nuttonson (1958) noted that rye has been used to suppress wild oats and many other weeds. Another report (Phillips and Young, 1973) indicated that a rye cropping program helped control both dandelions (*Taraxacum officinale* Weber.) and broad-leaved annual weeds. Robertson et al. (1976) found that rye residues suppressed weed growth when compared to sod or conventional tillage plots. These observations have provided the basis for more recent research involving allelopathy by rye.

A spring-planted cover crop of living rye reduced total weed biomass by 90% over unplanted controls (Table 16.1). A modified "stairstep" bioassay, where direct interplant competition for space and light was reduced, enabled testing for the effects of root leachates from living rye. Root leachates from two cultivars reduced lettuce (*Lactuca sativa* L.) shoot and total biomass by 25% and total biomass of tomatoes (*Lycopersicon esculentum* Mill.) by 18% (Barnes and Putnam, 1983). In a similar study, rye root leachates were found to reduce tomato root biomass by 20–32% and total biomass by 25–30%. The greatest reductions in tomato biomass occurred in treatments where rye was 10–30 days old when tomato was planted. These studies in quartz sand under greenhouse conditions indicated that rye root leachates were more inhibitory to the growth of tomato

TABLE 16.1.

Effect of spring-planted living rye cover crop on early season biomass of large crabgrass, common ragweed, and common lambsquarters in Spinks loamy sand

Cover Crop	Large Crabgrass (g/m²)	Common Ragweed (g/m²)	Common Lambsquarters (g/m²)	Total (g/m²)
No rye	12	21	165	265
MSU-13 rye	7	2	4	16

Source. Barnes and Putnam (1983).

than tomato root leachates were on themselves. Unconcentrated solutions of rye root leachates had no effect on the germination of several weed species (Barnes, 1981).

The interference observed by a living cover of rye might also arise from toxic compounds released by leachates of shoots. Shoots of different aged rye were misted and leachates were collected for bioassay. Shoot leachates had no effect on the germination of lettuce, barnyardgrass (*Echinochloa crusgalli* L. Beauv.), cress (*Lepidium sativum* L.), and tomato or seedling growth of tomatoes (Barnes, 1981).

Instances of poor germination and seedling growth were reported by growers in the Salinas Valley, California, where lettuce was planted too soon after turning in barley or rye cover crops. Patrick et al. (1963) initiated experiments to determine whether phytotoxicity could be detected when the cultural practice was repeated in the greenhouse with soil and plant residue obtained directly from the field. Lettuce root lengths and fresh weights were reduced in treatments where rye residue was present. Rye extracts with marked phytotoxicity were obtained after residues had been decomposed for 10–25 days. Toxicity of residues declined as the decomposition period increased until, by the 30th day, little or no phytotoxicity was observed.

Patrick and Koch (1958) found that decomposing rye residues inhibited respiration in tobacco (*Nicotiana tabacum* L.) seedlings. In a later experiment, Patrick (1971) sampled tobacco fields that contained decomposing rye residues previously plowed under as a green manure. He determined that extracts of decomposing residue fragments delayed germination and reduced the root growth of lettuce and tobacco. Similar phytotoxicity was exhibited with extracts of soil that were in contact with decomposing rye residues. No phytotoxicity was obtained with extracts of soil from which all recognizable decomposing rye residues had been removed. Ether extraction and gas chromatographic analysis identified several acids, including acetic, butyric, benzoic, and phenylacetic, to be implicated as the phytotoxic compounds. It was not determined whether phytotoxic substances were synthesized by soil microorganisms using the plant material as a substrate, or were the breakdown products inherent in the plant tissue. Both situations may have occurred, since phytotoxic compounds of significant potency were detectable only after the decomposing plant residue had been freed from most of the adhering soil, prior to extraction.

Chou and Patrick (1976) later identified nine acids from ether extracts of decaying rye residues in soil. Phenylacetic, 4-phenylbutyric, and salicyclic acids inhibited lettuce root growth at concentrations between 25 and 100 ppm. Other compounds identified included vanillic, ferulic, *p*-coumaric, *p*-hydroxybenzoic, *o*-coumaric, and salicyladehyde and were active at ca. 100 ppm. They assumed the toxic compounds in decomposing residues of rye were (1) released as volatiles, (2) polar in nature, and (3) of a phenolic origin. Therefore, their methods may have precluded identification of compounds other than phenolics.

The decomposition of rye residues, or any organic substrate, is a continuing process which requires rapid and sensitive assay methods to detect phytotoxic compounds during their short interval of production and disappearance (Patrick, 1971). This is evident from the various contradictory results often obtained during decomposition of similar plant residues. The ephemeral nature of such products, and their relatively rapid transition from one type of physiological activity to another, may explain why their occurrence is frequently missed. Plant injury is dependent on the frequency of a chance encounter with a growing root system, or with fragments of plant residues, at a time when decomposition is favorable for toxin production.

Sholte and Kupers (1977, 1978) investigated the causes of the lack of self-tolerance of winter rye grown on light sandy soils. In rotation studies, they found that the yield of rye seed and straw decreased up to 30% when rye followed itself, compared to following other crops. Nematodes and foot-rot fungi were excluded as possible causes. In a later study, they provided circumstantial evidence that the soil microflora was involved in the self-intolerance. This evidence, however, does not preclude allelopathy as the causative factor.

Kimber (1973) found that cold aqueous extracts of several grasses, including slightly green rye straw that had rotted for periods up to 21 days, inhibited the growth of wheat grown under aseptic conditions. Sterile conditions were used to eliminate the possible interactions of pathogens and microbial products. The degree of inhibition varied from one species to another, and also with the length of the rotting period. He found slightly green straw to be more toxic than fully matured residue. The most toxic materials were from extracts of rye straw that had rotted for 4 days.

Putnam and DeFrank (1979) screened numerous cover crops for weed suppressing activity. They found fall-planted cover crops to reduce both weed populations and biomass in the next growing season. Fall-killed "Balboa" rye reduced weed biomass by 84% over no-residue controls; spring-killed rye residues were less toxic. Additional field trials indicated that a variety of annual weed species could be suppressed consistently with residues of immature cereals [including rye, corn, barley, wheat, sorghum (*Sorghum bicolor* L.), and oats], whereas larger seeded vegetables, such as legumes, grew quite well in the no-till systems (Putnam and DeFrank, 1983).

TOXICITY OF RYE RESIDUES

Residues of 40-day-old greenhouse-grown rye reduced total weed biomass by 80% as compared to a nontoxic control mulch which reduced biomass by 44% (Table 16.2). Fall-planted, spring-killed rye residues also reduced total weed biomass by 68–95% when compared to controls with no residue (Figure 16.1; Barnes and Putnam, 1983). Weed biomass was reduced an additional 35% when compared to the control mulch, further indicating that allelopathy was one component of the interference by rye on other plant species.

Researchers in North Carolina have also examined the feasibility of using a rye mulch for weed suppression in no-till cropping systems (Shilling et al., 1985).

TABLE 16.2.
Weed suppression with surface residues of rye compared with *Populus* wood shavings

Residue	Weed Fresh Weight (g)
Populus wood shavings	224
Ryc	81
Control	398
LSD at $P < 0.05$	65

Source. Putnam and DeFrank (1983).

Figure 16.1 Weed regrowth in (a) control plot and (b) plot containing rye residues. Photograph taken 30 days after spraying both areas with glyphosate.

They found that fall-planted, spring-killed rye reduced the aboveground biomass of several weed species, including redroot pigweed (*Amaranthus retroflexus* L.), common lambsquarters (*Chenopodium album* L.), and common ragweed (*Ambrosia artemisiifolia* L.). Subsequent isolation of compounds with toxicity in aqueous extracts of field-grown rye resulted in the identification of two chemicals, B-phenyllactic acid (PLA) and B-hydroxybutyric acid (HBA). Both acids inhibited common lambsquarters root growth 20% at 2 mM. Redroot pigweed root growth was inhibited 59 and 39% at 2 mM by PLA and HBA, respectively.

Although there are numerous reports suggesting allelopathic activity by rye and its residues, few have followed the protocols for proof of allelopathic interference, as suggested by Fuerst and Putnam (1983). The aim of our recent research was to investigate the mechanism of interference by rye, with particular emphasis on crop and weed responses to its residues. The experimental plan was to identify the symptoms of the interference in the field and in bioassays, to isolate, identify, and quantify the toxicities of the most active chemicals from the residues, and to assess the importance of these chemicals in the interference.

A petri-dish bioassay of rye residues in soil was developed to allow for more accurate determination of effects on seed germination and seedling growth. The first series of experiments examined the effect of residue placement relative to seed. Rye residues had no effect on the germination of barnyardgrass and proso millet (*Panicum milaceum* L.) at any distance (Figure 16.2). In contrast, residues placed directly next to seed significantly reduced cress and lettuce germination. Overall, as distance between residue and seed increased, phytotoxicity decreased. The primary phytotoxic effect by residues on all species was the inhibition of root growth. Although soil diluted the phytotoxicity, residues still significantly reduced root growth of all species except proso millet when placed 15 mm from the seeds.

Lettuce was particularly sensitive to rye residues. When placed closest to residue, the seeds imbibed water but turned black and failed to grow. In cases where lettuce did germinate, the hypocotyls appeared swollen and the apical root meristems were discolored. Growth was subsequently inhibited. The symptoms of injury on lettuce were consistent with previous reports (Chou and Patrick, 1976). Rye shoot tissue was two- to threefold more inhibitory than root tissue. When root and shoot tissue were applied together, results implicated shoot tissue as the primary source of inhibition, although they also suggested that both root and shoot tissue can act together in the field to injure selected species.

Field studies have estimated rye surface residues at about 4900 kg/ha (Barnes, 1981). This was equivalent to 4.0 g tissue per petri dish, which resulted in 100% inhibition of germination of all species. Subsequently, lower rates were assayed to allow for comparative evaluation of treatments. Since 2.0 g shoot tissue almost totally inhibited germination of the more sensitive indicators (cress and lettuce), the rates of rye residue found under field conditions are probably sufficient to substantially reduce the germination and growth of many different plant species.

It is important to determine if phytotoxicity associated with residues in soil is of plant or microbial origin. In our studies there were no instances where the residues or plant extracts were more toxic in the presence of soil microorganisms (Figure 16.3). Using rye and relatively short-term bioassay durations, microbes (including possible pathogens) appeared to detoxify rather than enhance activity.

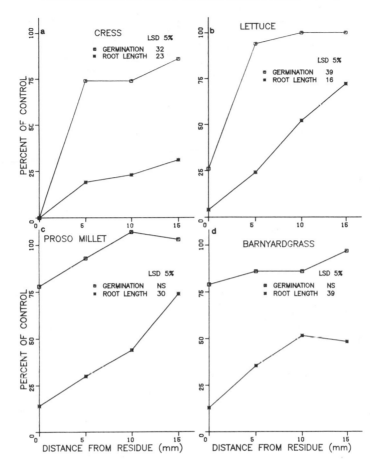

Figure 16.2   Effect of rye residue (1 0 g) placement relative to seed in soil on germination and root growth of four indicator species. (a) curly cress, (b) lettuce, (c) proso millet, and (d) barnyardgrass. *Source,* Barnes and Putnam (1986a).

ISOLATION AND IDENTIFICATION OF RYE ALLELOCHEMICALS

Extraction

In preliminary tests for extraction efficiency, water removed more compounds from dried rye shoot tissue than 50% methanol. From 50 g of dried tissue, 34% of the dried plant material was extracted with 50% alcohol in 24 hours, whereas extraction with H_2O yielded 49% of the initial weight. There was little difference in recovery between the 0.5- and 24-hour aqueous extractions. As water is the solvent of extraction in nature, and efficiently removes many compounds from rye, all subsequent separations were based on a 24-hour aqueous extraction.

After sequential partitioning against solvents of increasing polarity, all frac-

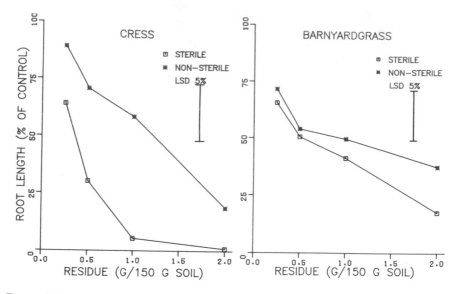

Figure 16.3 Toxicity of rye residues at concentrations of 0.5-2.0 g/150 g soil under sterile vs. nonsterile conditions to curly cress and barnyardgrass. *Source.* Barnes and Putnam (1986a).

tions exhibited some degree of activity in bioassay. Therefore, the yield, I_{50}, and unit activity of each fraction were determined and compared (Table 16.3). The I_{50} for the initial crude extract was approximately 1.90 mg or 1270 ppm. Based on the quantity of material recovered from the initial extraction, the aqueous crude extract contained 13,000 potential units (weight × mg I_{50}) of activity. After initial separations, 48% of the unit activity was associated with the protein precipitate and final aqueous extracts. Although these fractions accounted for

TABLE 16.3.
Activity of rye shoot tissue on cress (*Lepidium sativum* L.)

Fraction	Percent of Crude	I_{50} (mg)	Units[a]
Initial aqueous	—	1.90	13,000
Acetone precipitate	17.0	1.64	2,550
Hexane	0.1	0.68	47
Ethyl ether	1.0	0.15	1,580
Dichloromethane	0.2	0.38	144
Ethylacetate	0.6	0.29	498
Final aqueous	55.0	3.80	3,570

Source. Barnes et al. (1986b).
[a]Determined by weight of dried sample × quantity necessary for 50% inhibition.

72% of the crude weight and 97% of the recovered material, their specific activity was considerably less than those of the organic soluble fractions.

The ether fraction had the greatest specific activity with an I_{50} of 150 μg or 100 ppm (w/v). The organic fractions accounted for 18% of the activity, but only 1.9% of the initial crude weight. The ether fraction alone accounted for 12% of the activity and only 1.0% of the crude weight. Therefore, the ether fraction was chosen for further isolation work.

Separation and Identification of Active Compounds

The ether extracts were separated by thin-layer chromatography on silica gel plates and each Rf zone was bioassayed. Activity on silica gel plates was spread out in two zones. The more polar toxin (compound 1) found at Rf 0.20 was detected by UV, $FeCl_3$/HCl, $CeSO_4$/H_2SO_4, and vanillin/H_2SO_4. A second zone of toxicity was evident at Rf 0.66. This zone reacted with all previously mentioned detection reagents except $FeCl_3$.

The ether fraction was also separated on silica gel columns for purposes of identification and bioassay. All fractions with activity in the nonpolar range reacted blue with $FeCl_3$, suggesting that this compound may be a good indicator of activity. After sample concentration and storage at $-70°C$ overnight, compound 1 began to crystallize.

Compound 1 formed a blue complex with $FeCl_3$, which is specific for phenols and hydroxamic acids. The UV spectrum of compound 1 in methanol showed two major absorption bands at 281 and 254 nm. Upon addition of base (1 drop of 10% KOH in MeOH), the spectrum shifted to 301 and 217 nm, indicating an enolic or phenolic type compound. Results from NMR (^{13}C- and ^{1}H-) indicated that the compound had a base structure of C_8H_7-. The molecular weight was determined to be 181 via electron impact mass spectrometry (EIMS), which also indicated an odd number of nitrogens in the compound. A library search of compounds with similar fragmentation patterns resulted in the retrieval of a compound, 2(*3H*)-benzoxazolinone (BOA), structurely related to the unknown. The molecular formula was subsequently determined to be $C_8H_7NO_4$ and the compound was identified as 2,4-dihydroxy-1,4(*2H*)-benzoxazin-3-one or DIBOA (Figure 16.4).

DIBOA was originally identified primarily via nonspectral methods by Virtanen and Hietala (1960). The structure was subsequently confirmed through synthesis of the compound. These workers determined the major UV absorption bands of DIBOA in ethanol at 255 and 282 nm, which correspond closely to those obtained for compound 1 in MeOH (254 and 281 nm). Owing to the lack of availability of similar EIMS for DIBOA in the literature, the spectra of compound 1 was compared with that obtained for 2,4-dihydroxy-7-methoxy-1,4-benzoxazin-3-one (DIMBOA) by Klun et al. (1970). DIMBOA differs by the presence of a methoxyl group on carbon-7 of the aromatic ring, which should result in similar mass spectra patterns in which fragments differ by 30 molecular weight units. The major fragments of DIMBOA at m/z 211, 195, 193, and 165

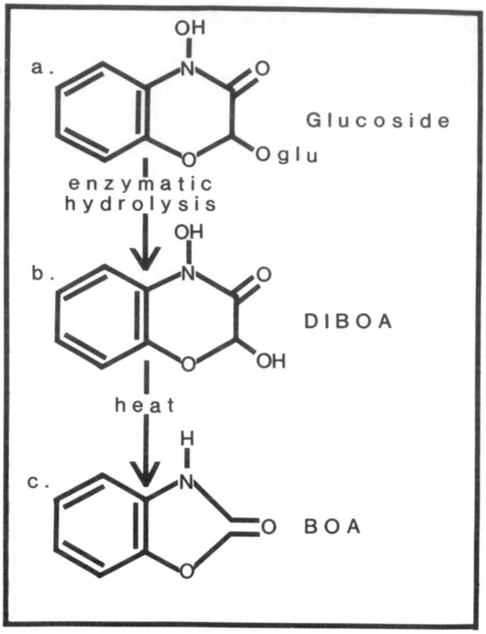

Figure 16.4 Structures for (a) DIBOA-glycoside, previously reported in noninjured plants; (b) compound 1 or DIBOA; and (c) compound 2 or BOA isolated from rye residues. *Source.* Barnes et al. (1986b).

do differ by 30 from the DIBOA ions of m/z 181, 165, 163, and 135, confirming the structure previously determined.

The pure compound was subsequently assayed for activity on cress. Increasing concentrations of DIBOA reduced both root and shoot length of cress, but showed little effect on seed germination at these concentrations. Cress root growth was more inhibited than shoot growth by DIBOA. The I_{50} values for root and shoot length by DIBOA were ca. 100 μg (67 ppm) and 190 μg (127 ppm), respectively.

Additional separation of the ether extract was directed toward isolation of the active, higher Rf (0.61–0.70) zone which was separated from compound 1 by TLC. This compound was further purified by column chromatography. Mass spectrometry of the most active components yielded a spectrum identical to that of BOA, a known compound believed to be a breakdown product of other benzoxazinones (Figure 16.4).

POSSIBLE ROLE OF HYDROXAMIC ACIDS IN RYE ALLELOPATHY

Relative Toxicity on Plants

Since the relative activities of alleged allelochemicals varies with the bioassay species tested, chemicals should be tested with several bioassays for comparison of activity. Shilling et al. (1985) implicated phenyllactic acid (PLA) and hydroxy-butyric acid (HBA) in the allelopathic activity of field-grown rye residues. For determination of relative activity, these compounds were compared with DIBOA and BOA in seed germination and seedling growth bioassays (Figure 16.5). Overall, DIBOA and BOA were consistently more inhibitory than PLA and HBA to germination and seedling growth of all weeds and crops tested. Monocotyledenous and dicotyledenous plants varied in their response and sensitivity to the four compounds tested.

Of the four chemicals, DIBOA was most active against the monocot species (Figure 16.5A). Although no chemical inhibited barnyardgrass germination, DIBOA significantly reduced germination of proso millet. Root growth of monocots was more sensitive than shoot growth, with greatest activity by DIBOA. The compound BOA reduced the growth of proso millet at high concentrations, but had little effect on barnyardgrass; PLA and HBA were active only at the highest concentrations on root growth.

In contrast, both germination and seedling growth of dicot species tested were significantly inhibited by the benzoxazinone compounds (Figure 16.5B). Here BOA was generally inhibitory to germination of all tested dicot species, whereas DIBOA showed significant reductions only at the higher concentrations. Both PLA and HBA had little effect on germination of the test species. Root and shoot growth of dicots were similarly affected by the chemicals tested; DIBOA and BOA were most toxic to seedling growth; BOA in particular caused lettuce root meristems to turn black. The necrosis of the apical root meristem closely resembled symptoms evident on lettuce germinating in the soil/residue petri-dish bioassays. Here PLA resulted in some significant reduction of dicot seedling growth, but was still less active than DIBOA and BOA. Once again, HBA had no effect on any growth parameter of dicots tested.

In general, the dicots were 30% more sensitive than the monocots to all rates of all chemicals tested. DIBOA and BOA were anywhere from 2 to 30 times more active than PLA and HBA. Of the two benzoxazinone compounds, BOA was more toxic to dicot germination and DIBOA had a greater effect on the germination of monocots. Both chemicals similarly reduced dicot seedling growth, but only DIBOA had any significant effect on seedling growth of monocots.

Queirolo et al. (1981) found that both DIMBOA and its glucoside, extracted from maize, inhibited both cyclic and noncyclic photophosphorylation in spinach (*Spinacea oleracea* L.) chloroplasts. Inhibition of coupled electron transport was attained at concentrations of 1 and 4 mM, respectively. Moreland and Hill (1963) proposed that the H-N-C=0 fragment found in many polycyclic urea herbicides was related to activity. This fragment is present in BOA and reduced forms of DIBOA. Chlorosis was a symptom of injury by rye residues on several indicators (Barnes and Putnam, 1983) and could result from the effects of DIBOA and BOA on photophosphorylation and electron transport.

Corn, wheat, rye, and Job's tears (*Coix lacryma jobi* L.) were all found to contain these compounds, but rice (*Oryza sativa*), barley, oats, and sorghum did not (Tang et al., 1975). Obviously, benzoxazinones cannot explain allelopathy with all cereal grains. It appears that many compounds are responsible.

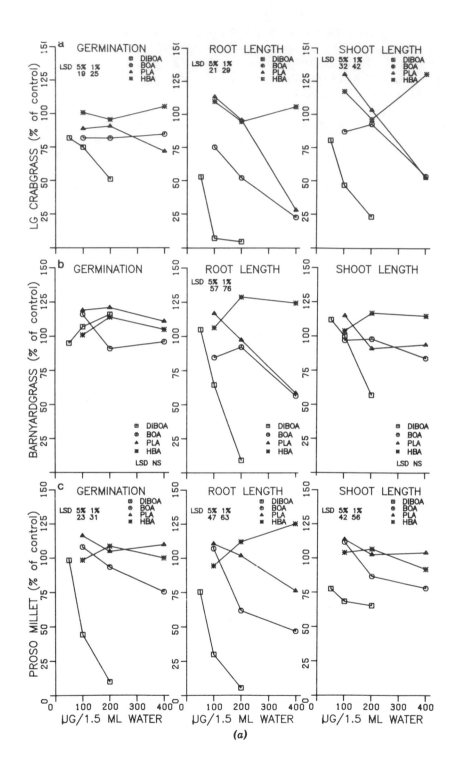

(a)

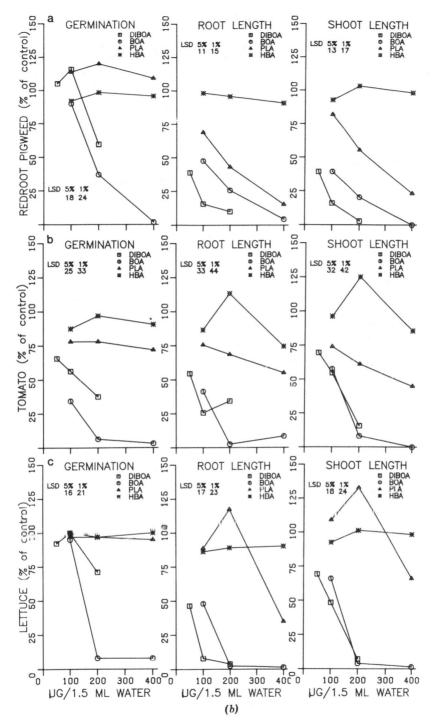

Figure 16.5 Response of (a) three monocot and (b) three dicot species to various concentrations of DIBOA, BOA, PLA, and HBA in petri-dish assays. *Source.* Barnes and Putnam (1986c).

283

Activity of DIBOA and BOA in Soil

The uptake of allelochemicals from rye residue may be by direct contact with tissue fragments or through the soil. Therefore, the pure compounds, DIBOA and BOA, were applied to soil for evaluation of activity on the emergence of three indicator species.

Two weeks after chemical application, the emergence of lettuce and cress was still inhibited completely by the highest rates of DIBOA and BOA (Table 16.4). Although still reduced relative to control, some cress and lettuce germinated and emerged in the lower rates of DIBOA after 14 days. Barnyardgrass seedlings that emerged appeared stunted and chlorotic.

The concentrations of benzoxazinones applied to soil in our studies may be considerably higher than the potential for production in field situations. Calculations of hydroxamic acids in residues of 35-day-old greenhouse grown (based on 4.7 MT biomass/ha) specimens indicate a potential for ca. 13.5 kg/ha from the shoots and ca. 0.8 kg/ha from the roots. None of the calculations account for the quantity of BOA present, which also shows activity. When the pure compounds were bioassayed at rates lower than 10 kg/ha, there was no inhibition or possible stimulation by the chemicals. There are no reliable data on concentrations in field-grown plants of various ages. According to the concentrations presented by Argandona et al. (1980), 34-day-old rye contains 1.9 m mole hydroxamic acids/1 kg fresh weight or a potential of 16.2 kg hydroxamic acids/ha. This calculation is somewhat greater than our estimate, but further supports the potential for high levels in rye. It must be remembered that these hydroxamic acids were but a fraction of the toxicity of the original aqueous extract.

TABLE 16.4.

Emergence of barnyardgrass (BYGR), cress, and lettuce (expressed as a percent of control) 14 days after spray applications of DIBOA and BOA to soil

Chemical	Rate (kg/ha)	Percent Control		
		BYGR	Cress	Lettuce
DIBOA	25	65	11	10
	50	19	0	0
	100	7	0	0
BOA	25	81	51	72
	50	61	9	16
	100	32	0	0
LSD	5%	29	19	24

Source. Barnes and Putnam (1986c).

While our major field observations for proof of allelopathy have been based on toxicity by residues of fall-planted, spring-killed rye, there is evidence for toxicity by spring-planted rye (Barnes, 1981). Residues of 40-day-old spring-planted rye reduced total weed density by 69% and total weed biomass by 32% when compared to a control mulch. As age may be an important variable and could be managed by timing the kill of rye, the maximum benzoxazinone concentration might be obtained. Additional studies that quantify and compare concentrations in both fall- and spring-planted rye killed at different ages would provide useful data. Ideally, the actual release of benzoxazinones into soil and uptake by plants, perhaps using radiolabeled material, should also be documented for conclusive proof of their role in allelopathy.

CONCLUSIONS

While toxicity by residues in nonsterile soil was still significant after incubation for 72 hours, benzoxazinones may be easily degraded by microorganisms. Undoubtedly, several compounds, in addition to the solvent extractable benzoxazinones, contribute toward toxicity by rye residues. This is because almost 50% of the initial aqueous extract activity was associated with the protein precipitate and final aqueous fractions. The identification of benzoxazinones and their subsequent implication in allelopathy by rye has been based on steps directed toward isolation of the most active chemicals. Although the ether extract accounted for only 12% of the potential toxicity, it was 11–25 times more active than the protein and final aqueous fractions.

Additional toxicity from residues may result from degradation products. Chou and Patrick (1976) found significant toxicity with residues of rye that had decomposed for up to 30 days. The compound BOA is the major chemical breakdown product of DIBOA and in our studies shows greater activity against the germination of all dicot species tested. Although the $FeCl_3$ method precludes quantitation of BOA and subsequent determination of its significance in rye residue toxicity, levels should presumably increase as DIBOA is broken down. Monitoring the breakdown and resultant toxicity in the soil over time should provide evidence for the persistence of the chemicals.

The importance of allelochemicals released in response to injury or infection is indicated by the benzoxazolinone series of compounds which occur naturally in many grass species (Beck and Reese, 1976). Their role and significance in resistance to insects and disease organisms have been investigated more thoroughly than their role in allelopathy. The greenhouse bioassays simulating no-tillage systems provided evidence for toxicity on whole plants. The similar injury symptoms obtained where both residues and pure compounds were applied support the premise of allelopathic interference by rye with benzoxazinone compounds playing a prominant role.

REFERENCES

Argandona, V. H., J. Luza, H. M. Niemeyer, and L. J. Corcuera. 1980. *Phytochem.* 19:1665.

Barnes, J. P. 1981. Exploitation of Rye (*Secale cereale* L.) and Its Residues for Weed Suppression in No-Tillage Cropping Systems. M. S. Thesis, Michigan State University. 128 pp.

Barnes, J. P. and A. R. Putnam. 1983. *J. Chem. Ecol.* 9:1045.

Barnes J. P. and A. R. Putnam. 1986a. *Weed Sci.* 34:384.

Barnes, J. P., A. R. Putnam, B. A. Burke, and A. J. Aasen, 1986. *Phytochemistry* (In Press.)

Barnes, J. P. and A. R. Putnam, 1986c, *J. Chem. Ecol.* (In Press).

Beck, S. D. and J. C. Reese. 1976. Insect–Plant Interactions: Nutrition and Metabolism. *In* J. W. Wallace and R. L. Mansel (eds.), *Recent Advances in Phytochemistry. Vol. 10: Biochemical Interactions Between Plants and Insects*. Plenum, NY, pp. 41–92.

Benoit, R. E., N. A. Willits, and W. J. Hanna. 1962. *Agron. J.* 54:419.

Blevins, R. L., D. Cook, S. H. Phillips, and R. E. Phillips. 1971. *Agron. J.* 63:593.

Chou, C. H. and Z. A. Patrick. 1976. *J. Chem. Ecol.* 2:369.

Cubbon, M. H. 1925. *Am. Soc. Agron.* 17:568.

Faulkner, E. H. 1943. *Plowman's Folly*. University of Oklahoma Press. Norman, OK. 156 pp.

Fuerst, E. P. and A. R. Putnam. 1983. *J. Chem. Ecol.* 9:937.

Hill, H. H. 1926. *J. Agric. Res.* 33:77.

Kimber, R. W. 1973. *Plant Soil.* 38:437.

Klun, J. A., C. L. Tipton, J. F. Robinson, D. L. Ostrem, and M. Beroza. 1970. *J. Agric. Food Chem.* 18:663.

Moreland, D. E. and K. L. Hill. 1963. *Weeds.* 11:284.

Nuttonson, M. Y. 1958. *In Rye Climate Relationships: On the Use of Phenology in Ascertaining the Thermal and Photo-thermal Requirements of Rye*. American Institute of Crop Ecology. 219 pp.

Osvald, H. 1953. On antagonism between plants. *Proc 7th Int. Congr. Bot.*, Stockholm. pp. 167–170.

Overland, L. 1966. *Am. J. Bot.* 53:423.

Patrick, Z. A. 1971. *Soil Sci.* 111:13.

Patrick, Z. A. and L. W. Koch. 1958. *Can. J. Bot.* 36:621.

Patrick, Z. A., T. A. Toussoun, and W. C. Snyder. 1963. *Phytopathology.* 53:152.

Phillips, S. H. and H. M. Young, Jr. 1973. No-tillage farming. Reiman Association, Milwaukee, WI, 224 pp.

Putnam, A. R. and J. DeFrank. 1979. Use of Allelopathic Cover Crops to Inhibit Weeds. *Proc. IX Int. Congr. Plant Protect.*, pp. 580–582.

Putnam, A. R. and J. DeFrank. 1983. *Crop Protect.* 2:173.

Queirolo, C. B., C. Andreo, R. H. Vallejos, H. M. Niemeyer, and L. D. Corcyera. 1981. *Plant Physiol.* 68:941.

Robertson, W. K., H. W. Lundy, G. M. Prine, and W. L. Currey. 1976. *Agron. J.* 68:271.

Scholte, K. and L.J.P. Kupers. 1977. *Netherlands J. Agric. Sci.* 25:255.

Scholte, K. and L.J.P. Kupers. 1978. *Netherlands J. Agric. Sci.* 26:250.

Shear, G. M. 1968. *Outlook Agric.* 5:247.

Shilling, D. G., R. A. Liebl, and A. D. Worsham. 1985. *In* A. C. Thompson (ed.), *The Chemistry of Allelopathy*. American Chemical Society, Washington, D.C., pp. 243–271.

Tang, C. S., S. H. Chang, D. Hoo, and K. H. Yanagihara. 1975. *Phytochem.* 14:2077.

Virtanen, A. I. and P. K. Hietala. 1960. *Acta Chem. Scand.* 14:499.

17

MICROBIALLY PRODUCED PHYTOTOXINS AS HERBICIDES – A PERSPECTIVE

STEPHEN O. DUKE

*United States Department of Agriculture, Agricultural Research Service,
Southern Weed Science Laboratory, Stoneville, Mississippi*

The need for more cost-effective, efficacious, selective, and environmentally safe herbicides is being increased by several current trends in agriculture and agricultural chemistry. The development of microbial toxins as herbicides or as sources of new herbicide chemistries promises to help fill this growing need.

A greater reliance on chemical weed control will result from the accelerating shift toward reduced- or no-till agriculture. Current herbicides are inadequate for weed control in some geographical areas with certain crops using reduced tillage practices. The increasing incidence of resistance of weeds to important herbicide classes such as s-triazines (Gressel, 1985) and dinitroanilines (Mudge et al., 1984) accelerates the need for new herbicides. Herbicide-resistant weeds are often resistant to several herbicide classes that have similar sites of action, creating a need for toxicants with new sites of action. Shifts in weed populations to species that are more closely related to the crop that they infest, for example, wild oats (*Avena fatua* L.) in oats, shattercane [*Sorghum bicolor* (L.) Moench] in sorghum (*Sorghum vulgare* Pers.), wild okra [*Abelmoschus esculentus* (L.) Moench] in cotton (*Gossypium hirsutum* L.), and red rice in rice (both *Oryza sativa* L.), are resulting in a greater need for more selective herbicides than are presently available, hence new strategies in herbicide development are required.

One of the more obvious causes of this change in strategy is that traditional approaches to the discovery of new herbicides are yielding diminishing returns. The number of chemicals that must be screened to develop a new herbicide is growing to levels that will soon make past methods economically unfeasible. Thus new, more carefully contrived methods of herbicide discovery that require evaluating significantly fewer compounds are required. Biorational approaches, such as the design of herbicides around specific biochemical sites of action, are options that are being pursued by many companies. An alternative is to seek new herbicides from naturally occurring products. Natural products offer exotic new chemistries that traditional pesticide chemists are unlikely to produce. In fact, many natural products are extremely difficult to synthesize after their structure is determined.

Registration of natural products as herbicides may be less expensive, in terms of time and money, than synthetic herbicides. The U.S. Environmental Protection Agency has reduced the registration requirements for certain natural compounds that it terms "biorational pesticides" (Charudattan, 1982; Hodosh et al., 1985). Although microbial phytotoxin herbicides may be acutely toxic to livestock or humans (some of the most toxic compounds known are of microbial origin), they are much less likely to cause environmental damage or alterations than traditional herbicides. The half-life of most natural compounds in the environment is relatively short, because microbes and plants have evolved to metabolize them for food sources or to detoxify them. This short half-life may be a problem in the development of some microbial toxins as herbicides.

Many microbially produced phytotoxins have ready-made selectivity because the producing microorganisms have co-evolved with host plant species. In fact, extreme selectivity may be an obstacle to the commercialization of host-specific toxins, except for those that affect the worst problem weeds.

Biocontrol (control of weeds with living microorganisms) (Charudattan and Walker, 1982) is not discussed in this chapter. Although microbial biocontrol agents are actively being developed and commercialized [e.g., DeVine® (*Phytopthora palmivora*) and Collego® (*Colletotrichum gloeosporioides* f. sp. *aeschynomene*)], microbially produced toxins have several advantages over the living pathogens. The phytotoxic compound generally has a longer shelf life and usually requires less storage space. The logistics of formulation and application are usually simpler. The possibility of disease spreading to nontarget species does not exist. Efficacy is easier to predict and is not as dependent on environmental conditions.

Microbially produced phytotoxins also offer several advantages over natural products from higher plants as herbicides. Allelochemicals from higher plants tend to have relatively little selectivity and, in many cases, are autotoxic to the producing species. Also, most are not very efficacious when compared to phytotoxic microbial toxins.

Herbicides of microbial origin have been reviewed previously (Cutler, 1984; Fischer and Bellus, 1983, McCalla and Haskins, 1964; Misato and Yamaguchi, 1984; Sekizawa and Takematsu, 1983) and pesticides of natural origin have been examined recently (Rice, 1983). In this chapter, examples of phytotoxic microbial toxins that are either being or not being developed as herbicides are discussed. This is not, however, meant to be an extensive review of all known microbial phytotoxins. Neither are extraction, purification, and identification discussed—a vast area of literature (Walton and Earle, 1984) partially covered by Cutler in Chapter 9 of this volume. Only microbial phytotoxins with known structures are mentioned.

HOST-SPECIFIC TOXINS

Host-specific toxins are those that affect only the species or, in some cases, the variety of a species infected by the toxin-producing pathogen. They are thought to be "pathogenicity factors" in the development of plant disease (Scheffer, 1983). Few of these toxins have been chemically characterized—those that have are listed in Table 17.1 and several of their structures are shown in Figure 17.1. To date, all toxins in this category are of fungal origin and are phytotoxic to a commercially valuable species—the reason for their discovery.

Little emphasis has been placed on the discovery of host-specific toxins affecting weed species, the likely reason that no host-specific toxins are known to affect weed species. Only recently has the search for host-specific phytotoxins for weeds begun (Robeson et al., 1984). Because of their extreme selectivity, such searches should focus on phytotoxins from pathogens that affect weed species with high economic impact such as velvetleaf (*Abutilon theophrasti* Medic.) or johnsongrass [*Sorghum halepense* (L.) Pers.].

Chemical manipulation of host-specific toxins might produce knowledge about their mechanisms of action and selectivity, about which little is known. This knowledge could be useful in designing molecules to extend the range of species affected by the toxin or its analogues and increasing its efficacy, thus increasing the potential of the toxin as a herbicide.

TABLE 17.1.
Fungal host-specific toxins for which structures are known

Toxin	Source	Species Affected
HC	*Helminthosporium carbonum* race 1	Maize
HS	*H. sacchari*	Sugarcane
HmT	*H. maydis* race T	Maize
AK	*Alternaria kikuchiana*	Japanese pear
AM	*A. mali*	Apple
AL	*A. alternaria f. lycopersici*	Tomato

Source. Scheffer (1983).

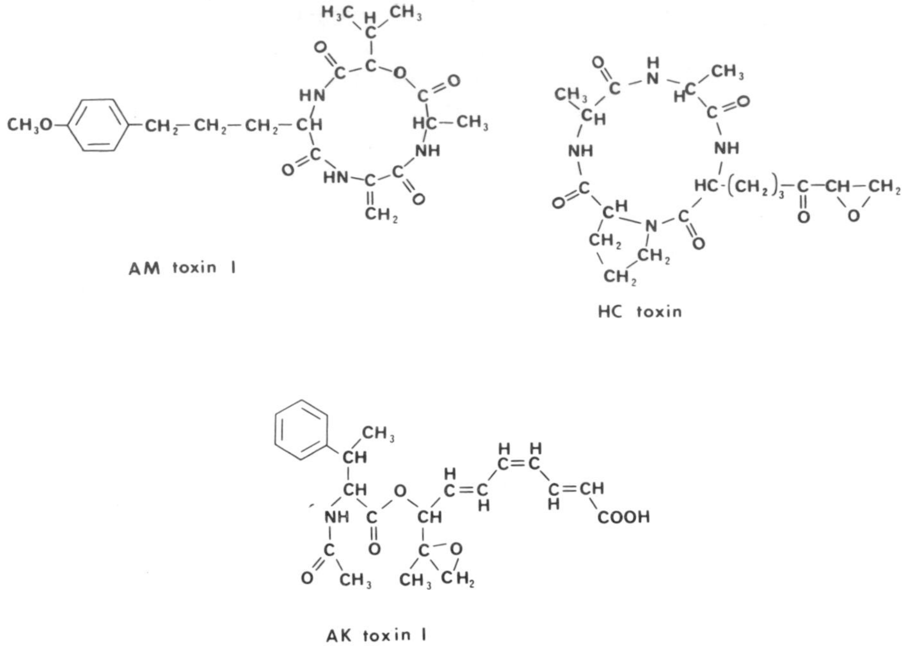

AM toxin I

HC toxin

AK toxin I

Figure 17.1 Structures of several host-specific toxins listed in Table 17.1.

NONSPECIFIC TOXINS

Non-host-specific toxins are defined by plant pathologists as those that are toxic to plant species that are not infected by the producing microorganism in nature. In fact, the producing microorganism need not be pathogenic. In a weed science context, however, many of the phytotoxins are quite selective (Barash et al., 1982; Cutler et al., 1984; Robeson and Strobel, 1984). All bacterially produced and many fungal phytotoxins belong in this category. They are the most characterized microbially produced phytotoxins from a physiological and biochemical standpoint. Several nonspecific phytotoxins that are the bases for registered herbicides or that show promise as microbially produced herbicides are discussed in detail.

Anisomycin and Related Compounds

The first microbial metabolite that resulted in a commercial herbicide was aniso-mycin (Figure 17.2), a product of *Streptomyces* sp. 638. Anisomycin was reported to have no activity toward tomato (*Lycopersicon esculentum* Mill.) or turnip (*Brassica rapa* L.), while being strongly phytotoxic toward barnyardgrass [*Echinochloa crus-galli* (L.) Beauv.] and crabgrass (*Digitaria* sp.) (Yamada et

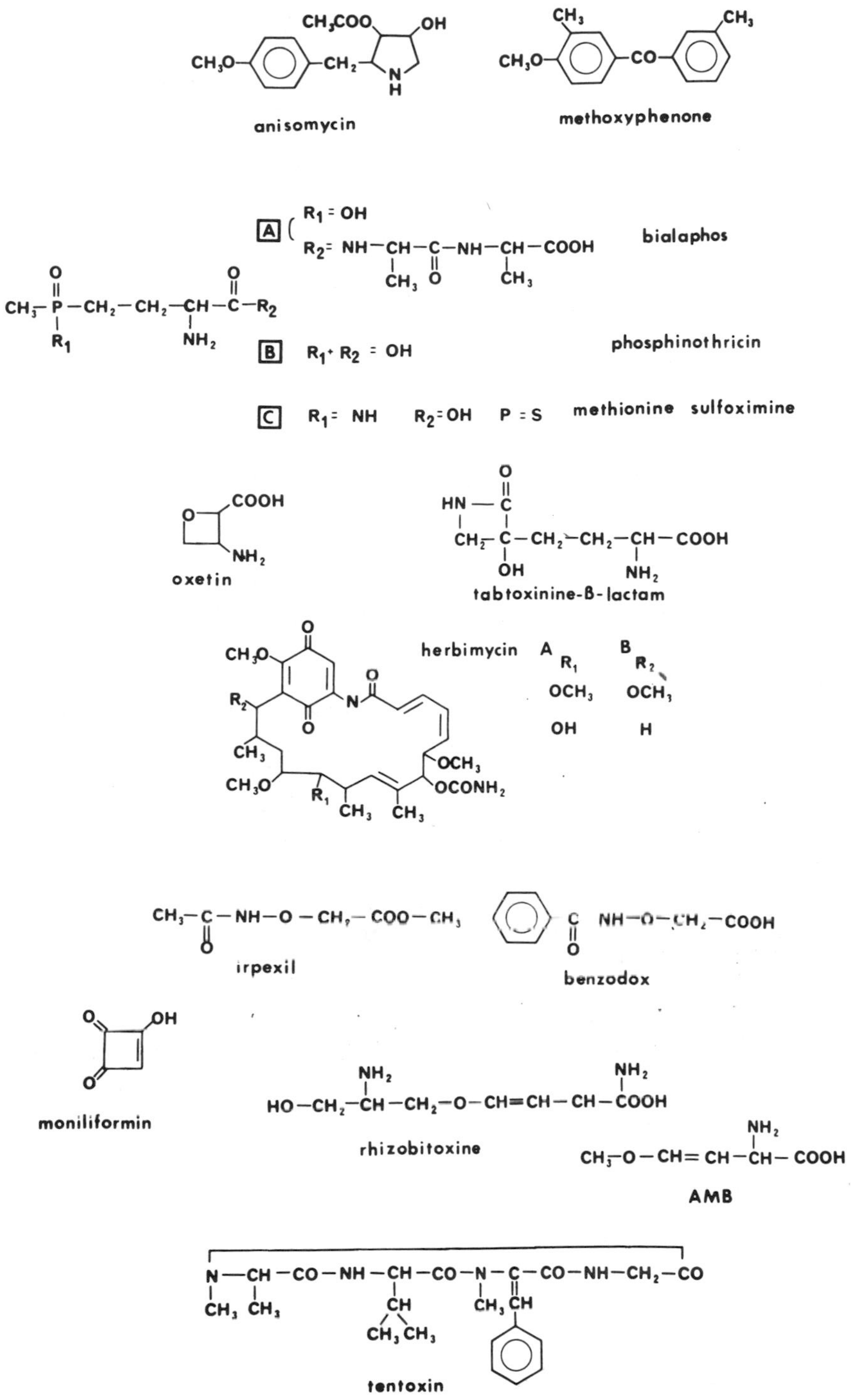

Figure 17.2 Structures of several non-host-specific toxins discussed in the text.

al., 1974a,b). This naturally occurring metabolite provided the chemical basis for the development of a new synthetic herbicide, methoxyphenone or NK-049 (Figure 17.2) (3,3'-dimethyl-4-methoxy benzophenone) (Munakata et al., 1973). This herbicide has been developed commercially in Japan for use in rice fields. At 4 kg/ha methoxyphenone causes extreme chlorosis in barnyardgrass without affecting rice. Its effect is apparently on chloroplast pigment synthesis rather than directly on photosynthesis (Ito et al., 1976). Methoxyphenone is readily degraded in soil, neither it nor its metabolites are toxic to rice (Munakata et al., 1973), and it is relatively easily synthesized. Many analogues of anisomycin have been screened for growth regulator and herbicidal activity (Yamada et al., 1974a,b). Several 4-methoxybenzophenones and 4-methoxydiphenylmethanes altered phototropic or gravitropic responses and/or caused chlorosis.

Bialaphos and Other Glutamine Synthetase Inhibitors

Bialaphos   (L-2-amino-4-[(hydroxy)(methyl)phosphinoyl]-butyrl-L-alanyl-L-alanine) (Figure 17.2) is currently marketed as a herbicide (Herbiace®) in Japan (Anonymous, 1984) and is being tested for use in the United States. It is a product of *Streptomyces viridochromogenes* with nonselective properties, killing both dicot and monocot species. It acts more rapidly than glyphosate (*N*-(phosphonomethyl)glycine), but has a persistent effect like glyphosate on perennials, inhibiting regrowth from subterranean plant organs (Sekizawa and Takematsu, 1983), and has a half-life of 20–30 days in soil (Mase, 1984). It also has a low mammalian toxicity.

Bialaphos is readily metabolized to phosphinothricin [L-2-amino-4-((hydroxy)(methyl)phosphinoyl)-butyic acid] (Figure 17.2) and L-2-amino-4-((hydroxy)(methyl)-phosphinoyl)-butyryl-L-alanine in higher plants (Sekizawa and Takematsu, 1983). Phosphinothricin is the herbicidal moiety of bialaphos and is readily translocated from foliage to subterranean organs. A synthetic version of phosphinothricin has been patented by Hoechst AG as a herbicide with the common name of glufosinate (HOE 39866) (Fischer and Bellus, 1983; Wild and Mandersheid, 1984). The half-life of glufosinate in soil is apparently much longer than that of bialaphos (Sekizawa and Takematsu, 1983). Bialaphos is somewhat selective in the greenhouse, having an LD_{50} range of < 0.125 to 8.5 kg/ha on seven different weed species (Ridley and McNally, 1985).

The mechanism of action of phosphinothricin (glufosinate) is well understood. It is an analogue of the potent synthetic glutamine synthetase (GS) inhibitor L-methionine sulfoximine (Figure 17.2). L-Methionine sulfoximine has long been known to inhibit GS (Leason et al., 1982) and was proposed as a herbicide in 1967 (Fischer and Bellus, 1983). It is apparently the toxic principle of the root bark of the plant *Cnestis glabra*, which was termed "glabrin" before it was identified (Jeannoda et al., 1985). The effectiveness of glufosinate as a GS inhibitor, however, is about 100 times greater than L-methionine sulfoximine (Leason et al., 1982; Wild and Mandersheid, 1984). Because GS is involved in ammonia assimilation, its inhibition results in the accumulation of toxic levels of ammonia. Ammonia production is enhanced by light-dependent reduction of nitrite to ammonia and by photorespiration, hence the efficacy of bialophos or glufosinate is enhanced by nitrogen fertilizers and light (Misato and Yamaguchi, 1984). Ammonia levels that accumulate in glufosinate-treated plants exceed those known to uncouple photophosphorylation (Sekizawa and Takematsu, 1983). No correlation could be found between the LD_{50} levels and sensitivities of glutamine synthetases of seven weed species (Ridley and McNally, 1985), indicating that selectivity is based on absorption, translocation, and/or detoxication.

Two other microbially-produced GS inhibitors that are glufosinate analogues are also known. The L-(N^5-phosphono)methionine-S-sulfoximine moiety of L-(N^5-phosphono)methionine-S-sulfoximinyl-L-alanyl-L-alanine product of an unclassified *Streptomyces* species is a potent GS inhibitor (Sekizawa and Takematsu, 1983). Likewise, tabtoxinine-β-lactam (Figure 17.2), a product of *Pseudomonas syringae* pv *tabaci* and closely related pathovars, is a strong, irreversible GS inhibitor (Langston-Unkefer et al., 1984; Thomas et al., 1983). The structurally unrelated oxetin (Figure 17.2), a product of *Streptomyces* sp. OM-2317, was recently found to be a GS inhibitor (Omura et al., 1984).

Herbimycin

The Japanese have discovered that two microbial products, herbimycin A and B (Figure 17.2), are efficacious herbicides against many monocot and dicot weeds without affecting rice at the same concentrations (Sekizawa and Takematsu, 1983). This herbicidal combination is generally more active as a preemergence rather than a postemergence treatment.

Irpexil and Related Compounds

Irpexil (Figure 17.2) is a herbicide produced by the basidiomycete *Irpex pachyodon*, which is structurally similar to the herbicide benzadox (Figure 17.2) (Fischer and Bellus, 1983). Analogues of benzadox that more closely resemble irpexil have been patented as herbicides. Extensive structure/activity studies of related compounds have been made for fungicidal and herbicidal activities.

The mechanism of action of irpexil is not known; however, it and benzadox are both translocated basipetally in plants and produce symptoms similar to those produced by glyphosate (Fischer and Bellus, 1983). This would suggest that their mechanism of action is the same. Benzadox is known to be metabolized by sensitive species to a potent inhibitor of pyridoxyl phosphate-requiring enzymes (aminooxyacetic acid) (Duke, 1985a). Aminooxyacetic acid also inhibits anthocyanin synthesis by inhibiting phenylalanine–ammonia lyase (Duke, 1985a,b) and irpexil inhibits anthocyanin synthesis (Fischer and Bellus, 1983).

Moniliformin and Its Analogues

Moniliformin (3-hydroxycyclobut-3-ene-2,3-dione) (Figure 17.2) is a potent phytotoxin produced by *Fusarium moniliforme* (Cole et al., 1973). It inhibits growth and causes chlorosis and necrosis in several species. Numerous analogues of this phytotoxin have been synthesized and some have strong, nonselective activity (Bellus et al., 1980; Fischer and Bellus, 1979). For instance, moniliformin O-dodecyl ether and a 3,4-dibutoxy analogue were more active than moniliformin.

Rhizobitoxine and Related Phytotoxins

Rhizobitoxine (Figure 17.2), a product of certain strains of *Rhizobium japonicum*, causes chlorosis and other phytotoxicity symptoms in many species. Rhizobitoxine is an analogue of cystathionine and inhibits β-cystathionase irreversibly (Giovanelli et al., 1971). β-Cystathionase converts cystathionine to homoserine, the immediate precursor of methionine. There may be other sites of action of rhizobitoxine (Gilchrist, 1983), such as inhibition of enzymatic production of ethylene from methionine (Lieberman, 1979). A herbicidal analogue of rhizobitoxine is 2-amino-4-methoxybut-3-enoic acid (AMB) (Figure 17.2), which is a product of *Streptomyces* and *Rhizobium* strains and is a weak herbicide (Fischer and Bellus, 1983). In higher plants, it irreversibly inhibits transaminases and other pyridoxal phosphate-requiring enzymes.

Tentoxin

The cylic tetrapeptide tentoxin (Figure 17.2), a product of *Alternaria alternata* (formerly *A. tenuis*, hence tentoxin), causes extreme chlorosis in a wide range of taxonomically unrelated weed species (Duke et al., 1980; Durbin and Uchytil, 1977), but does not affect several important crop species, including soybeans [*Glycine max* (L.) Merr.] and maize (*Zea* mays L.). It has been of interest as a possible herbicide because it is the only one known to kill johnsongrass in maize (Figure 17.3) and virtually every important weed in soybeans is susceptible to it. A number of analogues have been synthesized, but they are all less effective than tentoxin (Ballio, 1981).

Tentoxin was originally thought to cause chlorosis by inhibition of photophosphorylation (Steele et al., 1976), however, there is growing evidence that it interferes with the uptake of certain nuclear-coded proteins by the developing chloroplast (Vaughn and Duke, 1981; 1982; 1984b). Tentoxin does interfere with the energy transfer of coupling factor 1, thus inhibiting photophosphorylation (Steele et al., 1976), however, several effects of tentoxin on etioplasts and early chloroplast development cannot be explained by this mechanism (Duke et al., 1983; Vaughn and Duke, 1981; Wickliff et al., 1982). Tentoxin selectively inhibits the posttranslational processing and uptake of the nuclear-coded plastidic protein polyphenol oxidase (PPO) in susceptible species (Vaughn and Duke, 1984b). Although it is not clear whether chlorosis is caused by the lack of incorporation of PPO into the plastid, there is a strong correlation between the PPO effect and chlorosis. The functional role of PPO in the chloroplast is unknown

Figure 17.3   Effects of soil-incorporated tentoxin on johnsongrass in maize.

(Vaughn and Duke, 1984a). Susceptibility to tentoxin at all levels of action is plastome-coded and, thus, maternally inherited.

Other Non-Host-Specific Phytotoxins

A large number of non-host-specific toxins have been characterized and a few are listed in Table 17.2. The diversity in the structures of selected toxins from Table 17.2 is partially illustrated in Figure 17.4. As discussed earlier, the structures of many of these toxins are quite exotic and unlikely to be synthesized by traditional pesticide chemists. Some of these toxins are virulence factors of microbial plant pathogens, probably the result of millions of years of co-evolution of the host and pathogen. Others are from microorganisms that are not known to be plant pathogens.

Soil microorganisms have been the focus of several searches for phytotoxic compounds from microbes (DeFrank and Putnam, 1985; Heisey et al., 1985). Phytotoxins from nonpathogenic soil microflora are less likely to have extreme selectivity; however, they are more likely to be very nonselective, which could influence their agricultural uses.

TABLE 17.2.
Examples of several non host-specific phytotoxins produced by microorganisms

Microbial Toxin	Source	Effect	Reference
Acetylaranotin	*Aspergillis terreus*	Growth inhibition	Kamata et al. (1983)
Alteichin	*Alternaria eichorniae*	Necrosis	Robeson et al. (1984)
Altersolanol A	*Alternaria porri*	Growth inhibition	Suemitsu et al. (1984)
Botrydienol	*Botryotinia squamosa*	Growth inhibition	Kimata et al. (1985)
Brefeldin A	*Alternaria carthami*	Chlorosis	Tietjen et al. (1983)
Chaetoglobosin K	*Diplodia macrospora*	Growth inhibition	Cutler et al. (1980)
Cladosporin	*Cladisporium cladosporioides*	Growth inhibition	Springer et al. (1981)
	Aspergillus spp.		
	Eurotium spp.		
Colletotrichin	*Colletotrichum nicotianae*	Necrosis and growth inhibition	Gohbara et al. (1978)
Coronatine	*Pseudomonas syringae*	Chlorosis	Mitchell (1981)
Cyclopenin	*Penicillium cyclopium*	Growth inhibition and necrosis	Cutler et al. (1984)
Cytochalasins	*Phomopsis* spp.	Growth inhibition	Cole et al. (1981a,b); Cox et al. (1983); Wells et al. (1976)
Dehydrocurvularin	*Alternaria macrospora*	Necrosis	Robeson and Strobel (1985)
Desmethoxyviridiol	*Nodulisporium hinnuleum*	Necrosis monocots	Cole et al. (1975)

Dihydropergillin	*Aspergillus ustus*	Growth inhibition	Cutler et al. (1981)
Gougerotin	*Streptomyces* sp.	Growth inhibition	Murao and Hayashi (1983)
Hydroxyterphenyllin	*Aspergillus candidus*	Growth inhibition	Cutler et al. (1978)
Mevinolin	*Aspergillus terreus*	Growth inhibition	Bach and Lichtenthaler (1983)
Moniliformin	*Fusarium moniliforme*	Growth inhibition, necrosis, and chlorosis	Cole et al. (1973)
Monocerin	*Exserohilum turcicum*	Chlorosis and necrosis	Robeson and Strobel (1982)
Naramycin B	*Streptomyces griseus*	Growth inhibition	Berg at al. (1982)
Neosolaniol monoacetate	*Fusarium tricinctum*	Growth inhibition	Lansden et al. (1978)
Oosporein	*Chaetomium trilaterale*	Growth inhibition	Cole et al. (1974)
Ophiolbolin	*Helminthosporium oryzae* and *Cochliobolus miyabeanus*	Disrupts membrane function	Cocucci et al. (1983)
Orlandin	*Aspergillus niger*	Growth inhibition	Cutler et al. (1979)
Pergillin	*Aspergillus ustus*	Growth inhibition	Cutler et al. (1980)
Phomenone	*Phoma destructiva*	Wilt and growth inhibition	Capusso et al. (1984)
PR-toxin	*Penicillium roqueforti*	Growth inhibition	Capusso et al. (1984)
Prehelminthosporol	*Dreschlera sorokiana*	Growth inhibition, necrosis, and chlorosis	Cutler et al. (1982)
Radicinin	*Alternaria helianthi*	Necrosis	Tal et al. (1985)
Stemphylotoxin I	*Stemphylium botryosum*	Necrosis	Barash et al. (1982)
Toyocamycin	*Streptomyces toyocaensis*	Chlorosis	Yamada et al. (1972)
Viridiol	*Gliocladium virens*	Necrosis	Howell and Stipanovic (1984)
Zinniol	*Alternaria carthami*	Necrosis	Robeson and Strobel (1984); Tietjen et al. (1983)

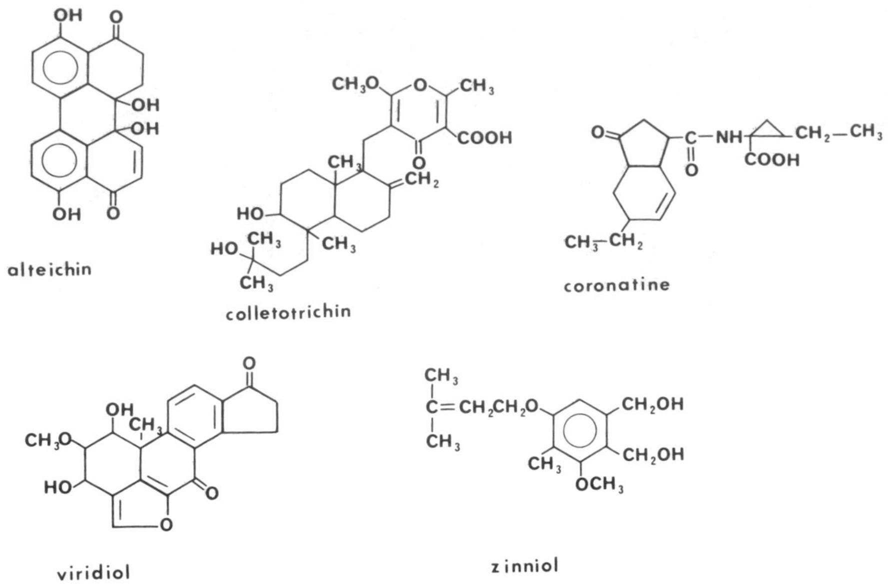

alteichin

colletotrichin

coronatine

viridiol

zinniol

Figure 17.4 Structures of several of the phytotoxins listed in Table 17.2.

MODES OF ACTION OF MICROBIALLY PRODUCED PHYTOTOXINS

Few of the potential biochemical sites of action have been identified for commercial herbicides (Table 17.3). However, the sites of a relatively large percentage are known when compared to microbially produced phytotoxins. Remarkably, little overlap exists between the known mechanisms of action of herbicides and microbial phytotoxins (Table 17.3). Thus the chemistries of these phytotoxins can be used as a basis for the exploitation of new sites of action for commercial herbicides.

Additionally, knowledge of the mechanism of action of these microbial phytotoxins has and will continue to be an important tool in developing a better understanding of plant metabolism and rational designs for herbicides.

PRODUCTION OF MICROBIAL PHYTOTOXINS

Several steps exist in the development of a naturally occurring compound as a herbicide which are not necessary in the development of conventional herbicides. These include culture of the producing organisms, extraction and purification of the compound, elucidation of the structure of the compound, and determination of whether biosynthesis or chemical synthesis of the compound will be least expensive. The last of these steps is the one most frought with difficulty. Many naturally occurring phytotoxins are relatively complex (Scheffer and Livingston, 1984) compared to chemically synthesized herbicides, and are correspondingly difficult to synthesize. Thus, before the development of a natural compound as a herbicide can be implemented, whether the compound can be economically produced by either biosynthetic or chemical means must be determined. In the case of the most publicized natural product-derived herbicide, one company chose biosynthesis (bialophos produced by fermentation) and another chose traditional organic synthesis (glufosinate). There is no general rule of thumb for making this decision. Development of new molecular engineering and biotechnological methods may, however, increasingly tip the balance toward biosynthesis.

The fermentation industry has developed considerable knowledge and expertise during the last few decades in the area of production of pharmaceutical microbial products. Both cultural and genetic manipulation of production for improved yields is extensively used. Examples of cultural manipulation to improve phytotoxin production are (1) enhancement of stemphylotoxin I with ferric chelates (Manulis et al., 1984) and (2) enhancement of tentoxin production with low phosphate (Brucker, 1983; Liebermann and Oertel, 1983). Also, simple strain selection techniques can greatly improve yields of tentoxin from *Alternaria alternata* (A. R. Lax, personal communication).

Depending on the complexity of the metabolic pathway, the precursors, and the genetics of biosynthesis of a compound, molecular genetics could be a useful tool for enhancing the production of herbicidal natural products through fermentation techniques. Microorganisms that do not normally produce the product might be engineered to produce phytotoxins either directly or by biotransformation. In fact, microorganisms might be superior vehicles for such synthesis of higher plant-derived compounds, provided conversion of the substrate to product involves few steps.

CONCLUSIONS

With the increasing need for new herbicides with greater efficacy, safer toxicological properties, improved selectivities, and new sites of action, interest in microbial products as herbicides is growing. In many cases these products offer novel chemistries, a high level of efficacy, new and more desirable selectivity, and favorable environmental properties. Their exotic structures, however, are an obstacle, in many instances, to economical production. Other problems are absorption (the pathogen normally synthesizes them in the host), short half-life, and often a very narrow range of selectivity. Nevertheless, microbially derived and microbially produced herbicides are beginning to be registered and sold. Many companies are now including research on microbial phytotoxins in their herbicide discovery plans.

TABLE 17.3.
Known sites of action of herbicides and microbial toxins

Affected Site (Enzyme)	Herbicide (Reference)	Microbial Toxin (Reference)
	Amino Acid Metabolism	
Methionine synthesis (β-Cystathionase)		Rhizobitoxine (Giovanelli et al., 1971)
Aromatic amino-acid synthesis (EPSP synthase)	Glyphosate (Duke, 1985a)	
Arginine synthesis (Ornithine carbamoyl transferase)		Phaseolotoxin (Gilchrist, 1983)
Histidine synthesis (Imidazoleglycerol phosphate dehydratase)	Amitrole (Duke, 1985a)	
Glutamine synthesis (Glutamine synthetase)	Methionine sulfoximine[a] (Duke, 1985a)	Tabtoxinine-β-lactam (Thomas et al., 1983) Bialaphos (Fischer and Bellus, 1983) Oxetin (Omura et al., 1984)
Valine and isoleucine synthesis (Acetolactate synthase)	Sulfonyl ureas (Duke, 1985a) Imidazolinones (Duke, 1985a)	
All Transaminases	Benzadox (Duke, 1985a)	
Glutamate synthesis (Aspartate aminotransferase)		Gostatin (Nishino and Murao, 1983)
	Plastid Functions	
Electron transport (32 kD protein)	Many (e.g., triazines) (Gressel, 1985)	
Energy transfer Uncoupling (CF_1)		Tentoxin (Steele et al., 1976)

Process	Herbicide	Toxin
Terpenoid synthesis		Mevinolin (Bach and Lichtenthaler, 1983)
Chlorophyll synthesis	Thiocarbamates (Duke, 1985a)	
Energy diversion	Oxadiazon (Duke, 1985a)	
Carotenoid synthesis	Bipyridiliums (Vaughn and Duke, 1983)	
Nuclear-coded protein uptake	Many (e.g., pyridazinones) (Duke, 1985a)	Tentoxin (Vaughn and Duke, 1984b)
Plasma Membrane		
K^+-ATPase activity		Fusicoccin (Gilchrist, 1983)
Alters membrane potential		Fusaric acid (D'Alton and Etherton, 1984)
Mitochondrion		
Coupled electron transport	Many (e.g., benzonitriles) (Moreland, 1985)	
Alternate oxidase	Ioxynil (Moreland, 1985)	
Other Sites		
Calmodulin function		Ophiobolin A (Leung et al., 1985)
Cellulose synthesis	Dichlobenil (Montezinos and Delmer, 1980)	
Folic acid synthesis	Asulam (Duke, 1985a)	
Lipid synthesis	Thiocarbamates (Duke, 1985a)	Cercosporin (Daub and Hangarter, 1983)
Singlet oxygen production	Diphenylethers (Kenyon et al., 1985) Rose bengal[a] (Knox and Dodge, 1984)	
Pyrimidine synthesis (Aspartate carbamoyl transferase)		AAL-toxin (Gilchrist, 1983)
Tubulin polymerization	Dinitroanilines (Hess and Bayer, 1977)	
Microsomal ATPase		CBT-toxin (Macri et al., 1983)

[a]Not a commercial herbicide.

REFERENCES

Anonymous, 1984. Meiji Seika Kaisha: Fermentation specialist. *Farm Chem.* Sept:52.

Bach, T. J. and H. K. Lichtenthaler. 1983. *Physiol. Plant.* 59:50.

Ballio, A. 1981. Structure–activity relationships. *In* R. D. Durbin, (ed.). *Toxins in Plant Disease.* Academic, New York, pp. 395–441.

Barash, I., G. Pupkin, D. Netzer, and Y. Kashman. 1982. *Plant Physiol.* 69:23.

Bellus, D., P. Martin, H. Sauter, and T. Winkler. 1980. *Helvetica Chim. Acta.* 63:1130.

Berg, D., M. Schedel, R. R. Schmidt, K. Ditgens, and H. Weyland. 1982. *Z. Naturforsch.* 37c:1100.

Brucker, B., I. Hanel, F. Hanel, and R. Troger. 1983. *Z. Allg. Mikrobiol.* 23:549.

Capusso, R., N. J. Iacobellis, A. Bottalico, and G. Randozzo. 1984. *Phytochemistry.* 23:2781.

Charudattan, R. 1982. Regulation of microbial weed control agents. *In* R. Charudattan and H. L. Walker, (eds.). *Biological Control of Weeds with Plant Pathogens.* Wiley Interscience, NY, pp. 175–188.

Charudattan, R. and H. L. Walker (eds.). 1982. *Biological Control of Weeds with Plant Pathogens.* Wiley Interscience, NY. 293 pp.

Cocucci, S. M., S. Morgotti, M. Cocucci, and L. Gianani. 1983. *Plant Sci. Lett.* 32:9.

Cole, R. J., J. W. Kirksey, H. G. Cutler, and E. E. Davis. 1974. *Agric. Food Chem.* 22:517.

Cole, R. J., J. W. Kirksey, H. G. Cutler, B. L. Doupnik, and J. C. Peckham. 1973. *Science.* 179:1324.

Cole, R. J., J. W. Kirksey, J. P. Springer, J. Clardy, H. G. Cutler, and K. H. Garren. 1975. *Phytochemistry.* 14:1429.

Cole, R. J., J. M. Wells, R. H. Cox, and H. G. Cutler. 1981a. *J. Agric. Food Chem.* 19:205.

Cole, R. J., D. M. Wilson, J. L. Harper, R. H. Cox, T. W. Cochran, H. G. Cutler, and D. K. Bell. 1981b. *J. Agric. Food Chem.* 30:301.

Cox, R. H., H. G. Cutler, R. E. Hurd, and R. J. Cole. 1983. *J. Agric. Food Chem.* 31:405.

Cutler, H. G. 1984. Biologically active natural products from fungi: templates for tomorrow's pesticides. Amer. Chem. Soc. Symp. Series No. 257, pp. 153–170.

Cutler, H. G., J. H. LeFiles, F. G. Crumley, and R. H. Cox. 1978. *J. Agric. Food Chem.* 26:632.

Cutler, H. G., F. G. Crumley, R. H. Cox, O. Hernandez, R. J. Cole, and J. W. Dorner. 1979. *J. Agric. Food Chem.* 27:592.

Cutler, H. G., F. G. Crumley, R. H. Cox, R. J. Cole, J. W. Dorner, J. P. Springer, F. M. Latterell, J. E. Thean, and A. E. Rossi. 1980. *J. Agric. Food Chem.* 28:139.

Cutler, H. G., F. G. Crumley, J. P. Springer, R. H. Cox, R. J. Cole, J. W. Dorner, and J. E. Thean. 1980. *J. Agric. Food Chem.* 28:989.

Cutler, H. G., F. G. Crumley, J. P. Springer, and R. H. Cox. 1981. *J. Agric. Food Chem.* 29:981.

Cutler, H. G., F. G. Crumley, R. H. Cox, E. E. Davis, J. L. Harper, R. J. Cole, and D. R. Sumner. 1982. *J. Agric. Food Chem.* 30:658.

Cutler, H. G., F. G. Crumley, R. H. Cox, J. M. Wells, and R. J. Cole. 1984. *Plant Cell Physiol.* 25:257.

D'Alton, A. and B. Etherton. 1984. *Plant Physiol.* 74:39.

Daub, M. E. and R. P. Hangarter. 1983. *Plant Physiol.* 73:855.

DeFrank, J. and A. R. Putnam. 1985. *Weed Sci.* 33:271.

Duke, S. O. 1985a. Effects of herbicides on nonphotosynthetic biosynthetic processes. *In* S. O. Duke, (ed.). *Weed Physiology, Vol. II. Herbicide Physiology.* CRC Press, Boca Raton, FL, pp. 91–112.

Duke, S. O. 1985b. Biosynthesis of phenolic compounds. Chemical manipulation in higher plants. Amer. Chem. Soc. Symp. Series No. 268, pp. 113–131.

Duke, S. O., R. N. Paul, and J. L. Wickliff. 1980. *Physiol. Plant.* 49:27.

Duke, S. O., J. L. Wickliff, K. C. Vaughn, and R. N. Paul. 1983. *Physiol. Plant.* 56:387.

Durbin, R. D. and T. F. Uchytil. 1977. *Phytopathology.* 67:602.

Fischer, H. and D. Bellus. 1979. *Chem. Abstr.* 90:198882p.

Fischer, H. P. and D. Bellus. 1983. *Pestic. Sci.* 334.

Gilchrist, D. G. 1983. Molecular modes of action. *In* J. M. Daly and B. J. Deverall, (eds.). *Toxins and Plant Pathogenesis.* Academic, New York, pp. 81–136.

Giovanelli, J., L. Owens, and S. Mudd. 1971. *Biochim. Biophys. Acta.* 227:671.

Gohbara, M., Y. Kosuge, S. Yamasaki, Y. Kimura, A. Suzuki, and S. Tamura. 1978. *Agric. Biol. Chem.* 42:1037.

Gressel, J. 1985. Herbicide tolerance and resistance: Alteration of site of activity. *In* S. O. Duke, (ed.). *Weed Physiology, Vol. II. Herbicide Physiology.* CRC Press, Boca Raton, FL, pp. 159–189.

Heisey, R. M., J. DeFrank, and A. R. Putnam. 1985. A survey of soil microorganisms for herbicidal activity. Amer. Chem. Soc. Symp. Ser. No. 268, pp. 337–349.

Hess, F. D. and D. E. Bayer. 1977. *J. Cell Sci.* 24:351.

Hodosh, R. J., E. M. Keough, and Y. Luthra. 1985. Toxicological evaluation and registration requirements for biorational pesticides. *In* M. B. Mandara (ed.). *Handbook of Natural Pesticides: Methods. Vol. I. Theory, Practice, and Detection.* CRC Press, Boca Raton, FL, pp. 231–272.

Howell, C. R. and R. D. Stipanovic. 1984. *Phytopathology.* 74:1346.

Ito, K., F. Futatsuya, K. Hibi, S. Ishida, O. Yamada, and K. Mumakata. 1976. Some aspects of the mode of action of new herbicide NK-049. *Proc. Asian-Pacific Weed Sci. Soc. Conf.* 5:159.

Jeannoda, V.L.R., J. Valisolalao, E. E. Creppy, and G. Dirheimer. 1985. *Phytochemistry.* 24:854.

Kamata, S., H. Sakui, and A. Hirota. 1983. *Agric. Biol. Chem.* 47:2637.

Kenyon, W. H., S. O. Duke, and K. C. Vaughn. 1985. *Pestic. Biochem. Physiol.* 24:240.

Kimata, T., M. Natsume, and S. Marumo. 1985. *Tetrahadron Lett.* 26:2097

Knox, J. P. and A. D. Dodge. 1984. *Plant Sci. Lett.* 37:3.

Langston-Unkefer, P. L., P. A. Macy, and R. D. Durbin. 1984. *Plant Physiol.* 76:71.

Lansden, J. A., R. J. Cole, J. W. Dorner, R. H. Cox, H. G. Cutler, and J. D. Clark. 1978. *J. Agric. Food Chem.* 26:246.

Leason, M., D. Cunliffe, D. Parkin, P. J. Lea, and B. J. Miflin. 1982. *Phytochemistry.* 21:855.

Liebermann, B. and B. Oertel. 1983. *Z. Allg. Mikrobiol.* 23:503.

Lieberman, M. 1979. *Annu. Rev. Plant Physiol.* 30:533.

Leung, P. C., W. A. Taylor, J. H. Wang, and C. L. Tipton. 1985. *Plant Physiol.* 77:303.

McCalla, T. M. and F. A. Haskins. 1964. *Bacteriol. Rev.* 28:181.

Macri, F., P. Dell'Antone, and A. Vianello. 1983. *Plant Cell Environ.* 6:555.

Manulis, S., Y. Kushman, D. Netzer, and I. Barash. 1984. *Phytochemistry.* 23:2193.

Mase, S. 1984. *Jpn. Pestic. Inf.* 45:27.

Misato, T. and I. Yamaguchi. 1984. *Outlook Agric.* 13:136.

Mitchell, R. E. 1981. Structure: Bacterial. *In* R. D. Durbin (ed.), *Toxins in Plant Disease.* Academic, NY, pp. 259–293.

Montezinos, D. and Delmer, D. 1980. *Planta* 148:305.

Moreland, D. E. 1985. Effects of herbicides on respiration. *In* S. O. Duke (ed.). *Weed Physiology. Vol. II. Herbicide Physiology.* CRC Press, Boca Raton, FL, pp. 37–61.

Mudge, L. L., B. J. Gossett, and T. R. Murphy. 1984. *Weed Sci.* 32:591.

Munakata, K., O. Yamada, S. Ishida, F. Futatsuya, K. Ito, and H. Yamamoto. 1973. NK-049: From natural products to new herbicides. *Proc. Asian-Pacific Weed Sci. Soc. Conf.* 4:215.

Murao, S. and H. Hayashi. 1983. *Agric. Biol. Chem.* 47:1135.

Nishino, T. and S. Murao. 1983. *Agric. Biol. Chem.* 47:1961.

Omura, S., M. Murata, N. Imamura, Y. Iwai, H. Tanaka, A. Furusaki, and T. Matsumoto. 1984. *J. Antibiot.* 11:1324.

Rice, E. L. 1983. *Pest Control with Nature's Chemicals.* University of Oklahoma Press, Norman, OK, 224 pp.

Ridley, S. M. and S. F. McNally. 1985. *Plant Sci. Lett.* 39:31.

Robeson, D. J. and G. A. Strobel. 1982. *Agric. Biol. Chem.* 46:2681.

Robeson, D. J. and G. A. Strobel. 1984. *Phytochemistry.* 23:1597.

Robeson, D. J. and G. A. Strobel. 1985. *J. Nat. Prod.* 48:139.

Robeson, D., G. Strobel, G. K. Matusumoto, E. L. Fischer, M. H. Chen, and J. Clardy. 1984. *Experientia.* 40:1248.

Scheffer, R. P. 1983. Toxins as chemical determinants of plant disease. *In* J. M. Daly and B. J. Deverall (eds.). *Toxins and Plant Pathogenesis.* Academic, NY, pp. 1-40.

Scheffer, R. P. and R. S. Livingston. 1984. *Science.* 112:17.

Sekizawa, Y. and T. Takematsu. 1983. How to discover new antibiotics for herbicidal use. *In* N. Takahashi, H. Yoshioka, T. Misato, and S. Matsunaka (eds.). *Pesticide Chemistry, Human Welfare and the Environment. Vol. 2. Natural Products.* Pergamon Press, Oxford, pp. 261-268.

Springer, J. P., H. G. Cutler, F. G. Crumley, R. H. Cox, E. E. Davis, and J. E. Thean. 1981. *J. Agric. Food Chem.* 29:853.

Steele, J. A., T. F. Uchytil, R. D. Durbin, P. Bhatnagar, and D. H. Rich. 1976. *Proc. Natl. Acad. Sci. USA.* 73:2245.

Suemitsu, R., Y. Yasumasa, T. Sano, and K. Yamashita. 1984. *Agric. Biol. Chem.* 48:2383.

Tal, B., D. J. Robeson, B. A. Burke, and A. J. Aasen. 1985. *Phytochemistry.* 24:729.

Thomas, M. D., P. J. Langston-Unkefer, T. F. Uchytil, and R. D. Durbin. 1983. *Plant Physiol.* 71:912.

Tietjen, K. G., E. Schaller, and U. Matern. 1983. *Physiol. Plant Pathol.* 23:387.

Vaughn, K. C. and S. O. Duke. 1981. *Physiol. Plant.* 53:421.

Vaughn, K. C. and S. O. Duke. 1982. *Protoplasma.* 110:48.

Vaughn, K. C. and S. O. Duke. 1983. *Plant Cell Environ.* 6:13.

Vaughn, K. C. and S. O. Duke. 1984a. *Physiol. Plant.* 60:106.

Vaughn, K. C. and S. O. Duke. 1984b. *Physiol. Plant.* 60:257.

Walton, J. D. and E. D. Earle. 1984. Isolation and bioassay of fungal phytotoxins. *In* I. K. Vasil (ed.). *Cell Culture and Somatic Cell Genetics of Plants, Vol. I, Laboratory Procedures and their Applications.* Academic Press, NY, pp. 598-607.

Wells, J. M., H. G. Cutler, and R. J. Cole. 1976. *Can. J. Microbiol.* 22:1137.

Wickliff, J. L., S. O. Duke, and K. C. Vaughn. 1982. *Physiol. Plant.* 56:399.

Wild, A. and R. Mandersheid. 1984. *Z. Naturforsch.* 39c:500.

Yamada, O., Y. Kuise, F. Futatsuya, S. Ishida, K. Ito, H. Yamamoto, and K. Munakata. 1972. *Agric. Biol. Chem.* 36:2013.

Yamada, O., S. Ishida, F. Futatsuya, K. Ito, H. Yamamoto, and K. Munakata. 1974a. *Agric. Biol. Chem.* 38:1235.

Yamada, O., S. Ishida, F. Futatsuya, K. Ito, H. Yamamoto, and K. Munakata. 1974b. *Agric. Biol. Chem.* 38:2017.

INDEX

AAL-toxin, 301
Abelmoschus esculentis, 287
Abscisic acid (ABA), 15, 175, 185
Abutilon theophrasti, 45, 179, 289
Acacia confusa, 67
Acacia spp., 88
Acetic acid, 49, 64, 273
8α-Acetoxy-β-cyclocostunolide, 215
8α-Acetoxyzaluzanin, 25
Acetylaranotin, 296
Achillin, 117
Acid phosphatase, 176
Acroptilin, 34, 213
Adaptation, role of allelopathy in, 70
Adaptive autoinhibition, 71
Additive inhibition, 172
Adenostema fasiculatum, 9, 190, 193
Ageratum conyzoides, 65
Aglycones, 58
Agroecosystems, allelopathy in, 43, 59
Agropyrene, 48
Agropyron repens, 11, 44, 47, 52
  decay products, 44, 183
  effects on nodulation, 47, 52
  toxicity to crops, 44, 181
  toxins:
    from herbage, 44, 47
    from rhizomes, 44
Agrostemma githago, 33, 45
Agrostemmin, 33
AK toxin, 289
Alantolactones, 213
Albizzia lophantha, 116
Aldehydes, 176, 178
Alfalfa, growth stimulation by, 33

Algae:
  allelopathy in succession of, 27
  plating bioassay, 135
  stimulation:
    by bacteria, 25
    by other plants, 26, 31, 32
Aliphatic acids, 70
Alkaloids, 81, 137, 252
Alkanes, 260
Allelochemicals:
  additive effects of, 7
  bioassays for, 6, 133
  burning to destroy, 85
  classes of, 4, 149, 203, 219, 245
  crop residues contribute, 49, 273
  decomposition products as, 47, 49, 61, 94
  definition of, 2, 57
  growth stimulation by, 23
  hormones affected by, 176
  identification methods, 115, 125
  Koala bears modify, 90
  mechanisms, 189
  membranes influenced by, 181, 184
  methods of release, 3
  mitochondrial effects, 177
  modes of action, 189, 298
  nitrogen availability effects, 61
  nutrient uptake effects, 180
  pesticidal value of, 16
  pest management uses, 13
  presence in marine habitat, 91
  quantitative determination of, 4, 127
  soil accumulation of, 107
  sphere of mung bean, 229
  terms to describe, 2, 57